# Approaches, Advances and Applications in Sustainable Development of Smart Cities

# Approaches, Advances and Applications in Sustainable Development of Smart Cities

Special Issue Editors

**Tan Yigitcanlar**
**Hoon Han**
**Md. (Liton) Kamruzzaman**

MDPI • Basel • Beijing • Wuhan • Barcelona • Belgrade

*Special Issue Editors*

Tan Yigitcanlar
Queensland University of Technology
Australia

Hoon Han
University of New South Wales
Australia

Md. (Liton) Kamruzzaman
Monash University
Australia

*Editorial Office*
MDPI
St. Alban-Anlage 66
4052 Basel, Switzerland

This is a reprint of articles from the Special Issue published online in the open access journal *Energies* (ISSN 1996-1073) from 2018 to 2019 (available at: https://www.mdpi.com/journal/energies/special_issues/sustainable_smart_city).

For citation purposes, cite each article independently as indicated on the article page online and as indicated below:

LastName, A.A.; LastName, B.B.; LastName, C.C. Article Title. *Journal Name* **Year**, *Article Number*, Page Range.

**ISBN 978-3-03928-012-4 (Pbk)**
**ISBN 978-3-03928-013-1 (PDF)**

© 2020 by the authors. Articles in this book are Open Access and distributed under the Creative Commons Attribution (CC BY) license, which allows users to download, copy and build upon published articles, as long as the author and publisher are properly credited, which ensures maximum dissemination and a wider impact of our publications.

The book as a whole is distributed by MDPI under the terms and conditions of the Creative Commons license CC BY-NC-ND.

# Contents

About the Special Issue Editors . . . . . . . . . . . . . . . . . . . . . . . . . . . . . . . . . . . . . . vii

Preface to "Approaches, Advances and Applications in Sustainable Development of Smart Cities" . . . . . . . . . . . . . . . . . . . . . . . . . . . . . . . . . . . . . . . . . . . . . . . . . . . . . . ix

**Tan Yigitcanlar, Hoon Han and Md. Kamruzzaman**
Approaches, Advances, and Applications in the Sustainable Development of Smart Cities: A Commentary from the Guest Editors
Reprinted from: *Energies* 2019, 12, 4554, doi:10.3390/en12234554 . . . . . . . . . . . . . . . . . . . . 1

**Tan Yigitcanlar, Jamile Sabatini-Marques, Cibele Lorenzi, Nathalia Bernardinetti, Tatiana Schreiner, Ana Fachinelli and Tatiana Wittmann**
Towards Smart Florianópolis: What Does It Take to Transform a Tourist Island into an Innovation Capital?
Reprinted from: *Energies* 2018, 11, 3265, doi:10.3390/en11123265 . . . . . . . . . . . . . . . . . . . 12

**Martin De Jong, Thomas Hoppe and Negar Noori**
City Branding, Sustainable Urban Development and the Rentier State. How Do Qatar, Abu Dhabi and Dubai Present Themselves in the Age of Post Oil and Global Warming?
Reprinted from: *Energies* 2019, 12, 1657, doi:10.3390/en12091657 . . . . . . . . . . . . . . . . . . . 44

**Can Bıyık**
Smart Cities in Turkey: Approaches, Advances and Applications with Greater Consideration for Future Urban Transport Development
Reprinted from: *Energies* 2019, 12, 2308, doi:10.3390/en12122308 . . . . . . . . . . . . . . . . . . . 70

**You Jin Kwon, Dong Kun Lee and Kiseung Lee**
Determining Favourable and Unfavourable Thermal Areas in Seoul Using In-Situ Measurements: A Preliminary Step towards Developing a Smart City
Reprinted from: *Energies* 2019, 12, 2320, doi:10.3390/en12122320 . . . . . . . . . . . . . . . . . . . 103

**Maurício José Ribeiro Rotta, Denilson Sell, Roberto Carlos dos Santos Pacheco and Tan Yigitcanlar**
Digital Commons and Citizen Coproduction in Smart Cities: Assessment of Brazilian Municipal E-Government Platforms
Reprinted from: *Energies* 2019, 12, 2813, doi:10.3390/en12142813 . . . . . . . . . . . . . . . . . . . 127

**Hoon Han, Sang Ho Lee and Yountaik Leem**
Modelling Interaction Decisions in Smart Cities: Why Do We Interact with Smart Media Displays?
Reprinted from: *Energies* 2019, 12, 2840, doi:10.3390/en12142840 . . . . . . . . . . . . . . . . . . . 145

**Raluca Suciu, Paul Stadler, Ivan Kantor, Luc Girardin and François Maréchal**
Systematic Integration of Energy-Optimal Buildings With District Networks
Reprinted from: *Energies* 2019, 12, 2945, doi:10.3390/en12152945 . . . . . . . . . . . . . . . . . . . 162

**Fatemeh Karimi Pour, Vicenç Puig and Gabriela Cembrano**
Economic Health-Aware LPV-MPC Based on System Reliability Assessment for Water Transport Network
Reprinted from: *Energies* 2019, 12, 3015, doi:10.3390/en12153015 . . . . . . . . . . . . . . . . . . . 200

**Robert Olszewski, Piotr Pałka, Agnieszka Wendland and Jacek Kamiński**
A Multi-Agent Social Gamification Model to Guide Sustainable Urban Photovoltaic Panels Installation Policies
Reprinted from: *Energies* **2019**, *12*, 3019, doi:10.3390/en12153019 . . . . . . . . . . . . . . . . . . . . **221**

**Debora Sotto, Arlindo Philippi, Jr., Tan Yigitcanlar and Md Kamruzzaman**
Aligning Urban Policy with Climate Action in the Global South: Are Brazilian Cities Considering Climate Emergency in Local Planning Practice?
Reprinted from: *Energies* **2019**, *12*, 3418, doi:10.3390/en12183418 . . . . . . . . . . . . . . . . . . . . **248**

**Richard Hu**
The State of Smart Cities in China: The Case of Shenzhen
Reprinted from: *Energies* **2019**, *12*, 4375, doi:10.3390/en12224375 . . . . . . . . . . . . . . . . . . . . **279**

# About the Special Issue Editors

**Tan Yigitcanlar** is Associate Professor at the School of Civil Engineering and Built Environment, Queensland University of Technology, Brisbane, Australia. He is also Honorary Professor at the Federal University of Santa Catarina, Florianopolis, Brazil. He has been responsible for research, teaching, training, and capacity-building programs in the fields of urban and regional planning, development, and management in esteemed Australian, Brazilian, Korean, Finnish, Japanese, and Turkish universities. The main foci of his research interests are clustered around the following inter-related and interdisciplinary themes: knowledge-based urban development and knowledge cities, sustainable urban development and sustainable cities, and intelligent urban technologies and smart cities. He has extensively published his research findings. These publications also include over 150 articles published in leading journals, and 13 key reference books published by esteemed international publishing houses. He is Editor-in-Chief of Elsevier's Smart Cities Book Series, and has senior editorial positions in 13 prominent academic journals. He is also the Chairman of the annual Knowledge Cities World Summit series, and has organized conferences in many global locations since 2007, including Monterrey (Mexico), Shenzhen (China), Melbourne (Australia), Bento Gonçalves (Brazil), Matera (Italy), Istanbul (Turkey), Tallinn (Estonia), Daegu (Korea), Vienna (Austria), Arequipa (Peru), Tenerife (Spain), and Florianopolis (Brazil).

**Hoon Han** is Associate Professor and Director of the City Planning program in the Faculty of Built Environment, University of New South Wales, Sydney, Australia. He has over 20 years of research experience in city planning and urban innovation. He uses a range of spatial and longitudinal research methods to understand complex relationships between urban form, technology, and human behavior. His recent publications have focused on smart-city planning by measuring the impact of new digital technologies (e.g., IoTs, ML, and AI) on people's adaptive behaviors as part of every-day living. He endeavors to augment current urban-planning studies with a specific focus on machine-learning and artificial-intelligence approaches to future cities, which would give city planners a leading edge in this area in the Fourth Industrial Revolution. He edited a Special Issue journal on 'Innovation and identity in next-generation smart cities' (2018) by City, Culture, and Society (Elsevier), and published a book, 'Open City I Open Data' (2019) by Palgrave Macmillan. He is currently Associate Editor of the City, Culture and Society (Elsevier) journal, and sits on the international editorial boards of Housing Studies (Taylors and Francis) and Spatial Information Research (Springer).

**Md. (Liton) Kamruzzaman** is Associate Professor of Urban Planning and Design at Monash University, Australia. He is also Honorary Associate Professor of Global, Urban, and Social Studies at RMIT University, Australia. He has a PhD in Transport Planning, an MSc (with distinction) in Geoinformation Science and Earth Observation, a Bachelor's degree in Urban and Regional Planning, and a Graduate Certificate in Academic Practice. His research interests are in three key areas of urban/transport planning: a) effectiveness of strategic urban policies, e.g., transit-oriented development (TOD), innovation precincts, and urban form and structure; b) behavioral socioeconomic and travel impact of transport infrastructure, e.g., bicycle-sharing schemes, light rail, and airports; and c) envisioning the future of cities, e.g., smart cities, autonomous vehicles, and climate vulnerability, such as the urban-heat-island effect. He has vast experience in teaching transport and land-use planning, GIS, and remote sensing. Prior to joining Monash University, he taught in three universities: the Queensland University of Technology, Australia; the University of Ulster, UK; and Jahangirnagar University, Bangladesh. He is Editorial Board Member of the Journal of Transport and Land Use, and Section Editor of the Sustainability journal. He has closely collaborated with numerous professional and research bodies, including the World Conference on Transport Research Society (WCTRS), the World Society for Transport and Land Use Research (WSTLUR), the Planning Institute of Australia (PIA), the Bangladesh Institute of Planners, and The Chartered Institute of Logistics and Transport (UK).

# Preface to "Approaches, Advances and Applications in Sustainable Development of Smart Cities"

Over the past decade, digital technologies, as part of the global smart-city agenda, have begun to form the backbone of our cities and to enhance service quality in urban infrastructure. It is widely argued that this approach will create smart cities that are efficient, technologically advanced, green, and socially inclusive. Along with this technocentric viewpoint, the sustainability ideology has had significant impact on the planning and development of smart cities in recent years—recoining the term as 'sustainable smart cities'. In other words, this envirocentric viewpoint has led to consolidated efforts in the conceptualisation of the sustainable development of (sustainable) smart cities. The marriage of technocentric and envirocentric views is seen as the only way to constitute the 21st century's ideal city form. It is also argued that, in this way, current and forthcoming severe global ecological, societal, economic, and governance challenges will be adequately addressed.

This book aims to contribute to the conceptual- and practical-knowledge pools in order to improve research and practices on sustainable smart cities by offering an informed understanding of the subject to scholars, policy-makers, and practitioners. The book contains contributions offering insight into sustainable smart cities by providing in-depth conceptual analyses, and detailed case-study descriptions and empirical investigations from across the globe. This book comprises a repository of relevant information, material, and knowledge to support research, policy-making, practices, and experience transferability to address the aforementioned challenges.

The scope of the book includes the following areas, with a particular focus on the approaches to, and advances and applications in sustainable smart cities: (a) theoretical underpinnings, and analytical and policy frameworks of sustainable smart cities; (b) methodological approaches for the evaluation of sustainable smart cities; (c) technological developments in the techno–enviro-nexus of sustainable smart cities; (d) emerging sustainability solutions and integrated actions from sustainable smart cities; (e) best-practice sustainable-smart-city case investigations from the Global North and South; (f) geodesign and applications concerning the desired urban outcomes of sustainable smart cities; and (g) prospects, implications, and impact concerning the future of sustainable smart cities.

Tan Yigitcanlar, Hoon Han, Md. (Liton) Kamruzzaman
*Special Issue Editors*

*Editorial*

# Approaches, Advances, and Applications in the Sustainable Development of Smart Cities: A Commentary from the Guest Editors

**Tan Yigitcanlar [1],\*, Hoon Han [2] and Md. Kamruzzaman [3]**

[1] School of Civil Engineering and Built Environment, Queensland University of Technology, 2 George Street, Brisbane QLD 4000, Australia
[2] City Planning Program, Faculty of the Built Environment, University of New South Wales, Sydney, NSW 2052, Australia; h.han@unsw.edu.au
[3] Faculty of Art, Design and Architecture, Monash University, 900 Dandenong Road, Caulfield East, VIC 3145, Australia; md.kamruzzaman@monash.edu
\* Correspondence: tan.yigitcanlar@qut.edu.au; Tel.: +61-7-3138-2428

Received: 13 November 2019; Accepted: 28 November 2019; Published: 29 November 2019

**Abstract:** Environmental externalities of the Anthropocene—mainly generated from population growth, rapid urbanization, high private motor vehicle dependency, the deregulated market, mass livestock production, and excessive consumerism—have placed serious concerns for the future of natural ecosystems, which we are a part of. For instance, global climate change—the biggest challenge we have ever faced—is directly impacting wellbeing, and even the existence of humankind, in the long run. During the last two decades, the notion of the smart city—particularly the sustainable development of smart cities—has become a popular topic not only for scholars, particularly in the fields of technology, science, urban and environmental planning, development, and management, but also for urban policymakers and professional practitioners. This was due to digital technologies becoming a powerful enabler in stimulating paradigmatic shifts in urban development-related visions, strategies, implementation, and learning. This paper offers a critical review of the key literature on the issues relating to approaches, advances, and applications in the sustainable development of smart cities. It also introduces contributions from the Special Issue, and speculates on the prospective research directions to place necessary mechanisms to secure a smart and sustainable urban future for all.

**Keywords:** smart city; sustainable smart city; smart infrastructure; smart urban technology; smart governance; sustainable city; sustainable urban development; knowledge-based urban development; climate change; urban informatics; urban policy

---

## 1. Background and Literature Review

The 21st century is recognised as the 'century of cities', as more than half of the world's population now live in urban settlements, and the importance of urban environments has become even greater over the recent decades [1]. It is also seen as the 'century of climate change' or 'century of planetary survival', as today, unexceptionally, all parts of the world are confronted with various environmental and/or socioeconomic crises—e.g., climate change, life-threatening natural disasters, loss of biodiversity, destruction of natural ecosystems, regional disparities, social polarization, and digital and knowledge divides [2]. These crises—the climate emergency being the biggest—are mainly caused by rapid population growth and the irreversible commitment of natural resources, combined with industrialization, urbanization, mobilization, globalization, agricultural intensification, and excessive consumption-driven lifestyles [3].

Due to the rising abovementioned concerns—about environmental deterioration such as increasing energy expenditure and climate change aroused from greenhouse gas emissions—the concept of 'sustainable development of cities' or 'sustainable urban development' has gained ever-increasing interest [3,4]. The widely accepted definition of sustainable urban development can be described as meeting the needs of the present without compromising the ability of future generations to meet their own needs, by achieving environmental, economic, and social sustainability [5,6]. As such, the underlying notion of sustainable urban development is closely aligned with the concept of smart cities, which encourages interactions between humans and technologies for a sustainable urban living environment [7–9].

Most smart city practices overlook the well-established notion of sustainable development [10]. For example, in an examination of the European Union's framing of the smart city concept, Haarstad [11] found that the smartness approach is strongly tied to innovation, technology, and economic entrepreneurialism, and sustainability is not a motivating driver. Nevertheless, the importance of the sustainable development of smart cities is gaining importance in the literature [12]. These studies share the view that the two concepts are not entirely separate, rather, they share many commonalities and thus need to be integrated. For example, it is found that the concept of smart cities includes the smart environment, economy, and people, which aim at environmental, economic, and social sustainability, respectively. This is also reflected by several recent definitions of smart cities, which often embrace the underlying notions of sustainable development [13]. Moreover, Bakıcı et al. [14] and Haarstad [11] claim that the important question to answer is how to strategically integrate the two concepts. In fact, Ahvenniemi et al. [15] even suggest that a more accurate term of 'smart sustainable cities' should be used while there are several other studies—e.g., [16–18]—also adopting the same term in their studies.

The general consensus about the importance of becoming smart and sustainable has resulted in an emergence of studies suggesting various technologies, strategies, and initiatives in order to achieve their aims [19]. Of these, Suciu et al. [20] suggest that the integration of a multi-energy network and low carbon resources would help to deal with the issues facing today's cities such as the imbalance between energy supply and demand. To this extent, their study is closely aligned with the concept of zero-energy building, which aims to achieve a higher level of building sustainability by having a balance in building energy consumption and production [21,22]. Additionally, Pour et al. [23] and Olszewski et al. [24] suggest that the utilization of renewable energy and efficient water transport network systems can contribute to more than better energy efficiency or environmental protection.

In addition to the above, several studies also highlight the importance of smart homes and buildings [25–27], smart transportation [28–31], smart energy and resource management [32,33], and smart media displays [34], which may boost the interaction between cities and their residents, and therefore, leads cities to become smarter and more sustainable. Findings of these studies are further supported by numerous other studies [35] suggesting that the implementation of various smart systems would foster the environmental, economic, and social development of smart cities. Indeed, as highlighted by Komeily and Srinivasan [36], having a balance among environmental, economic, and social aspects of sustainable urban development is particularly important for smart cities considering the concept of smart cities lies beyond simply taking advantage of various modern technologies for better convenience. Furthermore, Millar and Choi [37] highlight the importance of the development of knowledge resources to tackle the socioeconomic and environmental challenges of our time.

Meanwhile, there are also several studies discussing the obstacles for the sustainable development of smart cities. Most notably, Höjer and Wangel [8] present five challenges, namely, strategic assessment, mitigating measures, top-down and bottom-up, competence, and governance. It is noted that these challenges are inter-related to each other. For example, the strategic assessment and evaluation of the effects of information and communications technology (ICT) require competent governance models, as well as the adoption of well-balanced top-down and bottom-down approaches. These obstacles are also discussed by Kudva and Ye [38] where several obstacles including socioeconomic inequalities and the digital divide hinder cities becoming smart and sustainable. The findings of these also tend to agree with several other studies [39–41] showing the country or region-specific obstacles for implementing various smart technologies, strategies, and initiatives to make their cities more sustainable.

In line with the above, many studies also highlight the importance of policy implications to address and possibly overcome such obstacles. For example, Sotto et al. [42] claim that the continuous development of policies is essential for cities vulnerable to climate change. The importance of policy implementation is further highlighted by numerous studies [43–45]. For example, findings of Kim and Lim [46] imply that the development of both mandatory and voluntary regulations is recommended to effectively deal with the contemporary energy consumption and carbon emission issues aroused from our cities. Similarly, Kramers et al. [47] highlight that information and communication technology (ICT) policy implementations can contribute to cities to reduce their energy usage and to meet climate targets. Their studies share the views of Yigitcanlar and Kamruzzaman [48] and Yigitcanlar [49], suggesting that implementing proper smart cities policies and strategies can contribute to not only their environment, but also economic and social aspects of sustainable development.

While the technology dimension is the key identity of smart cities, technology adoption alone is not adequate to make a city smart [50]. Other critical qualities are also required. To be more precise, urban smartness encompasses a mix of human and intellectual capitals (e.g., skilled/talented labour force), infrastructural capital (e.g., high-tech telecommunication facilities), social capital (e.g., intense and open network of social linkages), entrepreneurial capital (e.g., creative and risk-taking business activities), relational capital (e.g., good governance through transparent and democratic institutions), and environmental capital (e.g., protection and enhancement of natural assets within and outside the city) [51]. This holistic view helps in determining policies that can increase the smartness levels of cities, and thus establish a blueprint for a new city model. For example, by diffusing sustainable and smart city discourses and through collaborations between private and public sector actors, Gothenburg represents a successful case of improving the performance of cities [52].

This new city model is widely referred to as 'sustainable smart cities' as numerous studies have indicated that unsustainable cities cannot be considered as smart [53–55]. In recent years, this consolidated sustainable smart cities concept has gained wider acceptance on the global scale. However, most of them focus on measuring the performance of smart cities [56,57]. Few have attempted to conceptualise the sustainable smart cities notion more clearly and comprehensively in a cause–effect model. Such conceptualization could form the basis for developing a thorough understanding, theoretically and practically, of designing smart cities for sustainable and balanced growth [58]. One of the frameworks that represent the abovementioned consolidated view is illustrated in Figure 1.

The conceptual framework (Figure 1) bases itself on an input-process-output-impact model—that also contains a 'system of systems' view—which is a widely used model in urban and regional planning [59]. Assets of a city are the main inputs of that city's smart urbanism endeavours. These assets are put into use through various processes. These processes include key drivers of technology, community, and policy. Given that assets and drivers of a locality (e.g., community, city, and region) are successfully operationalised, various desired outputs are expected to be achieved. The result of the successful execution of these processes is to generate sustainable and knowledge-based development outputs—i.e., in the economic, societal, environmental, and institutional development domains—to achieve desired outcomes. Given that the extent of desired outcomes—i.e., productivity, innovation, liveability, wellbeing, sustainability, accessibility, governance, and planning—are realised, the resulting impacts will transform the city into a smarter one.

This framework emphasises smart 'communities' as the essential ingredient of smart cities, positioning it as the critical driver of smart city development (Figure 1). This approach involves providing access to appropriate technologies, services, and platforms, and modifying the perceptions and behaviours of local communities via awareness campaigns and engagement projects [59]. The framework promotes the customization and development of local and culturally sensitive solutions by local residents and companies, not only to provide locally tailored/accepted solutions, but also to make contributions to the local knowledge-based economic development, sustainable urban development, and participatory governance practices. The framework emphasises the role of the wider urban community as users and developers of the smart city they live in. It also advocates the importance of providing necessary traditional and technology-enabled methods to engage the community in local smart city projects [59].

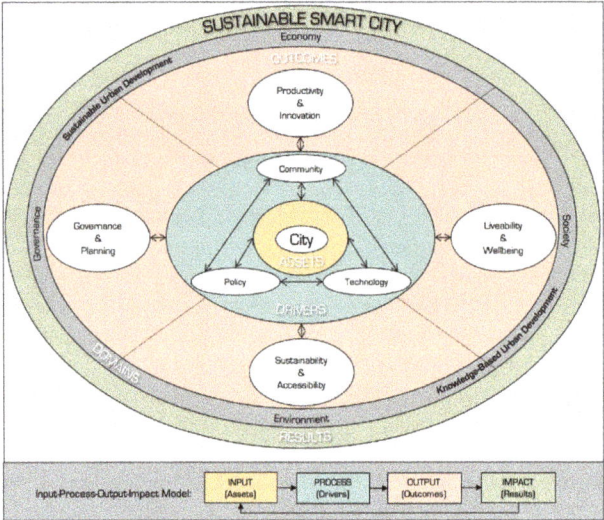

**Figure 1.** Sustainable smart city conceptual framework (derived from [59]).

In terms of 'technology', this framework, in parallel to the literature, considers a smart city as an organic whole, which is a networked and a linked system (Figure 1). While systems in industrial cities are mostly skeleton and skin, contemporary post-industrial cities, i.e., smart cities, are like organisms that develop an artificial nervous system, which enables them to behave in intelligently coordinated ways. The new intelligence of cities, then, resides in the increasingly effective combination of digital telecommunication networks (the nerves), ubiquitously embedded intelligence (the brains), sensors and tags (the sensory organs), and software (the knowledge and cognitive competences). In this way, the framework perceives urban technology only as a 'means' or an 'enabler' to an end—those ends being to achieve desired urban outcomes. It advocates the importance of a smart city as an organic whole of a network and a linked urban system that benefits from the technological offerings—but not solely dependent on or addicted to them [59].

This framework also highlights that the 'policy' context is vital to the understanding of the use of technology in appropriate ways (Figure 1). Hence, an innovative local government stresses the change in policies, as a government cannot innovate without a normative drive addressed in policy. Although innovation in technology for a smart city can be relatively easily observed and broadly agreed upon, subsequent changes in the policy context are more ambiguous. The policy context characterises institutional and non-technical urban governance issues. This policy, and governance, context creates conditions that can enable, or stymie, smart and sustainable urban development. The framework places urban policy at the heart of smart city development as a process that is critical to get it right. In this way, it frames technology as only one of the integral elements for a good policy and its implementation. It advocates the importance of developing competent strategies for the selection and adoption of technology or relevant solutions in appropriate ways [59].

Besides these drivers, the comprehensive conceptual view of the framework focuses on finding ways to achieve desired outcomes in the economy, society, environment, and governance domains. The desired outcomes or performance areas for smart cities consist of 'Productivity & Innovation', 'Liveability & Wellbeing', 'Sustainability & Accessibility', and 'Governance & Planning'. The integration of these desired smart city outcomes with smart city drivers is critical, and the framework emphasises this integration, or, in other words, intertwining [59].

The presented smart cities conceptual framework establishes a consolidated notion of smart cities, and seeks ways for achieving desired urban outcomes for an effective and efficient smart city transformation. The framework also offers the following consolidated definition of smart cities, which we believe will bring some clarity to what this report envisages a smart city as: "Smart city is an urban locality functioning as a healthy system of systems with sustainable and balanced practices of economic, societal, environmental, and governance activities generating desired outcomes and futures for all humans and non-humans" [59].

A review of the key literature finds that the majority of academic smart city research mainly interpret city smartness as technological solutions to the unsustainable development of cities, while issues such as governance and policymaking or community smartness in the traditional sense seem to be in neglect. As much as technology, the planning and development of sustainable smart cities require a comprehensive capital system—containing a mix of human and intellectual, infrastructural, social, entrepreneurial, relational, and environmental capitals. In other words, city smartness encompasses both modern urban production factors, in common frameworks utilizing advanced ICTs, and social and environmental capitals. It is these aspects together that form the competitive and sustainable cities of the information and knowledge age. Public officials commonly turn to smart urban technologies, for technology's sake, to funnel attention and funds to repair flailing urban systems in the absence of public funding and political action. Yet, urban smartness is not only about the technology adoption and use. Smartness—as a set of technologies, new sources of funding, and a branding strategy—helps local governments articulate pragmatic solutions in the immediate present to structurally thorny urban problems.

## 2. The Special Issue

Against the above literature background, it is possible to state that there has been growing, but still rather limited, research that systematically investigates cities from the angle of approaches, advances, and applications in the sustainable development of smart cities. Given that there is no silver bullet to unilaterally be applied in all urban environments to achieve sustainability and smartness, this Special Issue aims to gather diverse views and report progress towards the direction of sustainable smart cities. A fundamental objective of this Special Issue is to compile and present the cutting-edge work of researchers who focus on a joined-up thinking of themes—i.e., sustainability, smartness, and city. By doing so, we believe this Special Issue on "Approaches, Advances, and Applications in the Sustainable Development of Smart Cities" contributes to the knowledge pool in this area, particularly with new evidence driven from empirical research.

Following this guest editorial commentary, the Special Issue includes the following 11 case study, review, and research papers:

- 'The State of Smart Cities in China: The Case of Shenzhen' by Richard Hu [60];
- 'A Multi-Agent Social Gamification Model to Guide Sustainable Urban Photovoltaic Panels Installation Policies' by Robert Olszewski, Piotr Pałka, Agnieszka Wendland, and Jacek Kamiński [24];
- 'Economic Health-Aware LPV-MPC Based on System Reliability Assessment for Water Transport Network' by Fatemeh Karimi Pour, Vicenç Puig, and Gabriela Cembrano [23];
- 'Systematic Integration of Energy-Optimal Buildings with District Networks' by Raluca Suciu, Paul Stadler, Ivan Kantor, Luc Girardin, and François Maréchal [20];

- 'Modelling Interaction Decisions in Smart Cities: Why Do We Interact with Smart Media Displays?' by Hoon Han, Sang Ho Lee, and Yountaik Leem [7];
- 'Digital Commons and Citizen Coproduction in Smart Cities: Assessment of Brazilian Municipal E-Government Platforms' by Maurício José Ribeiro Rotta, Denilson Sell, Roberto Carlos dos Santos Pacheco, and Tan Yigitcanlar [9];
- 'Determining Favourable and Unfavourable Thermal Areas in Seoul Using In-Situ Measurements: A Preliminary Step towards Developing a Smart City' by You Jin Kwon, Dong Kun Lee, and Kiseung Lee [45];
- 'Smart Cities in Turkey: Approaches, Advances and Applications with Greater Consideration for Future Urban Transport Development' by Can Bıyık [35];
- 'City Branding, Sustainable Urban Development and the Rentier State. How Do Qatar, Abu Dhabi and Dubai Present Themselves in the Age of Post Oil and Global Warming?' by Martin De Jong, Thomas Hoppe, and Negar Noori [53];
- 'Aligning Urban Policy with Climate Action in the Global South: Are Brazilian Cities Considering Climate Emergency in Local Planning Practice?' by Debora Sotto, Arlindo Philippi, Jr., Tan Yigitcanlar, and Md Kamruzzaman [42];
- 'Towards Smart Florianópolis: What Does It Take to Transform a Tourist Island into an Innovation Capital?' by Tan Yigitcanlar, Jamile Sabatini-Marques, Cibele Lorenzi, Nathalia Bernardinetti, Tatiana Schreiner, Ana Fachinelli, and Tatiana Wittmann [61].

These articles focused on answering the three broad questions of this Special Issue—namely, what the approaches, advances, and applications in the sustainable development of smart cities are.

Four articles elaborate the approaches used to achieve the sustainable development of smart cities. The objective is to generate transferable knowledge-based diverse case studies. The work by Yigitcanlar et al. [61] entitled 'Towards Smart Florianópolis: What Does It Take to Transform a Tourist Island into an Innovation Capital?' demonstrates the processes used to transform the economic vulnerability of a tourist city, Florianópolis, the capital city of the Brazilian state of Santa Catarina, into a more sustainable economy through knowledge and innovation. Bıyık [35], in his work entitled 'Smart Cities in Turkey: Approaches, Advances and Applications with Greater Consideration for Future Urban Transport Development' describes the processes used to make a radical change in transport policy for the development of a smart transport vision for Turkey. De Jong et al. [53] present the processes used to introduce and operationalise sustainable branding of three middle-eastern cities in their article 'City Branding, Sustainable Urban Development and the Rentier State. How Do Qatar, Abu Dhabi and Dubai Present Themselves in the Age of Post Oil and Global Warming?'. Lastly, Hu [60] in his paper entitled 'The State of Smart Cities in China: The Case of Shenzhen' investigates the state of smart cities in the context of China by particularly focusing on Shenzhen. This paper provides lessons into China's fastest-growing experimental city that has adopted smart urbanization as a model for its development.

The second group of articles examine how the decision-making process can be advanced using technology—i.e., bringing smartness to city governance. Rotta et al. [9] examine how the implementation of the Municipal eGov Platform Assessment Model (MEPA) has enhanced citizen participation in Brazil in their study entitled 'Digital Commons and Citizen Coproduction in Smart Cities: Assessment of Brazilian Municipal E-Government Platforms'. While the results of this study are not promising, other avenues to enhance interaction still exist such as the use of smart media displays. Han et al. [7] present conditions for the effective use of such technologies using Sydney as a case in their article 'Modelling Interaction Decisions in Smart Cities: Why Do We Interact with Smart Media Displays?'

The remaining five articles both advance knowledge and demonstrate applications of specific technologies to bring smartness and sustainability to cities. Olszewski et al. [24] in their work, 'A Multi-Agent Social Gamification Model to Guide Sustainable Urban Photovoltaic Panels Installation Policies' present a model of social gamification to stimulate the photovoltaic panels installation process, and ultimately, to make cities more environmentally sustainable. A health-aware control approach for drinking water transport networks is proposed by Pour et al. [23] in their study entitled 'Economic Health-Aware LPV-MPC Based on System Reliability Assessment for Water Transport Network'. Kwon et al. [45] consider that thermal comfort (such as the urban heat island effect) is a public health issue and identify environmental factors that contribute to thermal comfort in cities. This work is entitled as 'Determining Favourable and Unfavourable Thermal Areas in Seoul Using In-Situ Measurements: A Preliminary Step towards Developing a Smart City'. Suciu et al. [20] in their paper entitled 'Systematic Integration of Energy-Optimal Buildings with District Networks' presents a method to combine multi-energy networks in order to reduce household dependency on fossil fuel. The last paper in this group by Sotto et al. [42] focuses on policy evaluation and examines the consistency between strategic policy objectives and policy implementation in terms of the commitment to reduce carbon emissions in Brazil. This work is presented under the title of 'Aligning Urban Policy with Climate Action in the Global South: Are Brazilian Cities Considering Climate Emergency in Local Planning Practice?'.

## 3. Concluding Remarks and Research Directions

Although the smart cities movement is not new, at present, there is not a single fully-fledged smart city example in the world [62]. Songdo from Korea is widely referred to as the most advanced smart city [63]. Nevertheless, this exemplar urban development project has also received heavy criticism for not being smart in terms of environmental and societal outcomes. Another popular smart city that is planned to be built from scratch is Google Sidewalk Labs' smart city project located at the waterfront area of Toronto. The project has also been criticised for using 'tech for tech's sake'—applying a complex technological solution to a situation that mostly does not need it. As evident in these two cases, while the smart city concept may be good in theory, in practice, there are numerous challenges in building truly smart cities [64–69]. These challenges can be grouped under the following categories:

- Technological and technical issues (e.g., technical barriers due to the size of the city and users, cyber security, privacy concerns, or over irrelevant or unnecessary technology offerings);
- Economic issues (e.g., requiring big-buck financial investments, particularly from the public sector, limited incentives and support to start-ups, incubators, accelerators, and so on);
- Societal issues (e.g., smart cities becoming enclaves for urban elites, gentrification and displacement, negative impacts of disruption, and the digital divide);
- Natural and built environmental issues (e.g., producing insignificant environmental sustainability outcomes or generating negative externalities, and establishing eco-human symbiosis);
- Governance or management issues (e.g., limited transparency, public participation, and bottom-up approach);
- Wider application of the smart city model (e.g., problematic nature of wide-scale development and the cost of retrofitting).

In order to adequately address these challenges, a holistic view on sustainable smart cities is needed [70,71]. We believe the aforementioned conceptual framework (Figure 1) along with other consolidated sustainable smart city research and practices will generate new insights into tackling these issues. In that very moment, it is necessary to place current and emerging approaches, advances, and applications in the sustainable development of smart cities under the microscope. In line with this necessity, the Special Issue generates new insights by investigating the sustainable smart cities from various disciplinary and contextual angles.

Against the literature review reported in this paper and the Special Issue, we compile the following sets of generic research questions focusing on the sustainable development of smart cities. We strongly believe that investigating these issues further in prospective research projects by scholars of this highly interdisciplinary field will shed light on the better conceptualization and practice of sustainable smart cities.

- What is a sustainable smart city supposed to be, and how can benchmarks be determined and set considering smartness is a vague term?
- What is the current status of cities and the inhibitors and threats on the way towards sustainable and smart urban development?
- What are the commonalities amongst cities that are moving towards smart and sustainable futures, and what are the factors of success and failure?
- How can sustainable smart city frameworks be developed and applied, recognizing that every city is unique, to the planning of cities?
- How can institutional and social capacities be developed and further enhanced for the formation of sustainable smart cities?
- How can sustainable smart cities be governed to make sure that existing high sustainability and smartness levels are maintained and improved over time?
- How can sustainable smart cities restore and tap the power of natural systems to enhance and protect urban life?
- How can sustainable smart city agendas contribute to the establishment of global eco-human symbiosis to avoid global ecocide?
- How can sustainable smart cities' climate innovations drive the next urban transformation to address the global climate crisis?
- How can sustainable smart city blueprints be developed for the next global transformation of cities to create carbon-free and adaptive futures for humanity?

**Author Contributions:** T.Y., H.H., and M.K. jointly prepared this Special Issue editorial that is written in the format of a literature review paper. All three authors read and approved the final manuscript.

**Funding:** This research did not receive any specific grant from funding agencies in the public, commercial or not-for-profit sectors.

**Acknowledgments:** We wish to thank the authors of the Special Issue papers for accepting our invitation and submitting and revising their manuscripts within a short time frame, and thank the referees for their thorough, constructive, and timely reviews. We also thank the journal management, for inviting us to serve as the Guest Editors of this Special Issue.

**Conflicts of Interest:** The authors declare no conflict of interest.

## References

1. Yigitcanlar, T.; Inkinen, T. *Geographies of Disruption: Place Making for Innovation in the Age of Knowledge Economy*; Springer: Cham, Switzerland, 2019.
2. Arbolino, R.; Carlucci, F.; Cirà, A.; Ioppolo, G.; Yigitcanlar, T. Efficiency of the EU regulation on greenhouse gas emissions in Italy: The hierarchical cluster analysis approach. *Ecol. Indic.* **2017**, *81*, 115–123. [CrossRef]
3. Mahbub, P.; Goonetilleke, A.; Ayoko, G.A.; Egodawatta, P.; Yigitcanlar, T. Analysis of build-up of heavy metals and volatile organics on urban roads in Gold Coast, Australia. *Water Sci. Technol.* **2011**, *63*, 2077–2085. [CrossRef] [PubMed]
4. Yigitcanlar, T.; Kamruzzaman, M.; Foth, M.; Sabatini-Marques, J.; Costa, E.; Ioppolo, G. Can cities become smart without being sustainable? A systematic review of the literature. *Sustain. Cities Soc.* **2019**, *45*, 348–365. [CrossRef]
5. UN. World Commission on Environment and Development. Our Common Future. 1987. Available online: https://sustainabledevelopment.un.org/content/documents/5987our-common-future.pdf (accessed on 5 November 2019).
6. Dur, F.; Yigitcanlar, T. Assessing land-use and transport integration via a spatial composite indexing model. *Int. J. Environ. Sci. Technol.* **2015**, *12*, 803–816. [CrossRef]
7. Han, H.; Lee, S.; Leem, Y. Modelling interaction decisions in smart cities: Why do we interact with smart media displays? *Energies* **2019**, *12*, 2840. [CrossRef]
8. Höjer, M.; Wangel, J. Smart Sustainable Cities: Definition and Challenges. In *ICT Innovations for Sustainability*; Hilty, L.M., Aebischer, B., Eds.; Springer: Cham, Switzerland, 2014; pp. 333–349.

9. Rotta, M.J.; Sell, D.; Pacheco, R.C.; Yigitcanlar, T. Digital commons and citizen coproduction in smart cities: Assessment of Brazilian municipal e-government platforms. *Energies* **2019**, *12*, 2813. [CrossRef]
10. Joss, S.; Sengers, F.; Schraven, D.; Caprotti, F.; Dayot, Y. The smart city as global discourse: Storylines and critical junctures across 27 cities. *J. Urban Technol.* **2019**, *26*, 3–34. [CrossRef]
11. Haarstad, H. Constructing the sustainable city: Examining the role of sustainability in the 'smart city' discourse. *J. Environ. Policy Plan.* **2017**, *19*, 423–437. [CrossRef]
12. Yigitcanlar, T.; Han, H.; Kamruzzaman, M.; Ioppolo, G.; Sabatini-Marques, J. The making of smart cities: Are Songdo, Masdar, Amsterdam, San Francisco and Brisbane the best we could build? *Land Use Policy* **2019**, *88*, 104187. [CrossRef]
13. Angelidou, M.; Psaltoglou, A.; Komninos, N.; Kakderi, C.; Tsarchopoulos, P.; Panori, A. Enhancing sustainable urban development through smart city applications. *J. Sci. Technol. Policy Manag.* **2018**, *9*, 146–169. [CrossRef]
14. Bakıcı, T.; Almirall, E.; Wareham, J. A smart city initiative: The case of Barcelona. *J. Knowl. Econ.* **2013**, *4*, 135–148. [CrossRef]
15. Ahvenniemi, H.; Huovila, A.; Pinto-Seppä, I.; Airaksinen, M. What are the differences between sustainable and smart cities? *Cities* **2017**, *60*, 234–245. [CrossRef]
16. Aina, Y.A. Achieving smart sustainable cities with GeoICT support: The Saudi evolving smart cities. *Cities* **2017**, *71*, 49–58. [CrossRef]
17. Bibri, S.E.; Krogstie, J. Smart sustainable cities of the future: An extensive interdisciplinary literature review. *Sustain. Cities Soc.* **2017**, *31*, 183–212. [CrossRef]
18. Yigitcanlar, T.; Kamruzzaman, M. Smart cities and mobility: Does the smartness of Australian cities lead to sustainable commuting patterns? *J. Urban Technol.* **2019**, *26*, 21–46. [CrossRef]
19. Yigitcanlar, T. Planning for smart urban ecosystems: Information technology applications for capacity building in environmental decision making. *Theor. Empir. Res. Urban Manag.* **2009**, *4*, 5–21.
20. Suciu, R.; Stadler, P.; Kantor, I.; Girardin, L.; Maréchal, F. Systematic integration of energy-optimal buildings with district networks. *Energies* **2019**, *12*, 2945. [CrossRef]
21. Han, J.H.; Kim, S.; Kim, J.H.; Lee, S.Y. A review of zero energy housing regulations for low-income households. *Int. J. Knowl.-Based Dev.* **2018**, *9*, 343. [CrossRef]
22. KiKylili, A.; Fokaides, P.A. European smart cities: The role of zero energy buildings. *Sustain. Cities Soc.* **2015**, *15*, 86–95. [CrossRef]
23. Pour, F.K.; Puig, V.; Cembrano, G. Economic health-aware LPV-MPC based on system reliability assessment for water transport network. *Energies* **2019**, *12*, 3015. [CrossRef]
24. Olszewski, R.; Pałka, P.; Wendland, A.; Kamiński, J. A multi-agent social gamification model to guide sustainable urban photovoltaic panels installation policies. *Energies* **2019**, *12*, 3019. [CrossRef]
25. Su, K.; Li, J.; Fu, H. Smart city and the applications. In Proceedings of the 2011 International Conference on Electronics, Communications and Control (ICECC), Ningbo, China, 9–11 September 2011; pp. 1028–1031.
26. Klein, C.; Kaefer, G. From smart homes to smart cities: Opportunities and challenges from an industrial perspective. In *NEW2AN 2008, Next Generation Teletraffic and Wired/Wireless Advanced Networking*; Balandin, S., Moltchanov, D., Koucheryavy, Y., Klein, C., Kaefer, G., Eds.; Springer: Berlin/Heidelberg, Germany, 2008; p. 260.
27. Plageras, A.P.; Psannis, K.E.; Stergiou, C.; Wang, H.; Gupta, B.B. Efficient IoT-based sensor big data collection–processing and analysis in smart buildings. *Future Gener. Comput. Syst.* **2018**, *82*, 349–357. [CrossRef]
28. Yigitcanlar, T.; Fabian, L.; Coiacetto, E. Challenges to urban transport sustainability and smart transport in a tourist city: The Gold Coast, Australia. *Open Transp. J.* **2008**, *2*, 29–46. [CrossRef]
29. Kyriazis, D.; Varvarigou, T.; Rossi, A.; White, D.; Cooper, J. Sustainable smart city IoT applications: Heat and electricity management & eco-conscious cruise control for public transportation. In Proceedings of the 2013 IEEE 14th International Symposium and Workshops, Madrid, Spain, 4–7 June 2013.
30. Menouar, H.; Guvenc, I.; Akkaya, K.; Uluagac, A.S.; Kadri, A.; Tuncer, A. UAV-enabled intelligent transportation systems for the smart city: Applications and challenges. *IEEE Commun. Mag.* **2017**, *55*, 22–28. [CrossRef]
31. Zawieska, J.; Pieriegud, J. Smart city as a tool for sustainable mobility and transport decarbonisation. *Transp. Policy* **2018**, *63*, 39–50. [CrossRef]

32. Bulkeley, H.; McGuirk, P.M.; Dowling, R. Making a smart city for the smart grid? The urban material politics of actualising smart electricity networks. *Environ. Plan A* **2016**, *48*, 1709–1726. [CrossRef]
33. Masera, M.; Bompard, E.F.; Profumo, F.; Hadjsaid, N. Smart (electricity) grids for smart cities: Assessing roles and societal impacts. *Proc. IEEE* **2018**, *106*, 613–625. [CrossRef]
34. Struppek, M. The social potential of urban screens. *Vis. Commun.* **2006**, *5*, 173–188. [CrossRef]
35. Bıyık, C. Smart Cities in Turkey: Approaches, Advances and Applications with Greater Consideration for Future Urban Transport Development. *Energies* **2019**, *12*, 2308. [CrossRef]
36. Komeily, A.; Srinivasan, R. Sustainability in smart cities: Balancing social, economic, environmental, and institutional aspects of urban life. In *Smart Cities: Foundations, Principles, and Applications*; Song, H., Srinivasan, R., Sookoor, T., Jeschke, S., Eds.; John Wiley & Sons: Hoboken, NJ, USA, 2017; pp. 503–534.
37. Millar, C.C.; Choi, C.J. Development and knowledge resources: A conceptual analysis. *J. Knowl. Manag.* **2010**, *14*, 759–776. [CrossRef]
38. Kudva, S.; Ye, X. Smart cities, big data, and sustainability union. *Big Data Cogn. Comput.* **2017**, *1*, 4. [CrossRef]
39. Mohammed, F.; Idries, A.; Mohamed, N.; Al-Jaroodi, J.; Jawhar, I. Opportunities and challenges of using UAVs for Dubai smart city. In Proceedings of the 6th International Conference on New Technologies, Mobility and Security, Dubai, UAE, 30 March–2 April 2014; pp. 1–4.
40. Maysoun, I.; Al-Nasrawi, S.; Adams, C.; El-Zaart, A. Challenges facing e-government and smart sustainable city: An Arab region perspective. In Proceedings of the 15th European Conference on e-government, University of Portsmouth, Portsmouth, UK, 18–19 June 2015; pp. 396–402.
41. Granier, B.; Kudo, H. How are citizens involved in smart cities? Analysing citizen participation in Japanese "smart communities". *Inf. Polity* **2016**, *21*, 61–76. [CrossRef]
42. Sotto, D.; Philippi, A.; Yigitcanlar, T.; Kamruzzaman, M. Aligning urban policy with climate action in the global south: Are Brazilian cities considering climate emergency in local planning practice? *Energies* **2019**, *12*, 3418. [CrossRef]
43. Visvizi, A.; Lytras, M.D.; Damiani, E.; Mathkour, H. Policy making for smart cities: Innovation and social inclusive economic growth for sustainability. *J. Sci. Technol. Policy Manag.* **2018**, *9*, 126–133. [CrossRef]
44. De Jong, M.; Joss, S.; Schraven, D.; Zhan, C.; Weijnen, M. Sustainable–smart–resilient–low carbon–eco–knowledge cities: Making sense of a multitude of concepts promoting sustainable urbanization. *J. Clean. Prod.* **2015**, *109*, 25–38. [CrossRef]
45. Kwon, Y.J.; Lee, D.K.; Lee, K. Determining favourable and unfavourable thermal areas in Seoul using in-situ measurements: A preliminary step towards developing a smart city. *Energies* **2019**, *12*, 2320. [CrossRef]
46. Kim, S.; Lim, B.T. How effective is mandatory building energy disclosure program in Australia? In Proceedings of the IOP Conference Series: Earth and Environmental Science, Banda Aceh, Indonesia, 26–27 September 2018; Volume 140, p. 12106. [CrossRef]
47. Kramers, A.; Höjer, M.; Lövehagen, N.; Wangel, J. Smart sustainable cities: Exploring ICT solutions for reduced energy use in cities. *Environ. Model. Softw.* **2014**, *56*, 52–62. [CrossRef]
48. Yigitcanlar, T.; Kamruzzaman, M. Does smart city policy lead to sustainability of cities? *Land Use Policy* **2018**, *73*, 49–58. [CrossRef]
49. Yigitcanlar, T. Technology and the city. In *Systems, Applications and Implications*; Routledge: New York, NY, USA, 2016.
50. Yigitcanlar, T.; Kamruzzaman, M.; Buys, L.; Ioppolo, G.; Sabatini-Marques, J.; Costa, E.; Yun, J. Understanding 'smart cities': Intertwining development drivers with desired outcomes in a multidimensional framework. *Cities* **2018**, *81*, 145–160. [CrossRef]
51. Yigitcanlar, T. Position paper: Benchmarking the performance of global and emerging knowledge cities. *Expert Syst. Appl.* **2014**, *41*, 5549–5559. [CrossRef]
52. Brorström, S.; Argento, D.; Grossi, G.; Thomasson, A.; Almqvist, R. Translating sustainable and smart city strategies into performance measurement systems. *Public Money Manag.* **2018**, *38*, 193–202. [CrossRef]
53. De Jong, M.; Hoppe, T.; Noori, N. City branding, sustainable urban development and the rentier state: How do Qatar, Abu Dhabi and Dubai present themselves in the age of post oil and global warming? *Energies* **2019**, *12*, 1657. [CrossRef]
54. Martin, C.J.; Evans, J.; Karvonen, A. Smart and sustainable? Five tensions in the visions and practices of the smart-sustainable city in Europe and North America. *Technol. Forecast. Soc. Chang.* **2018**, *133*, 269–278. [CrossRef]

55. Silva, B.N.; Khan, M.; Han, K. Towards sustainable smart cities: A review of trends, architectures, components, and open challenges in smart cities. *Sustain. Cities Soc.* **2018**, *38*, 697–713. [CrossRef]
56. Liu, Y.; Wang, H.; Tzeng, G.H. From measure to guidance: Galactic model and sustainable development planning toward the best smart city. *J. Urban Plan. Dev.* **2018**, *144*, 04018035. [CrossRef]
57. Bhattacharya, T.R.; Bhattacharya, A.; McLellan, B.; Tezuka, T. Sustainable smart city development framework for developing countries. *Urban Res. Pract.* **2018**, 1–13. [CrossRef]
58. Metaxiotis, K.; Carrillo, J.; Yigitcanlar, T. *Knowledge-Based Development for Cities and Societies: Integrated Multi-Level Approaches: Integrated Multi-Level Approaches*; IGI Global: Hersey, PA, USA, 2010.
59. Yigitcanlar, T. Smart city policies revisited: Considerations for a truly smart and sustainable urbanism practice. *World Technopolis Rev.* **2018**, *7*, 97–112.
60. Hu, R. The state of smart cities in China: The case of Shenzhen. *Energies* **2019**, *12*, 4375. [CrossRef]
61. Yigitcanlar, T.; Sabatini-Marques, J.; Lorenzi, C.; Bernardinetti, N.; Schreiner, T.; Fachinelli, A.; Wittmann, T. Towards smart Florianópolis: What does it take to transform a tourist island into an innovation capital? *Energies* **2018**, *23*, 3265. [CrossRef]
62. Alizadeh, T.; Irajifar, L. Gold Coast smart city strategy: Informed by local planning priorities and international smart city best practices. *Int. J. Knowl.-Based Dev.* **2018**, *9*, 153–173. [CrossRef]
63. Townsend, A.M. *Smart Cities: Big Data, Civic Hackers, and the Quest for a New Utopia*; WW Norton & Company: New York, NY, USA, 2013.
64. Han, H.; Hawken, S. Introduction: Innovation and identity in next-generation smart cities. *City Cult. Soc.* **2018**, *12*, 1–4. [CrossRef]
65. Alvarez, M.D. Creative cities and cultural spaces: New perspectives for city tourism. *Int. J. Cult. Tour. Hosp. Res.* **2010**, *10*, 171–175. [CrossRef]
66. McLaren, D.; Agyeman, J. *Sharing Cities: A Case for Truly Smart and Sustainable Cities*; MIT Press: Boston, MA, USA, 2015.
67. Plastrik, P.; Cleveland, J. *Life after Carbon: The Next Global Transformation of Cities*; Island Press: New York, NY, USA, 2018.
68. Heitlinger, S.; Foth, M.; Clarke, R.; DiSalvo, C.; Light, A.; Forlano, L. Avoiding ecocidal smart cities: Participatory design for more-than-human futures. In Proceedings of the 15th Participatory Design Conference: Short Papers, Situated Actions, Workshops and Tutorial-Volume 2, Hesselt/Genk, Belgium, 20–24 August 2018; ACM: New York, NY, USA, 2018; p. 51.
69. Clarke, R.; Heitlinger, S.; Foth, M.; DiSalvo, C.; Light, A.; Forlano, L. More-than-human urban futures: Speculative participatory design to avoid ecocidal smart cities. In Proceedings of the 15th Participatory Design Conference: Short Papers, Situated Actions, Workshops and Tutorial-Volume 2, Hesselt/Genk, Belgium, 20–24 August 2018; ACM: New York, NY, USA, 2018; p. 34.
70. Yigitcanlar, T. Smart cities: An effective urban development and management model? *Aust. Plan.* **2015**, *52*, 27–34. [CrossRef]
71. Heitlinger, S.; Bryan-Kinns, N.; Comber, R. The right to the sustainable smart city. In Proceedings of the 2019 CHI Conference on Human Factors in Computing Systems, Glasgow, UK, 4–9 May 2019; ACM: New York, NY, USA, 2019; p. 287.

© 2019 by the authors. Licensee MDPI, Basel, Switzerland. This article is an open access article distributed under the terms and conditions of the Creative Commons Attribution (CC BY) license (http://creativecommons.org/licenses/by/4.0/).

*Case Report*

# Towards Smart Florianópolis: What Does It Take to Transform a Tourist Island into an Innovation Capital?

**Tan Yigitcanlar [1],\*, Jamile Sabatini-Marques [2], Cibele Lorenzi [3], Nathalia Bernardinetti [4], Tatiana Schreiner [2], Ana Fachinelli [5] and Tatiana Wittmann [2]**

1. School of Civil Engineering and Built Environment, Queensland University of Technology, 2 George Street, Brisbane, QLD 4000, Australia
2. Department of Engineering and Knowledge Management, Federal University of Santa Catarina, Campus Universitario, Trindade, Florianopolis, SC 88040-900, Brazil; jamile@labchis.com (J.S.-M.); tati@labchis.com (T.S.); tatianaw@labchis.com (T.W.)
3. Prefeitura Municipal de Florianópolis, Rua Tenente Silveira, 60, Centro, Florianópolis, SC 88010-102, Brazil; cibele@pmf.sc.gov.br
4. Senac Santa Catarina, Rua Felipe Schmidt, 785, Andar, Centro, Florianópolis, SC 88010-002, Brazil; nathalia.bernardinetti@sc.senac.br
5. Department of Communication, University of Caxias do Sul, Rua Francisco Getulio Vargas 1130, Petrópolis, Caxias do Sul, RS 95070-650, Brazil; acfachin@ucs.br
* Correspondence: tan.yigitcanlar@qut.edu.au; Tel.: +61-7-3138-2418

Received: 7 October 2018; Accepted: 21 November 2018; Published: 23 November 2018

**Abstract:** During the last several decades, the diversification of economic activities has become a paramount policy for nations and cities with heavy dependence on a single economic driver. Particularly island economies, relying mainly on tourism income, are among the most vulnerable ones to the shocks of global financial crises. In the recent years, some of these tourist islands had attempts to diversify their economic activities by moving towards a knowledge and innovation economy. This paper places one of these islands—Florianópolis, the capital city of the Brazilian state of Santa Catarina—under the microscope to address the question of 'what it takes to transform a tourist island into an innovation capital'. In order to tackle this question, the study examines economic, social, spatial, and governance conditions and performances, along with the plans and processes of Florianópolis in moving towards an internationally recognized smart innovation island. The methodologic approach includes systematic review of the literature and qualitative analysis of the key development domains of Florianópolis through the lens of knowledge-based urban development. The results of this study provide insights into how to transform a resource-based economy into a knowledge-based one—by disclosing the transition journey of Florianópolis, including progress, challenges, and the new path creation processes. The findings are particularly useful for tourist islands that are aiming for an aspiring knowledge-based urban development and smart city transformation.

**Keywords:** tourist island; innovation hub; knowledge-based urban development; knowledge and innovation economy; smart city; urban branding; urban policy; economic resilience; Florianópolis; Brazil

## 1. Introduction

Today, in our highly globalized capitalist world, economic resilience has become an exceedingly critical issue not only for the economies of nations, but also for the regional and local economies [1]. Particularly, urban locations with a single-dominant economic sector and specific environmental sensitivities are the ones more vulnerable to the impacts of global financial crises [2–4]. Island territories are among those vulnerable locations—where tourism, in general, is the central economic activity, and high environmental sensitivities exist unexceptionally in all of them [5]. In other words, these islands, in general, pose the risk of high environmental vulnerabilities and economic dependencies [6].

For many tourist islands, the diversification of economic activities from tourism and service sectors to knowledge and innovation economy sectors have been a top government policy priority [7,8]. To name a few, Canary Islands, Cyprus, Florianópolis, Hawaii, Malta, and Puerto Rico are among these island territories. The rationale behind the diversification efforts is two-fold. Firstly, tourism has a tremendous externality on the environment and natural resources, and is not a sustainable economic activity in the long-run [9]. Secondly, knowledge and innovation economy activities produce higher value-addition with lesser environmental impact, and an upskilled workforce makes the economy more resilient to the prospective global financial crises [10,11].

Many island territories around the globe have been focusing on diversifying their mostly tourism-driven economies with knowledge and innovation economy sectors by adopting new development paradigms [12–14]. Knowledge-based urban development, and smart and sustainable urbanism are the most prominent paradigms considered for such adoption [15,16]. While there are specific differences between these two paradigms, their most common targets include: (a) searching for ways to achieve urban sustainability, and; (b) utilizing knowledge and technology to foster and/or attract innovative industries, businesses, and services [17–21]. This case report focuses on revealing the challenges confronted during the mostly organically emerging new pathway of Florianópolis. This pathway aspires to achieve prosperous knowledge-based urban development, and smart city transformation. With the main research question of 'What does it take to transform a tourist island into an innovation capital?' in mind, this study aims to examine economic, social, spatial, and governance conditions, and plans and processes of Florianópolis in moving towards an internationally recognized smart innovation island. In order to achieve the aim and addressing the research question, this case report profiles Florianópolis by exploring the knowledge-based urban development [22] conditions and performance, along with the plans and processes that support or hinder its transformation into a smart innovation island.

Following this brief introduction in Section 1 of the paper, the research design, methodologic approach, and the case study city context are introduced in Section 2. Next, in Section 3, the results of the qualitative study are presented under five sub-sections—the first four of them shedding light on the performance of Florianópolis in each of the knowledge-based urban development domains, and the last one revealing the knowledge-based urban development progress and challenges of the city. Lastly, Section 4 concludes the paper by highlighting the key findings, and concisely discussing their implications for the case study city Florianópolis and beyond.

## 2. Materials and Methods

### 2.1. Research Design and Methodology

This research applied a case study method for empirical investigation. The method was considered appropriate for this research, because it allows defining the topic more broadly (to explore what it takes to transform a tourist island into an innovation capital, or what it takes to transform a resource-based economy into a knowledge-based one) by considering contextual issues in the case, and relying on multiple sources of evidence [23]. The two most common methods of case study research include an inductive approach based on the grounded theory [24], and a deductive or testing approach [23]. The basic difference between these approaches is that while the grounded theory relies on data to generate new theories—there is no initial preconceived framework of concepts and hypotheses—the other approach develops a theory at the beginning of the research and focuses on testing and validating the theory in case settings. Another method proposed by Eisenhardt [25] lies in-between these two approaches that is inductive, but there are elements that follow a more planned approach. This research applies both the grounded theory and deductive approaches in a complementary way. In this study, the grounded theory approach was utilized in terms of a literature review to understand how the transformation took place in the selected case study by adopting a knowledge-based urban development perspective [26,27]. In contrast, this research also applied the deductive approach to test the perceptions of key actors on the opportunities and challenges the city faces today. This is accompanied by the likely solutions to the encountered or forthcoming major challenges.

*2.2. Case Study Context: Florianópolis in a Nutshell*

Florianópolis was selected as the case study to satisfy the replication logic of the deductive approach—commencing with generalizations and seeking to see if these generalizations apply to specific instances [28]. The reasons for the selection of the case city context include: (a) being a vibrant emerging economy city context with many developmental opportunities and also numerous challenges [29,30]; (b) achieving noteworthy progress in boosting its research, development, and innovation capabilities, despite being a mainly tourism-based economy [31]; (c) showing the signs of the Dutch disease—an increase in the development of the tourism sector and consequential decline in natural qualities—starting to impact the city in similar ways to the other tourist island destinations [32]; and (d) generated insights from the case investigation potentially being useful for other tourist cities or islands with similar local characteristics and policy ambitions.

The case study city, Florianópolis, is the state capital of Santa Catarina and one of the most developed cities in Brazil—a developing country that became an emerging market economy during the last decades [33]. The city is composed of a main island (97% of the territory), connected to the mainland by bridges, a small continental part, and surrounding uninhabited islands, as shown in Figure 1. The city contains 436.5 km$^2$ land and 453,285 inhabitants, while the metropolitan area has 1,096,476 people [34]. Florianópolis is a city with the third highest human development index score across all capital cities in Brazil (0.847), making it one of the most livable and safest cities in the nation [35].

**Figure 1.** Location of Florianópolis, drawn by authors on a Google Map base map [36].

The traditional inhabitants of the city were mainly from an indigenous community—the Carijós. Following the European arrival in the 16th century, a large proportion of the indigenous population died, and those who remained alive were held as slaves, and removed from the city. In 1738, Florianópolis started its urban development journey from the creation of the Captaincy of the Island of Santa Catarina (the main island where Florianópolis is located) due to its geostrategic military significance. This development was sped up by the Portuguese settlers exploiting opportunities for agriculture, manufacturing, and fishing. In 1823, Florianópolis, then named Desterro, became the capital city of Santa Catarina. In 1894, the capital city had its named changed to Florianópolis—in honor of the then president of Brazil, Floriano Peixoto. During the 20th century, Florianópolis initiated the process of modernization that contributed to the further development of the city, stimulating tourism, trade, innovation, and services [35].

During the last several decades, Florianópolis has become a popular international tourist destination, largely for South Americans. Besides its astonishing weather, beaches, and cuisine, Florianópolis also uses its colorful folklore to attract visitors. In 1985, the official city brand of 'Magic Island' was created, inspired by the old mythical tales [37]. The success of the tourism and service sectors generated a potential for the city to diversify its economic activities by moving towards a knowledge and innovation economy. A new city brand of 'Innovation Capital of Brazil' was coined in 2009. Today, Florianópolis is recognized as one of the premier innovation hubs of Brazil [38]. The most recent achievements of the city, at the national stage, have motivated urban administrators to think about turning Florianópolis into a globally recognized smart innovation island. Considering its current achievements and ambitions, a new brand is also suggested for the city of Florianópolis by the authors—that is 'Smart Florianópolis' or in short, 'Smart Floripa'. The salient characteristics of the city are presented in Appendix A; Table A1.

*2.3. Data Collection*

This case report employs two methodological approaches in its investigation, on the city of Florianópolis, to collect relevant data to analyze. These are: (a) reviewing the relevant literature systematically, and; (b) interviewing the key local actors thoroughly.

2.3.1. Literature Review

The study adopted a three-stage procedure as the methodologic approach of the systematic literature review. Highlighted by Bask and Rajahonka [39]: (Stage 1) planning stage contains objectives and review protocol for the review, defining sources and procedures for literature searches; (Stage 2) conducting the review stage consists of descriptive analysis; (Stage 3) reporting and dissemination stage includes analysis and synthesis of the results according to the established objectives.

In Stage 1 (planning stage), a research plan involving the research aim and question, keywords, and a set of inclusion and exclusion criteria was developed. The research aim was framed to identify the performance of Florianópolis under the four knowledge-based urban development domains—i.e., economy, society, environment, governance [40]. This helped to set the context for addressing the research question of: What does it take to transform a tourist island into an innovation capital? Therefore, 'Florianópolis', 'economic development', 'societal development', 'spatial development' and 'institutional development' were selected as the main search keywords (their Portuguese translations are also used). The query string used for database searches in English was: ((("Florianópolis") AND ("economy" OR "economic development") OR ("society" OR "societal development") OR ("environment" OR "spatial development") OR ("governance" OR "institutional development")). The Portuguese version was: ((("Florianópolis") E ("economia" OU "desenvolvimento econômico") OU ("sociedade" OU "desenvolvimento social") OU ("ambiente" OU "desenvolvimento espacial") OU ("governança" OU "desenvolvimento institucional")).

Given the limited literature coverage on the developmental issues of Florianópolis, both academic and grey literatures were targeted. Firstly, for the selection of the academic literature pieces, the Google

Scholar database was used. The following inclusion criteria were determined with the goal of the selection to relate to and help addressing the research aim and question. All publication types—i.e., journal articles, Procedia papers, books, book chapters, conference proceedings—in English and Portuguese were searched with the abovementioned keywords, but only the full-text and available online literature was considered. The exclusion criteria were determined as publications other than those mentioned in the inclusion criteria. Secondly, all Brazilian universities' websites were manually searched for relevant bachelor, master, and doctoral theses written in English or Portuguese. The same keywords were used in the search. In terms of the inclusion criteria, only the full-text theses, available online, and taking Florianópolis as a case study were considered. The exclusion criteria were kept the same. Thirdly, the grey literature consisting of government, private, and not-for-profit organizations' reports and websites were searched through the Google search engine by using the same keywords. In terms of the inclusion criteria, only full-text publications and websites in English and Portuguese were considered. The exclusion criteria were kept the same.

In Stage 2 (conducting the review stage), the search task of the relevant literature pieces was undertaken in April 2018. No starting publication date was introduced in the search, where the end date was when the search was conducted in April 2018. The keywords were directed to the titles and abstracts of the searched academic literature pieces. The abstracts of the selected articles were read. In the case that abstracts were found relevant, full-texts were read to decide whether to include the article in the review pool, considering the inclusion and exclusion criteria. As for the grey literature, the keywords were directed to the full-texts, and were read to decide whether to include the article in the review pool, considering the inclusion and exclusion criteria. Initially, the search returned in total 69 academic and grey literature pieces. All of them were 'eye-balled' for consistency and accuracy of the keyword search [41]. After evaluating the abstracts (or full-texts of the grey literature were skim read) against the research aim and also removing duplicates, this figure was brought down to 47 literature pieces. The full-texts of these initially screened literatures were then read carefully against the research aim. This resulted in the selection of the final 36 pieces, as shown in Appendix C, Table 2. Lastly, these pieces of literature were re-read, reviewed, and analyzed.

In Stage 3 (reporting and dissemination stage), the work focused on writing up and presenting the findings. At the write-up stage, other publications on the topic were also incorporated as additional supporting literature evidence to better analyze the topic and elaborate the overall findings. The findings of the review were used to support and elaborate the profiling of the development and transformation process of Florianópolis.

2.3.2. Interviews in Focus Group Meetings

Following the systematic review of the literature, the study adopted a semi-structured, interview-based approach to carry out the empirical investigations on the selected case study. The purpose here was to capture the perceptions of key actors on the opportunities and challenges the city faces, and the likely solutions to those issues. The interview exercise was designed to discuss various developmental issues with the key local actors of Florianópolis. The goal of these interviews was to capture the perspectives of these actors, that have a say in the city's policies and future development, on critical matters. In order to gain a holistic understanding, the perceptions of a range of key local actors were taken into consideration from public institutions, non-governmental agencies, and academia. Interviews took place in Florianópolis between May and August 2018 with 14 key local actors. These actors were high-ranked policymakers, and senior professional practitioners representing local (city and state) and national organizations, along with local and regional senior academics with expertise on the relevant areas. The interviewed actors were from the following organizations: State Government of Santa Catarina, Local Government of Florianópolis, Federation of Trade in Goods, Services and Tourism Santa Catarina, National Service of Commercial Learning Santa Catarina, Brazilian Association of Software Companies, Brazilian Institute of Zero Waste, Federal University of Santa Catarina, and University of Caxias do Sul, as shown in Table 1.

**Table 1.** Profiles of the interview participants.

| No | Institution | Position |
|---|---|---|
| 1 | State Government of Santa Catarina | Senior State Development Officer |
| 2 | State Government of Santa Catarina | Economic Development Officer |
| 3 | Local Government of Florianópolis | Senior Architect & Urban Planner |
| 4 | Local Government of Florianópolis | Senior Innovation Policy Officer |
| 5 | Federation of Trade in Goods, Services & Tourism Santa Catarina | Senior Economist & Social Scientist |
| 6 | Federation of Trade in Goods, Services & Tourism Santa Catarina | Senior Business Performance Analyst |
| 7 | National Service of Commercial Learning Santa Catarina | Tourism and Education Specialist |
| 8 | National Service of Commercial Learning Santa Catarina | Science and Technology Specialist |
| 9 | Brazilian Association of Software Companies | Director and Business Analyst |
| 10 | Brazilian Institute of Zero Waste | President and Environmental Scientist |
| 11 | Federal University of Santa Catarina | Professor of Urban & Regional Studies |
| 12 | Federal University of Santa Catarina | Professor of Innovation Studies |
| 13 | Federal University of Santa Catarina | Professor of Knowledge-Based Development |
| 14 | University of Caxias do Sul | Professor of Information Science |

The interviews were planned in the form of a set of focus group meetings. In total, four meetings were held in each month between May and August 2018. Each of these meetings lasted between three and four hours in an interactive workshop format. Most of the interviewees (or focus group members) were present in all of the four meetings. However, where an interviewee could not physically attend a workshop, alternative options were offered to capture their opinions, such as joining the workshop virtually or visiting them in their offices at a later day.

The study adopted the following procedures used in the research work of Pancholi et al. [42,43]: Firstly, as highlighted by Yin [44], in order to make a single case study method successful, it is necessary to utilize multiple sources of evidence as data. Hence, the data from the interviews were integrated with others collected from primary and secondary sources—i.e., literature review findings, and policy and plan documentation obtained from government organizations and other institutes. Secondly, other sources, such as field observations, photographs, physical plans, and maps, also contributed to the analysis as primary data sources in three ways [45]: (a) prior to interviews, to identify issues to form specific questions; (b) during interviews (or focus group meetings or workshops), to support the discussion, such as maps and other spatial data; (c) post-interviews, to confirm the findings. Thirdly, in order to carry out the analysis, a multidimensional conceptual framework for knowledge-based urban development, as shown in Figure 2, was adopted to integrate the diverse developmental perspectives together [46]. Fourthly, a content analysis, informed by phenomenographic methodology [47], is used to analyze findings and derive the attributes from the participants' perceptions. Lastly, the analysis is done through manual coding to identify the themes emerging in the form of significant concepts overarched by the guiding framework. Data from the different groups of interviewees contributed to different dimensions of the framework.

The workshops were chaired and moderated by a master of ceremony, and two project administrators were present to capture the key messages by manually recording them (taking notes in turns) and collecting the written material produced during the workshops. These notes were then transcribed into a document and converted into workshop meeting notes. After each meeting, these notes were circulated to the participants for confirmation of the minutes and the detailed discussions and highlights of the meetings. After the fourth workshop, all meeting notes were combined and forwarded to the participants. Participants were asked to put their consolidated views into a memo-style written piece representing their institutions and own perspectives on the knowledge-based urban development progress and challenges of the city. The meeting notes and memos were used for the analysis of local actors' perceptions on the development achievements, potentials, opportunities, and challenges of Florianópolis. By adopting this procedure, we meet the criterion of theoretical saturation, establishing the empirical validity of the data obtained [48].

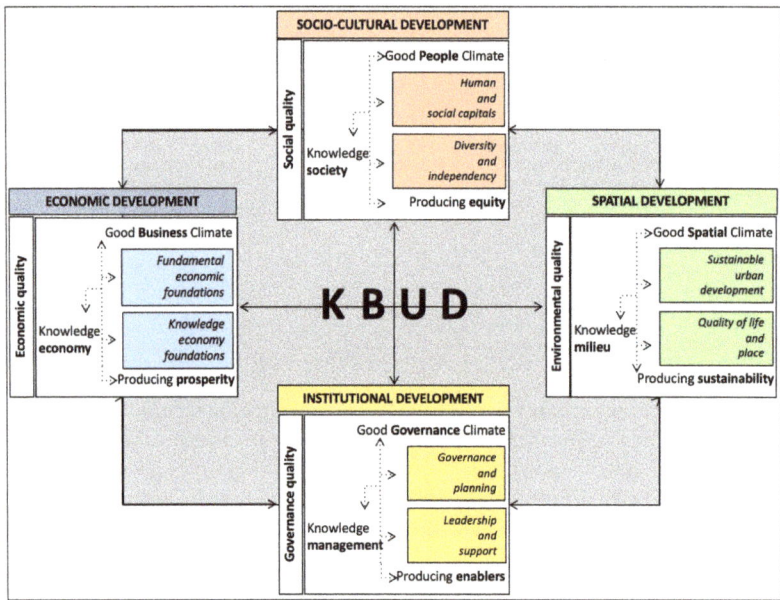

**Figure 2.** Conceptual framework of knowledge-based urban development (KBUD), derived from Reference [46].

The first workshop, that was held in May 2018, was used to introduce the project team and participants, and present the aims and objectives of the study. Conceptual and practical issues were also presented to the group and their roles and expectations were discussed. This helped us to determine some key priorities that local actors wanted the project to specifically focus on. The second and third workshops, that were held in June and July 2018, adopted the following format. In each workshop the project team, in the first hour, provided background to the study and summarized the highlights from the previous meeting. This was followed by, in the second hour, presenting a set of issues and questions to start the conversation between the participant key local actors. However, discussions were not limited to these issues, and in some instances, this led to inspiring debates on the topic and seldomly on its tangents. Table 2 lists the five-what and five-how questions directed to the participants. In the last hour, discussed issues were recapped and a consensus was sought in the case of disagreement. Additionally, participants were asked to put together their personal and institutional perspectives in a brief report (or memo) until the final meeting. The last workshop was held in August 2018. This workshop particularly focused on the presentation and discussion of the participant prepared memos and on establishing a consensus on, or recording the conflicting views on, various developmental related issues concerning Florianópolis.

**Table 2.** Five-what and five-how discussion starter questions.

| No | Question |
| --- | --- |
| 1 | What are the key economic development strengths and weaknesses of Florianópolis? |
| 2 | What are the key societal development strengths and weaknesses of Florianópolis? |
| 3 | What are the key spatial development strengths and weaknesses of Florianópolis? |
| 4 | What are the key institutional development strengths and weaknesses of Florianópolis? |
| 5 | What are the key milestones of Florianópolis' knowledge-based urban development and smart city journey? |
| 6 | How has the transformation of Florianópolis from a tourist island to smart innovation island been formulated? |
| 7 | How effective are the policy mechanisms in Florianópolis to achieve a thriving knowledge-based urban development? |
| 8 | How can encountered challenges of Florianópolis in the knowledge-based urban development journey be tackled? |
| 9 | How can the progress of Florianópolis be sustained given the downturn of the national economics and politics? |
| 10 | How can Florianópolis become an internationally recognized location for smart and sustainable development? |

## 3. Results

The findings of the study are reported in this section under the four categories based on the conceptual framework of knowledge-based urban development presented in Figure 2. These are economic, societal, spatial, and institutional development domains. Additionally, a sub-section on knowledge-based urban development challenges and opportunities is also included. The narrative of the findings is specially designed to disclose the developmental journey of the city as much possible in a chronological order. Moreover, a timeline of Florianopolis' developmental milestones is also developed, as shown in Appendix B, Figure 1. The reason for this approach was to form a comprehensive document to tell the story of Florianópolis and its new path creation efforts in detail. In turn, this may help inform the policy circles of the city and beyond.

*3.1. Economic Development*

The main economic activities in Florianópolis cluster around the commercial and service sectors—representing over 95%. Retail, public, hospitality-tourism, and technology-innovation sectors have almost a similar share in this figure—each representing about 20–30%. There also exists marginal levels of agricultural and maricultural (less than 1%) and limited light manufacturing (over 3%) activities in the city [34].

3.1.1. Tourism and Gastronomy

Despite the tourism potential of Florianópolis that was identified in the 1920s, the lack of infrastructure became an impediment [49]. Only in the 1970s, after the development of urban infrastructure and availability of public incentives and private investment, tourist lodging, sustenance, and entertainment projects were initiated [50]. From then onwards, tourism began to influence the growth directions of the city, advancing the development towards the beaches of the northern part of the island, as shown in Figure 3, and the main lagoon area—i.e., Lagoa da Conceição—through the highways [51].

Florianópolis Urban Development Plan in 1981 and Balneários Master Plan in 1985 were prepared with an aim of transforming Florianópolis into a tourist island [52]. In this period, publicity campaigns were intensified and the city was promoted as the South American Tourist Capital along with the official city brand of 'Magic Island'. The inauguration of the International Airport in 1976 and its expansion in 1988, allowed a greater flow of tourists into the city [53]. From the 1990s, seeking to circumvent the seasonality of tourism, Florianópolis extended its activities to strengthen business and event tourism—such as the opening of the Centro Sul Convention Centre. Today, the convention center is one of the most popular events centers in Brazil [54]. Some hotels in the city made alterations to meet the increased demand and (inter)national hotel chains began their operations from 1995 onwards [55]. According to the Brazilian Hotel Industry Association of Santa Catarina (ABIHSC), the city has 535 hotels with over 38,500 bed capacity. Along with the summer housing stock, Airbnb type accommodation options, and people self-renting out their homes while on vacation, the tourist bed capacity of Florianópolis reaches to well over 300,000 beds.

In addition to the sun-and-sea, business, and events tourism, other tourism segments were also strategically targeted—i.e., eco, party, sports, LGBT, high-end, gastronomic. Besides the Florianopolis Carnival, various annual off-peak tourism season events, such as the Ironman Triathlon (since 1998), Oysters Festival (since 1999), LGBT Parade (since 2005), and an out-of-season carnival (since 2006), are being held in the city [49].

Achievements in the gastronomic tourism area resulted in Florianópolis receiving the 'UNESCO Creative City of Gastronomy' recognition in 2014. Since then, actions have been taken to professionalize the gastronomy sector by improving the quality of dishes and services, encouraging the consumption of ingredients from local and regional producers, respecting the seasonality of the ingredients, and developing a Florianopolitan gastronomic identity—mainly based on authentic seafood dishes.

To further support the gastronomy sector, the production of seed oysters commenced in 1994 with the support of the Federal University of Santa Catarina (UFSC) [56].

**Figure 3.** A view of the Jurere International Beach, reproduced with permission from Eduardo Zappia (photo taken in February 2018).

### 3.1.2. Technology and Innovation

During recent years, Florianópolis has been making significant progress in knowledge-based economic activities. One of the primary reasons for that is its higher education system fostering human capital. The first higher education institution in the city was established in 1909, Federal Institute of Santa Catarina (IFSC). The subsequent institutions established were UFSC in 1960, University of South Santa Catarina (UNISUL) in 1964, and State University of Santa Catarina (UDESC) in 1965 [57].

At the federal-level, endeavors aiming to cultivate the Brazilian science and technology sector focused on the establishment of a number of key institutions. These include: National Council for Scientific and Technological Development (CNPq) in 1951; National Bank for Economic and Social Development (BNDES) in 1952; Southern Regional Development Bank (BRDE) in 1961; Brazilian Innovation Agency (FINEP) in 1967; and National Fund for Scientific and Technological Development (FNDCT) in 1971 [58]. The creation of Telebrás Research and Development Centre (CPqD) in 1974 boosted the telecommunication sector in the country. Florianópolis responded with the emergence of high-tech companies—e.g., Intelbrás in 1976, Dígitro in 1977, Eletrosul in 1978. These companies employed graduates of local universities and attracted specialized professionals to Florianópolis from other Brazilian cities [59].

At the state-level, during the last several decades, the technology sector has been targeted as the key driving force behind innovation. The establishment of the State Development Agency of Santa Catarina (BADESC) in 1975 generated the most needed support opportunities for the development of technology companies and initiated the innovation ecosystem formation in the state. In 1972, Santa Catarina Branch of Brazilian Institute of Management Assistance to Small and Medium Enterprises (IBAGESC) was created to promote training courses, facilitate access to financial services, stimulate cooperation between companies, and organize commerce fairs and business roundtables.

In 1991, IBAGESC was renamed as Santa Catarina Branch of Brazilian Micro and Small Business Support Service (SEBRAESC) [59].

At the local-level, the local government was concerned by the evident economic growth limitations of the city as an island. For example, as Florianópolis is an ecologically sensitive island, industrial developments are not permitted in its territory. In 1985, a municipal law (#2193/1985) was passed for the city to reinvent itself through knowledge economy drivers—basically to foster, attract, and retain talent and investment [60].

In the mid-1980s, the technology sector was established to generate high-value-added product and services in Florianópolis. In this direction, the first research and development (R&D) organization, Centers of Reference in Innovative Technology Foundation (CERTI), was created in 1984. This is followed by the Association of Technology Companies in Santa Catarina (ACATE) in 1986, gathering entrepreneurs to foster a technology and innovation ecosystem, and arousing the development of public policies in the state and the city. In the same period, the first incubator of Florianópolis, the Business Centre for Advanced Technologies Technology (CELTA), was created by CERTI.

At the federal-level, the mid-1980s marked the knowledge economy noteworthiness by the foundation of the Ministry of Science and Technology (MCTIC) in 1985 and the creation of the National Associations of Entities Promoting Innovative Enterprises (ANPROTEC) in 1987. ANPROTEC annually awards the best incubators and technology parks in the country. Thus far, Florianópolis is the only Brazilian city with two award-winning incubators (i.e., CELTA, MIDITEC). These incubators showcase the best practice incubation models in Brazil [38].

The efforts in creation of institutions to provide support for the development of an innovation ecosystem in Florianópolis was notable in the 1990s. Santa Catarina Research and Innovation Support Foundation (FAPESC), established in 1990, has made the collaboration between academia and companies possible and feasible. During this period, a number of successful software companies were created—e.g., Softplan in 1990, Reason/Alstom in 1991, Nexxera Group in 1992, Dot Group in 1996. In 1993, with the support of the state, ParqTec Alfa was created to accommodate growing numbers of technology-based enterprises that were surfacing from incubators and condominiums—equivalent to garage start-ups [61].

In 1994, the National Service for Industrial Training (SENAI), a network of not-for-profit secondary level professional schools, established and maintained by the Brazilian Confederation of Industry (CNI), inaugurated the Centre for Technology in Automation and Information Technology (CTAI). CTAI offers the industries of Santa Catarina a technological center capable of developing and transferring technology, with actions focused on quality and continuous improvement aiming at industrial competitiveness. In 1996, GERASUL, a state-owned energy company based in Florianópolis, was acquired by Belgian Tractebel. In 2015, its name changed to ENGIE—the only top-100 Brazilian company located in Florianópolis.

The Municipal Law (#4913/1996) offers tax exemption over services and land allocation for the companies established in the Industrial Computer Condominium and ParqTec Alfa. The growth of the technology sector pledged the development of another incubator in Florianópolis, MIDI Tecnológico, in 1998. It is managed by ACATE and maintained by SEBRAESC. The Network of Entities Promoting Technology Enterprises in Santa Catarina (RECEPETI) was founded in 2001 to improve the integration of incubators throughout the state. In 2013, its name was changed to the Catarinense Innovation Network.

In the late-2000s, Florianópolis started to be recognized both nationally and internationally. In the mid-2000s, the federal government took an important step towards promoting innovation and scientific and technological R&D by passing the Innovation Law (#10973/2004) that brings together companies and universities. The Good Law (#11196/2005) grants tax incentives to companies that carry out R&D of technological innovation [62]. Benefiting from these laws, in 2006, Sapiens Parque was conceived as an innovation park in Florianópolis. In 2013, it was included in the network of 13 innovation centers across the state.

Brazilian magazine Você S/A highlighted the city as one of the best Brazilian cities in which to work. Another national magazine, Exame, mentioned the city as one of the best cities to do business in Brazil. The Getúlio Vargas Foundation highlighted the city as the country's first capital in digital inclusion. Newsweek considered Florianópolis one of the top-ten most dynamic cities in the world. Florianópolis received mentions from BBC and Italian Corriere della Sera as the likely 'Silicon Valley of South America'.

In 2008, a program to promote innovative entrepreneurship—the Innovation Synapse Program—was conceived to offer financial resources, capabilities, and support to transform innovative ideas from different sectors of the knowledge economy into successful ventures for boosting the innovation ecosystem. In 2009, a new city brand, 'Innovation Capital of Brazil', was coined for Florianópolis by ACATE [63]. The joint effort constituted the positioning of the city, highlighting its innovative initiatives promoted not only by technology companies, but also universities, service companies, and the maricultural sector.

An 'Innovation Route' in Florianópolis was planned in 2013. The route aims to connect companies, academia, government, and community to leverage entrepreneurial and innovative potential in the city. An electric bus (locally developed technology) operates on the route—circulating between UFSC and the Sapiens Parque [64]. In 2017, two major companies were located on the innovation route—Peixe Urbano, one of the largest e-commerce companies in Latin America, and EMBRAER, the third largest aircraft manufacturer in the world. Figure 4 illustrates the location of high-tech company clusters on the innovation route.

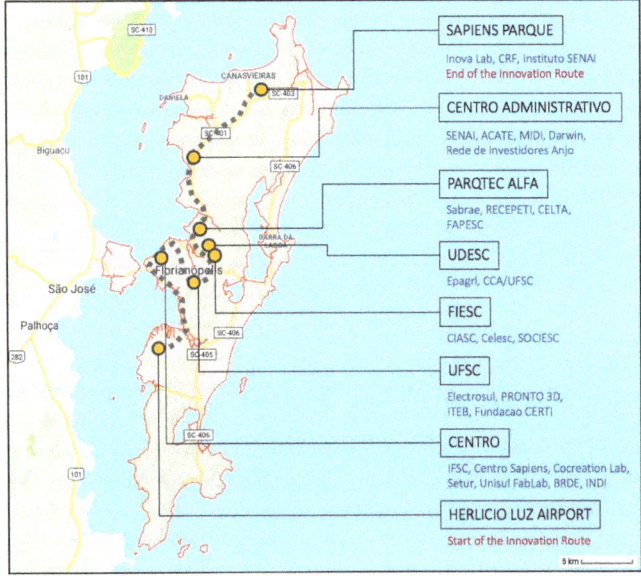

**Figure 4.** Location of the innovation route, drawn by authors on a Google Map base map [65]. SENAI: National Service for Industrial Training; ACATE: Association of Technology Companies in Santa Catarina; RECEPETI: Network of Entities Promoting Technology Enterprises in Santa Catarina; CELTA: Business Centre for Advanced Technologies Technology; FAPESC: Santa Catarina Research and Innovation Support Foundation; UDESC: State University of Santa Catarina; UFSC: Federal University of Santa Catarina; IFSC: Federal Institute of Santa Catarina; BRDE: Southern Regional Development Bank; CRF: Regional Council of Pharmacy; CCA: Center for Agrarian Sciences; FIESC: Federation of Industries of Santa Catarina; CIASC: Center of Informatics and Automation of the State of Santa Catarina; SOCIESC: Educational Society of Santa Catarina; ITEB: Rubber Technology Industry; INDI: Integrated Institute of Economic Development: CENTRO: City Centre.

Knowledge-based economic development endeavors of Florianópolis—through the evolution of human capital via its educational institutions and attractiveness, creativity in new product and service development, incentives, and legislative promotions—helped its gross domestic product (GDP) to be one of the highest among the Brazilian capital cities. The technology sector of Florianópolis makes more than R$4.3 billion per annum (about US$1 billion) and employs over 17,000 people. Florianópolis is ranked nationally: second most entrepreneurial city; second for access to venture capital; third for access to capital; third for innovation [66].

3.2. Societal Development

In order to support the growth of the innovation economy, numerous institutes contributed to the development of human capital on two fronts: training human resources and fostering entrepreneurs [67]. In terms of tertiary education, aforementioned institutes train prospective knowledge workers. According to the 2017 Census of Higher Education, over 45,000 students were enrolled in universities, where more than half of these resided in Florianópolis. In terms of skill development, National Service of Commercial Learning's (SENAC) Florianópolis branch delivers short-term technical qualification courses. In terms of fostering entrepreneurs, the Foundation for Research and Innovation Support of the State of Santa Catarina (FAPESC) supports research and innovation, education infrastructure development, scientific dissemination, and training [68]. Along with this, the 2018 federal regulation, introducing a Legal Framework for Science, Technology and Innovation, encourages collaboration between research centers, companies and public agencies.

Whilst the development of human capital in Florianópolis is promising, the same is not true for social capital and community development and cohesion. Florianópolis is a city of segregated settlements. This is due to the geographical and social conditions that generate a socially incohesive city form. In other words, socio-spatial segregation in the city is evident—the dominant social class in each neighborhood is clearly visible. Waterfront areas are generally occupied by higher-income groups and the environmentally fragile areas, such as hills and mangroves, by lower-income groups. Moreover, wealthier people reside in planned areas, where lower-income groups mostly in informal settlements, so-called 'favelas'—precarious human settlements originally resulting from the invasion of both private and public urban areas [69].

Over 80% of favelas are situated in the central city area—mostly in the hilly parts. In general, favelas neither have access to the main public services such as water, sewerage, electricity, street lighting, and garbage collection, nor to the facilities such as squares, schools, nurseries, and health centers. However, in the case of Florianópolis, most of these services are provided to the favelas. While the human development index score of the city is 0.847, the score goes down to 0.390 for the favela areas of the city [70]. There are only a few places that allow greater social interaction—such as parks and other public spaces. Beaches are the most democratic spaces of the city and the carnival time is the most inclusive time of the year, as shown in Figure 5.

According to the Ministry of Social Development, 10.16% of the households have income of up to half a minimum wage and 3.7% of the households receive government payments to move out of the extreme poverty level. According to the Atlas of Human Development in Brazil in 2013, the ratio of low-income groups in Florianópolis comes to a decline—9.63% in 1991, then 5.31% in 2000, and 1.35% in 2010. The ratio of the richest 20% to the poorest 20% fell from 21.92% in 1991 to 16.96% in 2010 [71]. However, these figures might be worsening in the near future as the economy took a downturn in recent years.

In line with a declining national economy since 2013, the unemployment rate in Florianópolis jumped from 3.8% in 2012 to 7.5% in 2017—still being well below the national average of 13.1%. In 2016, the average monthly salary was 4.7 minimum wages (15th in Brazil) and the proportion of employed persons in relation to the total population was 66.3% (17th in Brazil)—minimum monthly wage is around R$1000 or about US$240 [72]. However, given that in many national rankings the city places in the top five—technology company concentration, quality of life, human development index—in terms of these socioeconomic figures, the city is placed well behind many other national locations.

**Figure 5.** A view of the Florianópolis Carnival, reproduced with permission from Eduardo Zappia (photo taken in February 2018).

Community development and integration is supported with several public policies. Federal policies have been more effective in providing relative equality in education, health, and shelter areas. A social protection network was established in Florianópolis jointly by public and private entities to implement social assistance and community integration policies. The most active member of the network is Padre Vilson—a church-based organization. The public entities include all three tier governments and the most active private entities are Social Good Brazil and Greater Florianópolis Community Institute (ICOM). Public institutions house a number of reference centers for social assistance located in close proximity to the neediest communities [73].

*3.3. Spatial Development*

3.3.1. Built Environment and Infrastructure

The first master plan of Florianópolis was approved in 1955. This was followed by the second master plan in 1977. Following the downtown area's densification as shown in Figure 6, according to these plans, the urban expansion continued towards an area called Bacia do Itacorubi—particularly after the establishment of the UFSC university campus in the 1960s and Eletrosul enterprise (a major Brazilian energy company) in 1978. The municipality launched a master plan, Plano da Trindade, for the area in 1982. The plan, in addition to allocating land for housing for the new residents, also introduced zoning for environmental conservation—fully and partially preserved areas [53].

In 1985, the Balneários Master Plan offered for the first-time conditional development permission for the island's countryside. Although numerous new projects were developed in the northern area—e.g., Canasvieiras, Jurere International—most of the expansion was zoned as rural use and the development was informal. The areas with rural nature were occupied by informal housing—over

50.39% of the urban footprint, as shown in Figure 7. This plan defined the boundaries of the northern and southern preservation areas and only permitted a small neighborhood development at the countryside as a polycentric development zone. The central district master plan was revised in 1997 and the city master plan was approved in 2014. Despite planning regulations and development controls, informal land occupation and development continue to be a major problem in Florianópolis [74].

**Figure 6.** A view of downtown Florianópolis, reproduced with permission from Eduardo Trauer (photo taken in February 2018).

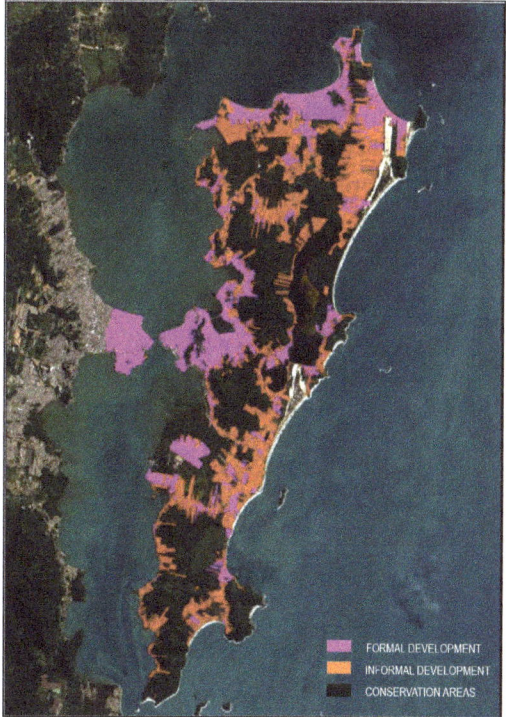

**Figure 7.** Formal and informal urban development, and conservation areas [73].

There is still quite a large part of the island that has not been subjected to exploitation. This is partly because of the existing planning mechanisms that do not allow rural zoned countryside land to be subdivided, because the former land ownership regulation did not permit individuals or corporations to own land on islands. However, the constitutional amendment (#46/2005) allows land ownership. This legislative change made legalization possible for informally occupied land and development. Furthermore, another recent federal land regularization legislation (#13465/2017) gave power to island municipalities to produce regularization policies to solve the informal occupation and development issue.

The risk still exists for further land exploitation as there is a general belief that if land is occupied eventually it will gain a legal status. Despite having large informal settlement areas, Florianópolis does not have too many favelas. This is due to most of the informal urban areas in the city, rather than low-income individuals, being occupied by middle-class households. Most of the favelas exist in the hills behind downtown—formed between 1970 and 2004, as shown in Figure 8. According to Oliveira-Musse et al. [75], in 2012 in Florianópolis, 13,231 households were residing in favelas. The surrounding municipalities house a larger number of favelas, where a considerable portion of the island's workforce lives.

**Figure 8.** A view of the favelas, reproduced with permission from Eduardo Trauer (photo taken in February 2018).

In order to solve the favela problem, the Social Housing Plan was approved in 2012. The plan provided directions to solve the problem in the city within the 15-year period by building new housing, constructing infrastructure in slum areas, and improving precarious houses [68]. However, federal funds provided to support the plan were not sufficient to carry out the required works. Interestingly, Florianópolis has about 24% unoccupied housing stock, one of the biggest rates in Brazil—13% used occasionally as summer house and 11% kept totally vacant.

Urbanization processes and lifestyle offerings generated rapid population increase and gentrification of the city. Amongst the residents, only less than half of them (47%) were born in the city. Accessibility and mobility are among the main problems in Florianópolis—particularly during the summer months when the island population almost doubles. According to Guerra et al. [35] (p. 216) "Florianópolis provides no concrete policy for promoting sustainable transport practices. This is evident in the robust growth rate of its car fleet, and hence a large increase in traffic levels". Fatal traffic accidents are part of daily life; 351 lives were lost on the highways between 2001 and 2017 [76].

While mobility is determined as one of the major challenges for the city to tackle, there are limited actions taken to better integrate land use and transport with sustainable considerations to the people and the environment of the city [77]. This also relates to the limited focus on sustainable urban development and climate change mitigation aspects in the planning circles of the city. The planning documents that emphasize sustainability issues are rather recent and have not found much implementation opportunity yet [71]. Likewise, despite the gaining popularity of the smart city notion [78] and the smart and resilient urban infrastructure concept [79] being widely pronounced and discussed by both state and local planning authorities; so far, no smart city strategy framework development or planning steps have been taken in the city or the state.

3.3.2. Natural Environment and Conservation

Population increase and informal land occupation problems are among the issues risking environmental integrity of Florianópolis. Environmental factors and quality of life attract more people—tourists, knowledge workers, and informal settlers [80]—but as ecosystems are under pressure by human activities, an increase in population leads to major environmental problems—and thus reversing the quality of life. The housing deficit in Florianópolis led to the growth of informal housing settlements to accommodate the growing population, especially among migrant workers who come to the area in search of high-season employment [81].

Another major environmental problem is the limited sewage system of the island. Only 62% of the Florianópolis population is provided with a sewage system. The rest of the population discharges their household and commercial wastewater into nature. The sea and lagoons have become highly vulnerable—particularly during the crowded peak tourism season [82]. In some popular beaches, seawater contains sewage-related biological waste that is causing both public and environmental health risks—e.g., in Canasvieiras.

Some attempts were made to address this issue. For instance, 'Floripa Se Liga Na Rede' is a program of the municipality, released in 2013 in partnership with the Catarinense Water and Sanitation Company (CASAN), which aims to promote the interconnection of real estate serviced by the public network of sewerage and to eliminate inadequacies in building installations [83]. The program targets 75% of the properties of the city—current coverage is 62%.

Another challenge relates to planning and regulation hiccups. The 2014 Master Plan used, for the first time, the concept of sustainability to identify the goals and tools to be used to create a sustainable city by programs and incentives. The main goal is to preserve natural habitats and promote energy efficient design and equipment [81]. However, the plan disagrees with the environmental legislation involving prevention of natural disasters and does not correspond with local needs linked to population growth [84].

On top of the discharge of domestic and industrial waste, deforestation and irregular land occupation are the main problematic areas that impact marine ecosystem and fishery productivity, intensifying the impoverishment of fishermen and food security in fishing communities [85]. In order to minimize the predatory effects of fishing on sea turtles, the Tamar Project launched in 2005 its base in the region of Barra da Lagoa Beach, 25 km from the center of Florianópolis [86].

Florianópolis is the first city in Brazil to officially aim to become zero-waste. This movement provoked civil society and organizations—such as Interinstitutional Group of Solid Waste (GIRS), Commercial and Industrial Association of Florianópolis (ACIF), universities, civil entities and companies—to start a process of implementation of internal programs. Increasing sustainability awareness led society to demand from the government the establishment of a goal for the city, which was established to achieve a 90% diversion of landfill by 2030. Further raising this awareness is critical, as particularly the sustainability of the water environment of the city—given its island and lagoon nature—is under already serious threat [87].

*3.4. Institutional Development*

Following tourism success, during the last years the innovation and technology sector has been performing quite well in Florianópolis. Actors in the city were aware that innovation emerges as a result of co-creation between companies, citizens, universities, and government, in a context marked by the existence of partnerships, collaborative networks, and symbiotic relationships [88]. Despite this quadruple-helix model partnership being frequently pronounced in policy circles, 'walking the talk' is still a big challenge.

A limitation of knowledge-based urban development of Florianópolis is the lack of the innovation sector's political power and say on the city's governance and planning. This is a dichotomous situation as innovation and technology activities are major tax generators, but only a small portion of this tax is kept locally in Florianópolis. This is due to the governance system of Brazil, leaving municipalities with limited financial resources. State government, however, has been active in providing support for knowledge-based urban development of the city—e.g., funding and incentives for innovation centers and enterprises.

Universities have their own dynamics and are generally more receptive than active. IFSC and UFSC have considerably more access to funds for R&D, but depending on the administration, at times there can be resistance in interacting with the private sector. Companies also receive guidance from business and industrial associations. The oldest association has been active for more than a century—Florianópolis Commercial and Industrial Association (ACIF). National associations for retail (CDL), accommodation (ABIHSC), entertainment (ABRASELSC), technology (ACATE), trade (FECOMERCIO), and industry (FIESC) opened their headquarters in Florianópolis.

Non-governmental organizations (NGOs) have been imprinting a culture of understanding on the city's problems. In 2005, FloripAmanhã was founded by a group of individuals aiming to contribute to strategies for sustainable development and building citizenship and social wellbeing. At the same time, ICOM was founded to promote community development by mobilizing, articulating, and supporting investors and social organizations. Since 2008, ICOM has been publishing an annual report called Sinais Vitais to inspire community participation and public debates.

A major governance challenge is providing safety and security. In parallel to elsewhere in Brazil, safety and security issues are at the forefront as they are highly impacting the lifestyle of everyone. The downward trend of the Brazilian economy since 2013 has had a tremendous impact on the crime rates. Florianópolis takes is share from this—although the city is still relatively much safer than elsewhere in Brazil. According to the State Department of Public Security, in Florianópolis, the homicide rate per 100,000 inhabitants is 11.7—an outcome of the bursting drug use and trafficking issue.

In Brazil, governments play a role in the creation of employment in the public sector. The reason for this is that providing employment (mostly low-income earners) would take those people from working for crime syndicates. While it serves the purpose of crime control, this creates a productivity problem for the public sector. Due to budgetary restrictions, the local government of Florianópolis faces serious difficulties in the generation of new jobs. Furthermore, according to the Ministry of Labor, 1510 formal jobs were lost in the city in 2017 as a result of the still incipient economic recovery of the services sector—that is, the city's flagship.

The city, however, still maintains one of the lowest unemployment rates among the national capitals due to higher-level income generating jobs in the technology sector, socioeconomic profile of a median-income city, and seasonal tourism sector jobs. In the first half of 2018, the average net monthly per capita income received was R$3225 (about US$775)—sixth among capitals, behind Vitória, Brasília, Porto Alegre, São Paulo, and Curitiba. This is explained by the high number of skilled knowledge workers—over 15% of employees have a university degree.

Observatories for urban mobility, social and social innovation, connected to federal and state universities, have been created to provide open and collaborative platforms to strengthen such respective areas. Despite the importance of creating indicators and making the city's diagnosis and planning, thus, producing relevant enablers for good governance, there is still a trend to articulate these initiatives towards specific and segmented guidelines. On the other hand, public service is dealing with the recent idea of developing a culture of horizontal engagement between citizen and government, demanding a new public service.

Aligned with this concept, the Network of Public Spaces (REP) is a recent initiative from Florianópolis Urban Planning Institute (IPUF), aiming to integrate the strategies of planning, intervention, and management of the city's public spaces, building partnerships with the community. However, there is a major challenge for institutional effectiveness of projects. The ineffectiveness issue is not solely unique to local governments; political discontinuity, misconduct, and corruption have their roots in all three-tier governments. Every four years, when elections are due, the city replaces an average of 700 senior positions.

Corruption and misconduct are major problems for the whole of Brazil, hampering its socioeconomic development. The misconduct issue is also evident in Florianópolis. One example is the Hercílio Luz Bridge. This engineering masterpiece was built in 1926 and closed in 1991 for safety concerns. After almost three decades, the restoration works of the bridge—through a partnership of the federal, state, and local governments—is still in progress, as shown in Figure 9. The bridge has been the symbol of the city for two reasons—for 'high aesthetics' and 'poor governance' of the city.

**Figure 9.** A view of the Hercílio Luz Bridge, reproduced with permission from Eduardo Trauer (photo taken in February 2018).

*3.5. Knowledge-Based Urban Development Progress and Challenges*

Its subtropical climate, island nature, pristine beaches, gourmet cuisine, and local culture transformed Florianópolis into a popular tourist destination in the 1990s. During recent years, an increasing concentration of high-tech companies provided an opportunity to turn Florianópolis into a smart innovation hub for Brazil. Today, Florianópolis is a prominent Brazilian city with ambitions to achieve thriving knowledge-based urban development and smart city transformation. Thus far, the city is recognized as an Innovation Capital of Brazil, and for some has the potential to become the 'Silicon Valley of South America' or 'Smart Floripa'.

The real challenge for the city, however, is finding a way to move the recognition of the city from a national level to international by placing it on the international map of innovation locations [89]. The city is willing to create a new path for a smart innovation island transformation to receive international attention. Its unique characteristics and achievements so far provide a possibility for developing such a pathway. However, these strengths and opportunities also come with counter weaknesses and challenges. For instance, Florianópolis has: (a) pristine but rapidly degrading natural environment; (b) existing but draining human capital; (c) increasing but inequitable income-levels; (d) diverse but incoherent community; (e) relative safety but raising serious security concerns; (f) numerous but conflicting planning schemes and regulations; and (g) established institutions but performing mostly ineffective governance.

The path Florianópolis takes in its journey from a tourist island to a smart innovation island is neither clear nor without major hurdles. Nevertheless, Florianópolis is still determined to achieve knowledge-based urban development and smart city transformation to realize such transformation. A priority seems to be so far given to making space and place for innovation companies and communities in the city. This approach is widely supported in the literature [90,91]. Likewise, some attempts also exist to create a smart innovation ecosystem—e.g., a municipal innovation system is currently being discussed—however, the complexity of governance mechanisms in such a smart innovation ecosystem is delaying the development process [92].

The findings of this study revealed the main prospects (or progress) and constraints (or challenges) in Florianópolis' knowledge-based urban development and smart city transformation journey. These findings are presented under the economic, societal, spatial, and governance domains in Tables 3 and 4. Additionally, the salient characteristics of the city in these development domains are presented in Appendix A, as shown in Table A1, and a timeline of Florianópolis' developmental milestones is illustrated in Appendix B, as shown in Figure 1.

**Table 3.** Florianopolis' knowledge-based urban development progress.

| Economy | |
|---|---|
| Condition | Implication |
| • Federal University of Santa Catarina (UFSC) research on maricultural and government policy on tourism and gastronomy | • Development of gastronomic tourism along with the sun-sand-sea tourism |
| • UNESCO Creative City of Gastronomy recognition | • Strong motivation in and reputation of the city |
| • Construction of event centers | • Expansion of tourism to off-season |
| • Tourism and gastronomy support institutions and workforce | • Tourism performance improvement in the recent years resulting in international recognition |
| • Technology and innovation companies that are locally fostered or nationally attracted | • Extension of tourism activities to business and event tourism |
| • Higher education system and several universities of the city | • Development of human capital and attraction of diverse student community |
| • Federal, state, and municipal legislation | • Promotion of innovation and technology sectors |
| • Entrepreneurs and early technology adopters | • Blue Ocean Strategy formulation |
| • Government incentive programs | • Emerging innovation ecosystem |
| • Unique nature and quality of life | • Attraction of visitors and talented knowledge workers |
| Society | |
| Condition | Implication |
| • Training availability generated employment opportunities in tourism sector for unskilled residents | • Positive contribution to employment levels resulting in improved social texture |

Table 3. *Cont.*

|  | Condition |  | Implication |
|---|---|---|---|
| ■ | Social protection networks provide help to the neediest—even though the support is limited it brings people above the poverty line | ■ | Slight improvement in the conditions of people living under poverty line, but the magnitude of the problem requires radical solutions |
| ■ | The city is an attractive location to live and work particularly for Brazilians that are moving to the city from other states for various reasons | ■ | Increasing inflow of migration from rest of the country bringing talent that is escaping from worsening safety conditions of elsewhere in Brazil |
| ■ | The city has several social organizations committed to help the most vulnerable—even though the funds are limited their existence is crucial | ■ | The magnitude of social problems is huge; however, social organizations are making positive changes in some individuals' lives |

**Environment**

| Condition | Implication |
|---|---|
| ■ Perfect climate and natural beauty of the city | ■ Attracting visitors and inflow migration |
| ■ Smart city has been pronounced in the city as a near future direction to follow—particularly there is a growing interest in smart city technologies | ■ There have been some attempts in consideration of smart city technology, however, nothing concrete developed |
| ■ Sustainability issue is at the forefront of policy discussions as it is critical for the future | ■ A zero-waste strategy has been adopted, however, it is too early to see the outcomes |
| ■ There are attempts to formalize some of the informal development—such as the UFSC campus—through new planning schemes | ■ While legalizing some of the informal buildings it could lead future development to follow formal procedures and standards |

**Governance**

| Condition | Implication |
|---|---|
| ■ Specialized industry associations working diligently to address main issues | ■ Improvements in company networking and clustering along with access to training |
| ■ Despite budgetary limitations local government is trying hard to deliver basic municipal services—however, at times there occur some interruptions | ■ Urban services and amenities are being delivered, even though they might not be timely and high quality at times |
| ■ Brazilian motto of 'order and progress' highlighting where the attention needs to be paid— safety-security and socioeconomic progress—is adopted by the city | ■ In both order and progress issues the city is ahead of many other locations in Brazil—despite the achievements, they are not enough |
| ■ Public sector was successful in creating and promoting the 'Magic Island' brand for the city, this was followed by the private sector initiative of the 'Innovation Capital of Brazil' brand | ■ Brand helped tourist perceptions and sector to develop and grow; however, a similar support has not been officially provided for the 'Innovation Capital of Brazil' brand |
| ■ Innovation route was a successful approach by the local government to link key innovation hubs | ■ The route increased networking and face-to-face interaction of the knowledge workers |

Table 4. Florianopolis' knowledge-based urban development challenges.

**Economy**

| Condition | Implication |
|---|---|
| ■ Low public investment and support in science, technology, innovation, and innovation ecosystem | ■ Hampering the growth potentials and keeping the reputation at the national level |
| ■ Lack of an adequate policy to support talented foreigner workforce | ■ The city mostly being able to attract talent from Brazil and Latin America |
| ■ Lot of work to be done for actors joining forces towards common goals, including sustainable economic activities | ■ Lack of a commonly agreed roadmap to achieve international reputation |
| ■ Lack of a big picture view thinking and required actions to turn city into a global innovation hub | ■ City reputation as an innovation island to remain in the national context only |
| ■ Delays in establishing and putting an effective innovation ecosystem in place | ■ Development not being planned and unable to foster, attract, and retain talent and investment |
| ■ The link between economic success and smart-sustainable development has not been clearly placed into practice | ■ Economy is advancing with ad hoc policies and practices, rather than with a conceptual framework of planned smart and sustainable economic progress |

**Society**

| Condition | Implication |
|---|---|
| ■ Socio-spatial segregation and inequality due to lack of social development and cohesion | ■ Decreasing community values and increasing inequality |

Table 4. Cont.

| Society | |
|---|---|
| Condition | Implication |
| - Socio-spatial segregation and inequality due to lack of social development and cohesion | - Decreasing community values and increasing inequality |
| - Inflow of low-income groups and not offering world-class education and skill development opportunities | - Contributing to the social disintegration problem and relatively lower education and training outcomes |
| - Limited support to the social protection networks | - Demand for further support or radical solutions |
| - Gentrification and inflow of middle/high income groups | - Weakening of cohesive community |
| - Worsening social conditions, including crime levels and economic recession | - Brain drain to overseas and very low levels of brain gain back |
| - While diversity in the community can be seen as an advantage, a clear socioeconomic gap exists between them | - Social cohesion is far from a reality, only during the carnival and football games and at beaches does society look united and cohesive |

| Environment | |
|---|---|
| Condition | Implication |
| - High population growth rate with lack of world-class urban infrastructure | - Adding on to the existing infrastructure and sustainability challenges |
| - Informal and favela developments | - Development with low standards and quality |
| - Increasing urbanization rate | - Threat to the natural ecosystems |
| - High inoccupancy housing rates despite of high housing demand | - Unsafe neighborhoods and rapidly weakening social functions |
| - Significant difference between peak summer tourism season and the rest of the year | - Overcapacity of roads, water, power, and sewerage during the peak season |
| - Informal land occupation | - Undermined environmental conservation/protection |
| - Housing deficit due to unavailability of providing land for demand | - Growth of informal housing settlements in risky or environmentally significant areas |
| - Control of deforestation and irregular occupation due to poor conservation practice | - At odds with existing environmental legislation—showing weakness at control and protection |
| - Lack of basic sanitation practice | - Pollution of rivers, sea, lagoons, and beaches |
| - Lack of efficient public transport system and high accident levels in a private motor vehicle dependent city | - Serious mobility and accessibility problems, high casualty accidents, unsafe city for cycling |
| - Poor and unsustainable transport infrastructure that brings traffic to standstill during the peak hours and peak tourism season | - Highly limits the mobility in the city, even including emergency services; hampers tourism as tourists cannot easily explore the island |

| Governance | |
|---|---|
| Condition | Implication |
| - Limited public, private, academia, community partnerships—quadruple-helix partnership | - Decreased effectiveness of plans and actions requiring broader collaboration |
| - Limited interdisciplinary and transdisciplinary research of local universities | - Limited innovativeness and university-industry collaboration |
| - Top university of the city lacks international reputation, education and skill development systems are below world-class | - UFSC's international university ranking is only 751 and has no research collaboration with global top-50 universities |
| - Lack of transparency, red tape, corruption in government practice and lobbying of influential organizations and individuals | - Limited ability of government to address local issues adequately due to poor, ineffective, and non-transparent governance |
| - Downturn of the economy since the end of the boom in early 2010s | - Increasing poverty and resulting unrest and sky rocketing crime rates |
| - Limited state government financial support to science, technology, and innovation | - Growth in the innovation sector is at a slow pace and lack of substantial incentives |
| - Lack of an official city innovation ecosystem and roadmap | - Planning of actions for the innovation sector growth is at best ad-hoc |
| - Lack of an adequate government policy to support talented foreigner workforce | - Highly limited attraction rate of foreign talent and resulting limited foreign direct investment |
| - Lack of strategic vision in the planning mechanism and plan and policy implementation | - Having ineffective plans and regulations, at times conflicting with each other |
| - Lack of development control and demolition power of the informal structures | - Encouraging informal developments and land speculation |

The study discloses three key policy priorities for each of the development domains. In terms of economic development, the top three priorities are: (a) investing in science, technology, and innovation; (b) creating an effective and efficient urban innovation ecosystem; (c) fostering sustainable and knowledge-based economic activities. In terms of societal development, the top three priorities are: (a) promoting social development and community cohesion; (b) establishing social protection networks particularly for the neediest; (c) improving the quality of the education, learning, and skill development system. In terms of spatial development, the top three priorities are: (a) conserving the pristine ecosystems of the island; (b) building world-class smart urban infrastructure; (c) establishing sustainable development and particularly sustainable mobility. In terms of institutional development and governance, the top three priorities are: (a) building effective, participatory, and transparent governance practice and support mechanisms; (b) planning strategically and acting upon for a timely implementation; (c) enabling quadruple-helix partnerships in key projects.

In the light of the research findings, the study also proposes Florianópolis to adopt a locally customized framework of knowledge-based urban development—where it can be used as a powerful instrument to facilitate an urban turnaround, bring prosperity and sustainability to the city, and form the geographies of disruption—such as smart cities and innovation districts [93]. This framework builds on the generic knowledge-based urban development framework presented in Figure 2, and specifically identifies the critical attributes—i.e., the top priorities in each domain—that are particularly required to put Florianópolis on the way to a thriving transformation towards a smart innovation island. Figure 10 illustrates this locally-customized framework.

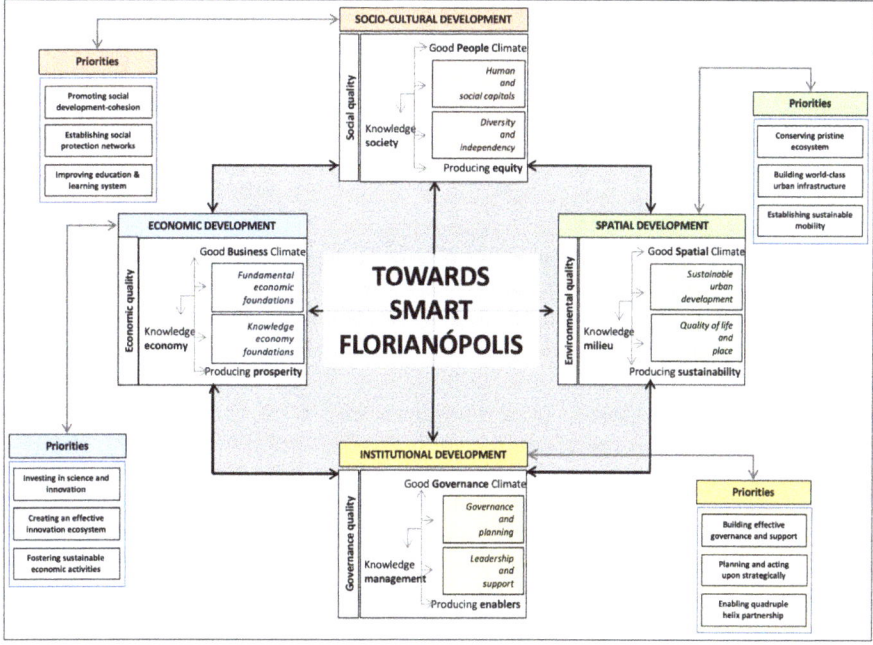

**Figure 10.** A framework for knowledge-based development of Florianópolis, derived from Reference [46].

## 4. Discussion and Conclusions

This paper focused on addressing the question of 'what it takes to transform a tourist island into an innovation capital' by placing Florianópolis—a.k.a. Smart Floripa—under the knowledge-based urban development microscope. The results of the study acknowledged the recent achievements of the city in establishing a promising innovation hub/capital for the nation, and by doing so diversifying the local economy—with inclusion of newly created knowledge-based activities to the traditionally dominant tourism and public service ones. Even though the accomplishments are to a level commendable, as shown in Table 3, the study also highlighted a large number of major challenges and their potential implications that the city needs to tackle without much delay, as shown in Table 4.

In sum, the findings suggested that particularly the municipal planning for the near-term needs to seek solutions to the most urgent problems—such as those related to environmental conservation, public security, urban mobility, and infrastructural improvements—and they need to be novel and sustainable solutions. The city has also a willingness to adopt technological tools as part of the smart solution. However, given the budgetary limitations and ongoing deficits in governmental agencies, urban administrators should keep in mind that smart city technology does not necessarily need to be new to be effective. Particularly in the developing country cities—such as Florianópolis—the most effective solutions often involve innovative uses of existing technologies at low cost [94]. The findings of the study also pointed out the serious governance hiccups of the nation that infiltrate into the state and municipal level politics as the biggest challenge to deal with.

While the knowledge-based activities are promising in Florianópolis, the same cannot be said for Brazil. For instance, the Global Innovation Index 2018 ranked Brazil as the 64th nation out of 126 [93]. This figure gets even worse when Brazil is compared with OECD countries—37th out of 37, the weakest performance. A closer look at the indicators of the index highlights the areas where performance is particularly suffering. In the following areas Brazil is ranked outside the top-80 nations: (a) political environment; (b) business environment; (c) tertiary education; (d) general infrastructure; (e) credit; (f) investment; (g) knowledge impact; (h) creative goods and services [95]. Unfortunately, Florianópolis also shows similar weaknesses in most of these areas—indicating a national path dependency.

The challenges are great for Florianópolis, but the path to solutions, from the identification of bottlenecks to the specifications of policy directions, has already been found. Putting them into action with a good governance practice is now required, which is the greatest challenge in any locality in Brazil. Brazil's National Museum's recent fire disaster in Rio de Janeiro is an example of poor management/governance practices of the authorities. Almost the entire collection of Latin America's largest museum and 200 years of work was burned to ashes. The collection was not insured, no smoke detectors were installed, and fire hydrants ran out of water. Nevertheless, Florianópolis is willing to address the governance challenges and showcase an exemplar performance. Is this willingness enough? Only time will tell.

Florianópolis achieved its goal of becoming a tourist island in the 1990s. In recent years, it has become the innovation capital of Brazil. Recently for the city, becoming a smart and sustainable city seems to be the next goal. However, this is the most challenging target the city has ever chosen for itself. The reason for this is simple; becoming smart and sustainable requires the entire city to meet this target. In other words, more or less only having natural beauty could lead to a success story in tourism. Similarly, in isolated exclusive locations of a city innovation activities could be performed and this could also lead to a recognition or claiming success in this area. However, a smart and sustainable city is a different animal. The achievements in this type of city require success in all knowledge-based urban development domains—economy, society, environment, and governance. Besides, these cities are inclusive and resilient [96,97]. Furthermore, the recent election of Brazil's new far-right president—Jair Bolsonaro—will likely inflate the sustainability related problems across the nation (and even globally). For instance, his views on the exploitation of the Amazon rainforest—often called the 'Lungs of the World' as it produces more than 20% of the world's oxygen—has already created serious concerns. The top-level political climate change denial rhetoric, and associated actions and malpractices will have inevitable consequences on Florianópolis' progress as well as the entire nation.

The results of this case report disclosed that despite some knowledge-based urban development achievements, Florianópolis' is currently neither 'smart' nor 'sustainable'. In the light of this finding, we suggest the use of a more realistic brand for the city: 'Towards Smart Florianópolis' or 'Towards Smart Floripa'. Moreover, adoption of a customized knowledge-based urban development framework—such as in Figure 10—would also be useful to the city to strategically plan its priorities for development. The findings also revealed insights on the new path creation processes and attempts of Florianópolis. In other words, the recipe of what it takes to transform a tourist island into an innovation capital has been disclosed. These insights are useful for local authorities and stakeholders of Florianópolis to consider and develop new strategies and find implementation grounds for a smart transformation. Furthermore, the insights might also be helpful for other developing country or emerging economy locations—aiming for aspiring knowledge-based urban development and smart city transformation along with generating socioeconomic and spatial resilience and good governance. Tourist cities or islands with similar characteristics could particularly benefit the most from the experiences of Florianópolis—even though the journey to a smart city has only just began, and there is a long and tough road ahead for Florianópolis.

Overall, this is a pioneering study, as it scrutinized the development journey of an ambitious city—that is constantly seeking to progress further—thoroughly from the lens of a knowledge-based urban development paradigm by engaging key local actors in the investigation. This paper revealed invaluable insights particularly into: (a) how to determine the strengths and weaknesses of a city by using a knowledge-based urban development framework; (b) how to engage key actors to discuss issues around the future development of their city as a smart, sustainable, and knowledge-based one; and (c) how to develop a customized framework for the knowledge-based development of a city to guide its future policy and actions. The learnings from the experiences of Florianópolis, and abovementioned insights, can inform the policymaking circles in Florianópolis, and other cities.

The following limitations of the study should also be considered while interpreting the findings: (a) the literature search could only find a moderate number of writings on the city and its relevant developmental issues; (b) the analysis conducted is descriptive and qualitative in nature, no thorough statistical elaborations were carried out; (c) there might be some possible unintended bias in conducting interviews with focus group members, the process of data analysis, and the method of data collection; (d) while public and academic sectors' inputs were collected directly, private sector's voices were only captured indirectly through the non-governmental organizations that participated in the study, and community perceptions were not considered; (e) a relatively moderate but acceptable number of local actors participated in the study, their potential unconscious bias might have affected the results. Our prospective studies will focus on addressing these limitations, and conducting further empirical investigations—by using both qualitative and quantitative techniques to measure the innovation and smartness performance of the city in comparative studies. Furthermore, our future research will also concentrate on developing a comprehensive 'smart city strategy' for the city to guide Florianópolis' smart and sustainable urbanism endeavors.

**Author Contributions:** T.Y. designed and led the research; T.Y., J.S.-M., A.F., C.L., N.B., T.W., and T.S. jointly prepared the manuscript; T.Y. and A.F. revised the manuscript. All authors read and approved the final version of the manuscript.

**Funding:** This research is an outcome of a project entitled 'Smart Floripa: Strategizing the Transformation of Florianópolis into a Smart Innovation Island'. The project received financial and in-kind support from the following project partners: Governo do Estado de Santa Catarina, Prefeitura Municipal de Florianópolis, Fecomércio Santa Catarina, Senac Santa Catarina, Associação Brasileira das Empresas de Software, Instituto Lixo Zero Brasil, Federal University of Santa Catarina, University of Caxias do Sul, and Queensland University of Technology.

**Acknowledgments:** The authors cordially thank: abovementioned project partners for the financial and in-kind support; focus group participants for sharing their invaluable knowledge and time; Eduardo Trauer and Eduardo Zappia for taking photographs used in the manuscript; Susan Yigitcanlar for native speaker English editing of the manuscript; the managing editor of the journal and four anonymous referees for their constructive comments on an earlier version of the manuscript.

**Conflicts of Interest:** The authors declare no conflict of interest.

## Appendix A

**Table A1.** Salient characteristics of Florianópolis, collected from various data sources.

| Domain | Characteristic | Description | Value | Unit |
|---|---|---|---|---|
| Geo-demographic | Population | Total resident population | 453,285 | ppl |
| | Area | Total area | 436.5 | km² |
| | Density | Population density | 1038.45 | ppl/km² |
| | Age | Median age | 33 | years |
| | Climate | Climate category | subtropical | type |
| | Economy | Dominant economic sector | service | type |
| Economic development | Gross domestic product | Gross domestic product (GDP) per capita | 39,678 | R$ |
| | Major companies | Number of top-100 Brazilian companies hosted | 1 | count |
| | Foreign direct investment | Ratio of international share in foreign direct investments | 0.375 | ratio |
| | Urban competitiveness | Urban competitiveness index ranking | 9th | national rank |
| | Innovation economy | City ranking in innovation economy | 3rd | national rank |
| | Research and development | Ratio of research and development expenditure in GDP | 0.002 | ratio |
| | Patent applications | Patent cooperation treaty patent applications per million inhabitants | 0.110 | ratio |
| | Knowledge worker pool | Ratio of professionals and managers in all workers | 0.152 | ratio |
| Societal development | Education investment | Ratio of public spending on education in GDP | 0.045 | ratio |
| | Professional skill base | Ratio of 18+ years old people with tertiary education (bachelor degree or above) | 0.163 | ratio |
| | University prestige | Ranking of the local top ranked university (UFSC) | 751st/9th | inter/national rank |
| | Mobile broadband | Ratio of mobile broadband subscribers in 18+ years old population | 0.546 | ratio |
| | Cultural diversity | Ratio of people that were born abroad | 0.008 | ratio |
| | Social cohesion and equality | Level of income inequality (Gini coefficient) | 0.547 | ratio |
| | Socio-economic dependency | Ratio between the elderly population and the working age (15-64 years) | 0.340 | ratio |
| | Unemployment level | Unemployment rate | 0.075 | ratio |
| Spatial development | Smart city formation | City ranking in smart city achievements | 6th | national rank |
| | Sustainable transport use | Ratio of sustainable transport mode use in commuting population | 0.010 | ratio |
| | Environmental impact | Per capita carbon dioxide emissions | 2.17 | tCO$_2$e/ppl |
| | Urban form | Ratio of informal buildings in all buildings | 0.504 | ratio |
| | Quality of life | City ranking in quality of life | 2nd | national rank |
| | Cost of living | City ranking in cost of living | 8th | national rank |
| | Housing affordability | Ratio of GDP per capita to median dwelling price | 0.073 | ratio |
| | Personal safety | City ranking in personal safety | 70th | national rank |
| Institutional development | Government effectiveness | Level of government effectiveness in achieving knowledge-based urban development | low | Likert-scale |
| | Electronic governance | Level of online applications for public services (vote, tax, payment, renewal) | low | Likert-scale |
| | Strategic planning | Level of knowledge-based urban development strategies in strategic development plans | none | Likert-scale |
| | City branding | Level of deliberately created city branding and resulting reputation | medium | Likert-scale |
| | Effective leadership | Level of institutional leadership in overseeing knowledge-based urban development | none | Likert-scale |
| | Strategic partnership | Level of effective triple-helix and public-private-partnerships | medium-low | Likert-scale |
| | Community engagement | Level of institutional mechanisms for community development and participation | medium-low | Likert-scale |
| | Infrastructure and amenities | Level of effective infrastructure and amenity development and maintenance | medium-low | Likert-scale |

B.

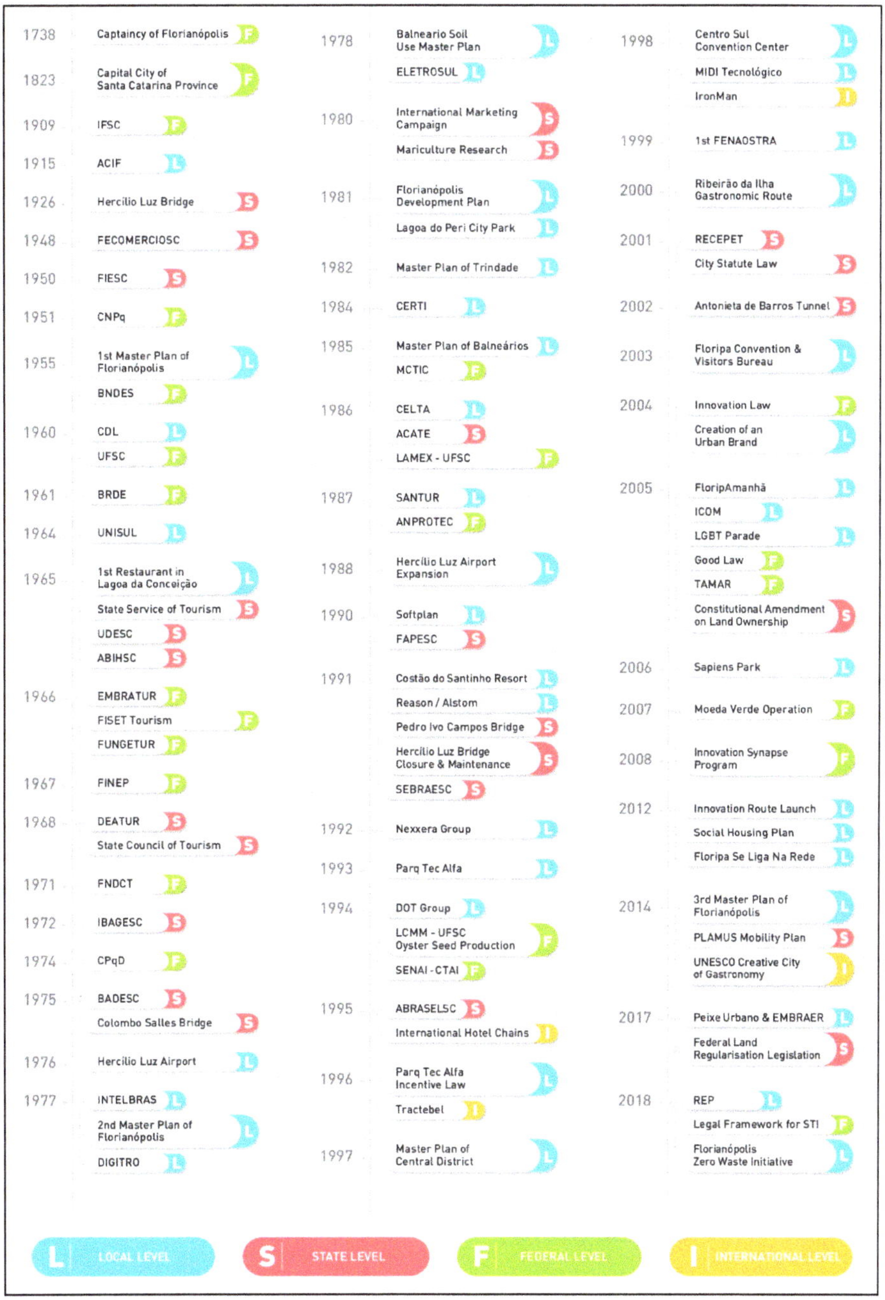

| Year | Milestone | Level |
|---|---|---|
| 1738 | Captaincy of Florianópolis | F |
| 1823 | Capital City of Santa Catarina Province | F |
| 1909 | IFSC | F |
| 1915 | ACIF | L |
| 1926 | Hercílio Luz Bridge | S |
| 1948 | FECOMERCIOSC | S |
| 1950 | FIESC | S |
| 1951 | CNPq | F |
| 1955 | 1st Master Plan of Florianópolis | L |
| | BNDES | F |
| 1960 | CDL | L |
| | UFSC | F |
| 1961 | BRDE | F |
| 1964 | UNISUL | L |
| 1965 | 1st Restaurant in Lagoa da Conceição | L |
| | State Service of Tourism | S |
| | UDESC | S |
| | ABIHSC | S |
| 1966 | EMBRATUR | F |
| | FISET Tourism | F |
| | FUNGETUR | F |
| 1967 | FINEP | F |
| 1968 | DEATUR | S |
| | State Council of Tourism | S |
| 1971 | FNDCT | F |
| 1972 | IBAGESC | S |
| 1974 | CPqD | F |
| 1975 | BADESC | S |
| | Colombo Salles Bridge | S |
| 1976 | Hercílio Luz Airport | L |
| 1977 | INTELBRAS | L |
| | 2nd Master Plan of Florianópolis | L |
| | DIGITRO | L |
| 1978 | Balneario Soil Use Master Plan | L |
| | ELETROSUL | L |
| 1980 | International Marketing Campaign | S |
| | Mariculture Research | S |
| 1981 | Florianópolis Development Plan | L |
| | Lagoa do Peri City Park | L |
| 1982 | Master Plan of Trindade | L |
| 1984 | CERTI | L |
| 1985 | Master Plan of Balneários | L |
| | MCTIC | F |
| 1986 | CELTA | L |
| | ACATE | S |
| | LAMEX - UFSC | F |
| 1987 | SANTUR | L |
| | ANPROTEC | F |
| 1988 | Hercílio Luz Airport Expansion | L |
| 1990 | Softplan | L |
| | FAPESC | S |
| 1991 | Costão do Santinho Resort | L |
| | Reason / Alstom | L |
| | Pedro Ivo Campos Bridge | S |
| | Hercílio Luz Bridge Closure & Maintenance | S |
| | SEBRAESC | S |
| 1992 | Nexxera Group | L |
| 1993 | Parq Tec Alfa | L |
| 1994 | DOT Group | L |
| | LCMM - UFSC Oyster Seed Production | F |
| | SENAI - CTAI | F |
| 1995 | ABRASELSC | S |
| | International Hotel Chains | I |
| 1996 | Parq Tec Alfa Incentive Law | L |
| | Tractebel | I |
| 1997 | Master Plan of Central District | L |
| 1998 | Centro Sul Convention Center | L |
| | MIDI Tecnológico | L |
| | IronMan | I |
| 1999 | 1st FENAOSTRA | L |
| 2000 | Ribeirão da Ilha Gastronomic Route | L |
| 2001 | RECEPET | S |
| | City Statute Law | S |
| 2002 | Antonieta de Barros Tunnel | L |
| 2003 | Floripa Convention & Visitors Bureau | L |
| 2004 | Innovation Law | F |
| | Creation of an Urban Brand | L |
| 2005 | FloripAmanhã | L |
| | ICOM | L |
| | LGBT Parade | L |
| | Good Law | F |
| | TAMAR | F |
| | Constitutional Amendment on Land Ownership | S |
| 2006 | Sapiens Park | L |
| 2007 | Moeda Verde Operation | F |
| 2008 | Innovation Synapse Program | F |
| 2012 | Innovation Route Launch | L |
| | Social Housing Plan | L |
| | Floripa Se Liga Na Rede | L |
| 2014 | 3rd Master Plan of Florianópolis | L |
| | PLAMUS Mobility Plan | S |
| | UNESCO Creative City of Gastronomy | I |
| 2017 | Peixe Urbano & EMBRAER | L |
| | Federal Land Regularisation Legislation | S |
| 2018 | REP | L |
| | Legal Framework for STI | F |
| | Florianópolis Zero Waste Initiative | L |

L: LOCAL LEVEL  S: STATE LEVEL  F: FEDERAL LEVEL  I: INTERNATIONAL LEVEL

**Figure 1.** Timeline of Florianopolis' developmental milestones, authors' own compilation.

C.

Table 2. Reviewed literature.

| No | Literature |
|---|---|
| [8] | Zouain, D.M.; Plonski, G.A. Science and technology parks: Laboratories of innovation for urban development-an approach from Brazil. *Triple Helix* **2015**, *2*, 7, doi:10.1186/s40604-015-0018-1. |
| [29] | Barufi, A.M.; Haddad, E.A.; Nijkamp, P. Industrial scope of agglomeration economies in Brazil. *Ann. Reg. Sci.* **2016**, *56*, 707–755, doi:10.1007/s00168-016-0768-3. |
| [35] | Guerra, J.; Knabben, J.; Fernandez, F.; Bailey, C.; Barbosa, S.; Neiva, S. Reprint of: The adoption of strategies for sustainable cities: A comparative study between Newcastle and Florianópolis focused on urban mobility. *J. Clean. Prod.* **2017**, *163*, S209–S222, doi:10.1016/j.jclepro.2017.05.142. |
| [37] | Michelmann, A.C. Franklin Cascaes, a Divulgação Turística de Florianópolis e a Invenção da 'Ilha da Magia'. Unpublished Bachelor Thesis, Federal University of Santa Catarina, Florianópolis, Brazil, 2017. |
| [38] | Azevedo, I.; Teixeira, C.; Teixeira, M. CELTA e MIDI Tecnológico: Um estudo de caso das incubadoras de Florianópolis. In Proceedings of the International Congress on Research & Development, Curitiba, Brazil, 21–23 September 2016; pp. 347–363. |
| [49] | Lenzi, M.H. A Invenção de Florianópolis Como Cidade Turística: Discursos, Paisagens e Relações de Poder. Unpublished Doctoral Thesis, University of São Paulo, São Paulo, Brazil, 2016. |
| [50] | Pereira, E.M. A importância de conceitos modernistas no planejamento urbano de Florianópolis. *Seminário de Historia da Cidade e do Urbanismo* **2000**, *6*, 1–15. |
| [51] | Machado, E.V. Florianópolis: Um Lugar em Tempo de Globalização. Unplublished Doctoral Thesis. Federal University of Santa Catarina, Florianópolis, Brazil, 2000. |
| [52] | Barreto, M.; Burgos, R.; Frenkel, D. *Turismo, Políticas Públicas e Relações Internacionais*; Papirus: Campinas, Brazil, 2003; ISBN 8530807154. |
| [53] | Oliveira, A.P. *A História do Turismo em Florianópolis Narrada por Quem a Vivenciou (1950–2010)*; Palavracom Editora: Florianópolis, Brazil, 2011; ISBN 9788564034051. |
| [54] | De Luca Filho, V. A Geografia das Feiras de Negócios em Santa Catarina: Origem, Evolução e Dinâmica das Transformações. Unpublished Doctoral Thesis, Federal University of Santa Catarina, Florianópolis, Brazil, 2014. |
| [55] | Santos, F.M.; Pereira, R.M. A rede hoteleira no núcleo urbano central de Florianópolis: Expansão urbana e turismo. In Proceedings of the IV Mercosur Tourism Research Seminar, Caxias do Sul, Brazil, 7–8 July 2006; pp. 1–15. |
| [56] | Paulilo, M.I. Maricultura e território em Santa Catarina, Brasil. *Geosul* **2002**, *17*, 87–112. |
| [57] | Azevedo, I.; Teixeira, C. Florianópolis: Uma análise evolutiva do desenvolvimento do desenvolvimento inovador da cidade a partir do seu ecossistema de inovação. *REAVI* **2017**, *6*, 108–121, doi:10.5965/2316419006092017108. |
| [59] | Xavier, M. *Polo Tecnológico de Florianópolis: Origem e Desenvolvimento*; Insular: Florianópolis, Brazil, 2010; ISBN 9788574744957. |
| [64] | Tarachucky, L. Sistematização da Aplicação do Brand DNA Process No Design de Marca de Cidades Criativas: Caso Projeto Rota da Inovação. Unpublished Master Thesis. Federal University of Santa Catarina, Florianópolis, Brazil, 2015. |
| [67] | Ueno, A. A Concepção de um Modelo de Empreendedorismo Inovador Baseado em Conhecimento: Um Estudo de Caso do Programa Sinapse da Inovação. Unpublished Master Thesis, Federal University of Santa Catarina, Florianópolis, Brazil, 2011. |
| [68] | Depine, A.C. Fatores de Atração e Retenção da Classe Criativa: O Potencial de Florianópolis Como Cidade Humana Inteligente. Unpublished Master Thesis. Federal University of Santa Catarina, Florianópolis, Brazil, 2016. |
| [69] | Fernandes, E. The legalisation of favelas in Brazil: Problems and prospects. *Third World Plan. Rev.* **2000**, *22*, 167, doi:10.3828/twpr.22.2.f765q34rm4600773. |
| [70] | Prefeitura Municipal de Florianópolis (PMF). *Municipal Plan of Housing of Social Interest of Florianópolis*; PMF: Florianópolis, Brazil, 2012. |
| [71] | Prefeitura Municipal de Florianópolis (PMF). *Florianópolis Sustainability Action Plan*; PMF: Florianópolis, Brazil, 2017. |
| [73] | Prefeitura Municipal de Florianópolis (PMF). *Florianópolis Urban Growth Study*; PMF: Florianópolis, Brazil, 2017. |
| [74] | Prefeitura Municipal de Florianópolis (PMF). *Florianópolis Mitigation and Climate Change Study*; PMF: Florianópolis, Brazil, 2017. |
| [75] | Oliveira-Musse, J.; Homrich, A.S.; Mello, R.; Carvalho, M.M. Applying backcasting and system dynamics towards sustainable development: The housing planning case for low-income citizens in Brazil. *J. Clean. Prod.* **2018**, *193*, 97–114, doi:10.1016/j.jclepro.2018.04.219. |
| [81] | Bridges, A. Leveraging Amenity-Led Growth and Collective Action for Sustainable Development in Florianópolis, 1965–2016. Unpublished Doctoral Thesis, Rutgers University, New Brunswick, NJ, USA, 2015. |
| [82] | Scherer, M.E.; Asmus, M.L. Ecosystem-based knowledge and management as a tool for integrated coastal and ocean management: A Brazilian initiative. *J. Coast. Res.* **2016**, *75*, 690–694, doi:10.2112/SI75-138.1. |
| [83] | Prefeitura Municipal de Florianópolis (PMF). *Floripa Se Liga Na Rede*; PMF: Florianópolis, Brazil, 2013. |
| [84] | Figueiroa, A.; Scherer, M. Para onde estamos indo? Uma avaliação do plano diretor do Município de Florianópolis para o entorno da Estação Ecológica de Carijós. *Desenvolvimento e Meio Ambiente* **2016**, *38*, 283–301, doi:10.5380/dma.v38i0.47110. |
| [85] | Bastos, G. Análise Financeira das Pescarias de Pequena escala No Município de Florianópolis. Unpublished Doctoral Thesis, University of São Paulo, São Paulo, Brazil, 2009. |
| [86] | Stahelin, G.D.; Fiedler, F.N.; Lima, E.P.; Sales, G.; Wanderlinde, J. Projeto Tamar's station in Florianópolis, State of Santa Catarina, Southern Brazil. *Mar. Turt. Newslett.* **2012**, *133*, 23. |
| [98] | Lara, A.P.; Marques, J.S.; Santos, N.; Costa, E.M. Projeto Florip@ 21: A construção de uma região inteligente na cidade de Florianópolis, Brasil. In Proceedings of the XV Latin Iberian-American Congress on Management of Technology, Porto, Portugal, 27–31 October 2013; pp. 1673–1691. |
| [99] | Lorenzetti, J.; de Lima Trindade, L.; Pires de Pires, D.E.; Souza Ramos, F.R. Tecnologia, inovação tecnológica e saúde: Uma reflexão necessária. *Texto & Contexto Enfermagem* **2012**, *21*, 432–439. |
| [100] | Martins, C. O Papel das Incubadoras de Empresas do Polo Tecnológico de Florianópolis No Desenvolvimento do Processo de Empreendedorismo Inovador. Unpublished Master Thesis, Universidade do Sul de Santa Catarina, Florianópolis, Brazil, 2013. |
| [101] | Sarquis, A.B.; Fiates, G.G.; Hahn, A.K.; Cavalcante, F.R. Empreendedorismo inovador no polo tecnológico de Florianópolis. *REEN* **2014**, *7*, 228–255, doi:10.19177/reen.v7e32014228-255. |
| [102] | Gaspar J.V.; Menegazzo, C.; Fiates, J.E.; Teixeira, C.S.; Gomes, L.S. A revitalização de espaços urbanos: O case do Centro Sapiens em Florianópolis. *Revista Livre de Sustentabilidade e Empreendedorismo* **2017**, *29*, 183–205. |
| [103] | Menezes, A.G.; Lezana, Á.G.; de Abreu-Ronconi, L.F.; de Oliveira-Menezes, E.C.; de Melo, É.N. A pesquisa-ação como estratégia de avaliação da inovação social: Estudo de uma entidade educacional do município de Florianópolis. *NAVUS Revista de Gestão e Tecnologia* **2016**, *6*, 93–105. |

## References

1. Simmie, J.; Martin, R. The economic resilience of regions: Towards an evolutionary approach. *Camb. J. Reg. Econ. Soc.* **2010**, *3*, 27–43. [CrossRef]
2. Briguglio, L.; Cordina, G.; Farrugia, N.; Vella, S. Economic vulnerability and resilience: Concepts and measurements. *Oxf. Dev. Stud.* **2009**, *37*, 229–247. [CrossRef]
3. Yigitcanlar, T. Planning for smart urban ecosystems: Information technology applications for capacity building in environmental decision making. *Theor. Empir.* **2009**, *4*, 5–21.
4. Yigitcanlar, T.; Dizdaroglu, D. Ecological approaches in planning for sustainable cities: A review of the literature. *Glob. J. Environ. Sci. Manag.* **2014**, *1*, 159–188. [CrossRef]
5. Ghaderi, Z.; Som, A.P.; Henderson, J.C. Tourism crises and island destinations: Experiences in Penang, Malaysia. *Tour. Manag. Perspect.* **2012**, *2*, 79–84. [CrossRef]
6. Scheyvens, R.; Momsen, J. Tourism in small island states: From vulnerability to strengths. *J. Sustain. Tour.* **2008**, *16*, 491–510. [CrossRef]
7. Hadjimanolis, A.; Dickson, K. Development of national innovation policy in small developing countries: The case of Cyprus. *Res. Pol.* **2001**, *30*, 805–817. [CrossRef]
8. Zouain, D.M.; Plonski, G.A. Science and technology parks: Laboratories of innovation for urban development-an approach from Brazil. *Triple Helix* **2015**, *2*, 7. [CrossRef]
9. Jones, C.; Munday, M. Exploring the environmental consequences of tourism: A satellite account approach. *J. Travel Res.* **2007**, *46*, 164–172. [CrossRef]
10. Cooke, P.; Leydesdorff, L. Regional development in the knowledge-based economy: The construction of advantage. *J. Technol. Transf.* **2006**, *31*, 5–15. [CrossRef]
11. Evers, H.D.; Gerke, S.; Menkhoff, T. Knowledge clusters and knowledge hubs: Designing epistemic landscapes for development. *J. Knowl. Manag.* **2010**, *14*, 678–689. [CrossRef]
12. Cassingena-Harper, J.; Georghiou, L. The targeted and unforeseen impacts of foresight on innovation policy: The eFORESEE Malta case study. *Int. J. Foresight. Innov. Policy* **2005**, *2*, 84–103. [CrossRef]
13. Cabrer-Borras, B.; Serrano-Domingo, G. Innovation and R&D spillover effects in Spanish regions: A spatial approach. *Res. Pol.* **2007**, *36*, 1357–1371. [CrossRef]
14. Chertow, M.R.; Ashton, W.S.; Espinosa, J.C. Industrial symbiosis in Puerto Rico: Environmentally related agglomeration economies. *Reg. Stud.* **2008**, *42*, 1299–1312. [CrossRef]
15. Yigitcanlar, T.; Inkinen, T.; Makkonen, T. Does size matter? Knowledge-based development of second-order city-regions in Finland. *DISP* **2015**, *51*, 62–77. [CrossRef]
16. Yigitcanlar, T. *Technology and the City: Systems, Applications and Implications*; Routledge: New York, NY, USA, 2016; ISBN 9781317575696.
17. Baum, S.; O'Connor, K.; Yigitcanlar, T. The implications of creative industries for regional outcomes. *Int. J. Foresight. Innov. Policy* **2009**, *5*, 44–64. [CrossRef]
18. Millar, C.C.; Ju Choi, C. Development and knowledge resources: A conceptual analysis. *J. Knowl. Manag.* **2010**, *14*, 759–776. [CrossRef]
19. Lönnqvist, A.; Käpylä, J.; Salonius, H.; Yigitcanlar, T. Knowledge that matters: Identifying regional knowledge assets of the Tampere region. *Eur. Plan. Stud.* **2014**, *22*, 2011–2029. [CrossRef]
20. Pancholi, S.; Yigitcanlar, T.; Guaralda, M. Urban knowledge and innovation spaces: Concepts, conditions and contexts. *Asia Pac. J. Innov. Entrep.* **2014**, *8*, 15–38.
21. Cowley, R.; Caprotti, F. Smart city as anti-planning in the UK. *Environ. Plan. D* **2018**. [CrossRef]
22. Yigitcanlar, T.; Metaxiotis, K.; Carrillo, F.J. (Eds.) *Building Prosperous Knowledge Cities: Policies, Plans and Metrics*; Edward Elgar: Cheltenham, UK, 2012; ISBN 9780857936035.
23. Yin, R.K. *Case Study Research, Design and Method*; Sage: Newbury Park, CA, USA, 2009; ISBN 9781412960991.
24. Glaser, B.G.; Strauss, A.L. *The Discovery of Grounded Theory*; Aldine: Chicago, IL, USA, 1967; ISBN 0202302601.
25. Eisenhardt, K.M. Building theories from case study research. *Acad. Manag. Rev.* **1989**, *14*, 532–550. [CrossRef]
26. Carrillo, F.J.; Yigitcanlar, T.; García, B.; Lönnqvist, A. *Knowledge and the City: Concepts, Applications and Trends of Knowledge-Based Urban Development*; Routledge: New York, NY, USA, 2014; ISBN 1317931378.
27. Yigitcanlar, T.; Edvardsson, I.; Johannesson, H.; Kamruzzaman, M.; Ioppolo, G.; Pancholi, S. Knowledge-based development dynamics in less favoured regions: Insights from Australian and Icelandic university towns. *Eur. Plan. Stud.* **2017**, *25*, 2272–2292. [CrossRef]

28. Hyde, K.F. Recognising deductive processes in qualitative research. *Qual. Mark. Res. Int. J.* **2000**, *3*, 82–90. [CrossRef]
29. Barufi, A.M.; Haddad, E.A.; Nijkamp, P. Industrial scope of agglomeration economies in Brazil. *Ann. Reg. Sci.* **2016**, *56*, 707–755. [CrossRef]
30. Barufi, A.M. Services that add value in the city: The rise of the modern economy in Brazil. *Cities* **2018**, *78*, 39–51. [CrossRef]
31. Mais, I.; de Carvalho, L.C.; Mohamed, A.; Hoffmann, M.G. The role of network relationships in innovation and internationalization of technology-based companies. *RAI* **2010**, *7*, 41. [CrossRef]
32. Capo, J.; Font, A.R.; Nadal, J.R. Dutch disease in tourism economies: Evidence from the Balearics and the Canary Islands. *J. Sustain. Tour.* **2007**, *15*, 615–627. [CrossRef]
33. Dalmarco, G.; Hulsink, W.; Blois, G.V. Creating entrepreneurial universities in an emerging economy: Evidence from Brazil. *Technol. Forecast. Soc. Chang.* **2018**. [CrossRef]
34. Instituto Brasileiro de Geografia e Estatísticas (IBGE). *Santa Catarina e Florianópolis*; IBGE: Rio de Janeiro, Brazil, 2014.
35. Guerra, J.; Knabben, J.; Fernandez, F.; Bailey, C.; Barbosa, S.; Neiva, S. Reprint of: The adoption of strategies for sustainable cities: A comparative study between Newcastle and Florianópolis focused on urban mobility. *J. Clean. Prod.* **2017**, *163*, S209–S222. [CrossRef]
36. Google Maps. Available online: https://goo.gl/maps/DycdnkEoDev (accessed on 14 August 2018).
37. Michelmann, A.C. Franklin Cascaes, a Divulgação Turística de Florianópolis e a Invenção da 'Ilha da Magia'. Unpublished Bachelor Thesis, Federal University of Santa Catarina, Florianópolis, Brazil, 2017.
38. Azevedo, I.; Teixeira, C.; Teixeira, M. CELTA e MIDI Tecnológico: Um estudo de caso das incubadoras de Florianópolis. In Proceedings of the International Congress on Research & Development, Curitiba, Brazil, 21–23 September 2016; pp. 347–363.
39. Bask, A.; Rajahonka, M. The role of environmental sustainability in the freight transport mode choice: A systematic literature review with focus on the EU. *Int. J. Phys. Distrib. Manag.* **2017**, *47*, 560–602. [CrossRef]
40. Sarimin, M.; Yigitcanlar, T. Towards a comprehensive and integrated knowledge-based urban development model: Status quo and directions. *Int. J. Knowl.-Based Dev.* **2012**, *3*, 175–192. [CrossRef]
41. Yin, R.K. Discovering the future of the case study: Method in evaluation research. *Eval. Pract.* **1994**, *15*, 283–290. [CrossRef]
42. Pancholi, S.; Yigitcanlar, T.; Guaralda, M. Governance that matters: Identifying place-making challenges of Melbourne's Monash Employment Cluster. *J. Place Manag. Dev.* **2017**, *10*, 73–87. [CrossRef]
43. Pancholi, S.; Yigitcanlar, T.; Guaralda, M. Attributes of successful place-making in knowledge and innovation spaces: Evidence from Brisbane's Diamantina knowledge precinct. *J. Urban Des.* **2018**, *23*, 693–711. [CrossRef]
44. Yin, R.K. *Qualitative Research from Start to Finish*; Guilford Press: London, UK, 2011; ISBN 1606239783.
45. Pancholi, S.; Yigitcanlar, T.; Guaralda, M. Place making in knowledge and innovation spaces: The Australia experience. *Technol. Forecast. Soc. Chang.* **2017**. [CrossRef]
46. Yigitcanlar, T. Position paper: Benchmarking the performance of global and emerging knowledge cities. *Expert Syst. Appl.* **2014**, *41*, 5549–5559. [CrossRef]
47. Pancholi, S.; Yigitcanlar, T.; Guaralda, M. Societal integration that matters: Place making experience of Macquarie Park Innovation District, Sydney. *City Cult. Soc.* **2018**, *13*, 13–21. [CrossRef]
48. Constantinou, C.S.; Georgiou, M.; Perdikogianni, M. A comparative method for themes saturation (CoMeTS) in qualitative interviews. *Qual. Res.* **2017**, *17*, 571–588. [CrossRef]
49. Lenzi, M.H. A Invenção de Florianópolis Como Cidade Turística: Discursos, Paisagens e Relações de Poder. Unpublished Doctoral Thesis, University of São Paulo, São Paulo, Brazil, 2016.
50. Pereira, E.M. A importância de conceitos modernistas no planejamento urbano de Florianópolis. *Seminário de Historia da Cidade e do Urbanismo* **2000**, *6*, 1–15.
51. Machado, E.V. Florianópolis: Um Lugar em Tempo de Globalização. Unplublished Doctoral Thesis, Federal University of Santa Catarina, Florianópolis, Brazil, 2000.
52. Barreto, M.; Burgos, R.; Frenkel, D. *Turismo, Políticas Públicas e Relações Internacionais*; Papirus: Campinas, Brazil, 2003; ISBN 8530807154.
53. Oliveira, A.P. *A História do Turismo em Florianópolis Narrada por Quem a Vivenciou (1950–2010)*; Palavracom Editora: Florianópolis, Brazil, 2011; ISBN 9788564034051.

54. De Luca Filho, V. A Geografia das Feiras de Negócios em Santa Catarina: Origem, Evolução e Dinâmica das Transformações. Unpublished Doctoral Thesis, Federal University of Santa Catarina, Florianópolis, Brazil, 2014.
55. Santos, F.M.; Pereira, R.M. A rede hoteleira no núcleo urbano central de Florianópolis: Expansão urbana e turismo. In Proceedings of the IV Mercosur Tourism Research Seminar, Caxias do Sul, Brazil, 7–8 July 2006; pp. 1–15.
56. Paulilo, M.I. Maricultura e território em Santa Catarina, Brasil. *Geosul* **2002**, *17*, 87–112.
57. Azevedo, I.; Teixeira, C. Florianópolis: Uma análise evolutiva do desenvolvimento do desenvolvimento inovador da cidade a partir do seu ecossistema de inovação. *REAVI* **2017**, *6*, 108–121. [CrossRef]
58. Sabatini-Marques, J.; Yigitcanlar, T.; Costa, E. Incentivizing innovation: A review of the Brazilian federal innovation support programs. *Asia Pac. J. Innov. Entrep.* **2015**, *9*, 31–56.
59. Xavier, M. *Polo Tecnológico de Florianópolis: Origem e Desenvolvimento*; Insular: Florianópolis, Brazil, 2010; ISBN 9788574744957.
60. Esmaeilpoorarabi, N.; Yigitcanlar, T.; Guaralda, M. Place quality and urban competitiveness symbiosis? A position paper. *Int. J. Knowl.-Based Dev.* **2016**, *7*, 4–21. [CrossRef]
61. Yigitcanlar, T.; Sabatini-Marques, J.; Kamruzzaman, M.; Camargo, F.; Costa, E.; Ioppolo, G.; Palandi, F. Impact of funding sources on innovation: Evidence from Brazilian software companies. *R&D Manag.* **2018**, *48*, 460–484. [CrossRef]
62. Yigitcanlar, T.; Sabatini-Marques, J.; Costa, E.M.; Kamruzzaman, M.; Ioppolo, G. Stimulating technological innovation through incentives: Perceptions of Australian and Brazilian firms. *Technol. Forecast. Soc. Chang.* **2017**. [CrossRef]
63. Techflier. 9 Tech Startups from Florianópolis You Need to Know about in 2016. Available online: https://www.techflier.com/2016/09/03/9-tech-startups-from-florianopolis-you-need-to-know-about-in-2016/ (accessed on 6 September 2018).
64. Tarachucky, L. Sistematização da Aplicação do Brand DNA Process No Design de Marca de Cidades Criativas: Caso Projeto Rota da Inovação. Unpublished Master Thesis, Federal University of Santa Catarina, Florianópolis, Brazil, 2015.
65. Google Maps. Available online: https://goo.gl/maps/EiLAFigrR1K2 (accessed on 14 August 2018).
66. Endeavor Brazil. Índice de Cidades Empreendedoras. Available online: http://info.endeavor.org.br/ice2017 (accessed on 28 June 2018).
67. Ueno, A. A Concepção de um Modelo de Empreendorismo Inovador Baseado em Conhecimento: Um Estudo de Caso do Programa Sinapse da Inovação. Unpublished Master Thesis, Federal University of Santa Catarina, Florianópolis, Brazil, 2011.
68. Depine, A.C. Fatores de Atração e Retenção da Classe Criativa: O Potencial de Florianópolis Como Cidade Humana Inteligente. Unpublished Master Thesis, Federal University of Santa Catarina, Florianópolis, Brazil, 2016.
69. Fernandes, E. The legalisation of favelas in Brazil: Problems and prospects. *Third World Plan. Rev.* **2000**, *22*, 167. [CrossRef]
70. Prefeitura Municipal de Florianópolis (PMF). *Municipal Plan of Housing of Social Interest of Florianópolis*; PMF: Florianópolis, Brazil, 2012.
71. Prefeitura Municipal de Florianópolis (PMF). *Florianópolis Sustainability Action Plan*; PMF: Florianópolis, Brazil, 2017.
72. Instituto Brasileiro de Geografia e Estatísticas (IBGE). Florianópolis Panorama. Available online: https://cidades.ibge.gov.br/brasil/sc/florianopolis/panorama (accessed on 6 September 2018).
73. Prefeitura Municipal de Florianópolis (PMF). *Florianópolis Urban Growth Study*; PMF: Florianópolis, Brazil, 2017.
74. Prefeitura Municipal de Florianópolis (PMF). *Florianópolis Mitigation and Climate Change Study*; PMF: Florianópolis, Brazil, 2017.
75. Oliveira-Musse, J.; Homrich, A.S.; Mello, R.; Carvalho, M.M. Applying backcasting and system dynamics towards sustainable development: The housing planning case for low-income citizens in Brazil. *J. Clean. Prod.* **2018**, *193*, 97–114. [CrossRef]

76. ND Online. SC-401 em Números: A Rodovia de Florianópolis é a Líder em Mortes e Acidentes. Available online: https://ndonline.com.br/florianopolis/especiais/sc-401-em-numeros-a-rodovia-de-florianopolis-e-a-lider-em-mortes-e-acidentes (accessed on 1 August 2018).
77. Yigitcanlar, T.; Kamruzzaman, M. Investigating the interplay between transport, land use and the environment: A review of the literature. *Int. J. Environ. Sci. Technol. (Tehran)* **2014**, *11*, 2121–2132. [CrossRef]
78. Yigitcanlar, T.; Kamruzzaman, M.; Buys, L.; Ioppolo, G.; Sabatini-Marques, J.; Costa, E.; Yun, J. Understanding 'smart cities': Intertwining development drivers with desired outcomes in a multidimensional framework. *Cities* **2018**, *81*, 145–160. [CrossRef]
79. Yigitcanlar, T. Smart cities: An effective urban development and management model? *Aust. Plan.* **2015**, *52*, 27–34. [CrossRef]
80. Metaxiotis, K.; Carrillo, J.; Yigitcanlar, T. (Eds.) *Knowledge-Based Development for Cities and Societies: An Integrated Multi-Level Approach*; IGI Global: Hersey, PA, USA, 2010; ISBN 1615207228.
81. Bridges, A. Leveraging Amenity-Led Growth and Collective Action for Sustainable Development in Florianópolis, 1965–2016. Unpublished Doctoral Thesis, Rutgers University, New Brunswick, NJ, USA, 2017.
82. Scherer, M.E.; Asmus, M.L. Ecosystem-based knowledge and management as a tool for integrated coastal and ocean management: A Brazilian initiative. *J. Coast. Res.* **2016**, *75*, 690–694. [CrossRef]
83. Prefeitura Municipal de Florianópolis (PMF). *Floripa Se Liga Na Rede*; PMF: Florianópolis, Brazil, 2013.
84. Figueiroa, A.; Scherer, M. Para onde estamos indo? Uma avaliação do plano diretor do Município de Florianópolis para o entorno da Estação Ecológica de Carijós. *Desenvolvimento e Meio Ambiente* **2016**, *38*, 283–301. [CrossRef]
85. Bastos, G. Análise Financeira das Pescarias de Pequena escala No Município de Florianópolis. Unpublished Doctoral Thesis, University of São Paulo, São Paulo, Brazil, 2009.
86. Stahelin, G.D.; Fiedler, F.N.; Lima, E.P.; Sales, G.; Wanderlinde, J. Projeto Tamar's station in Florianópolis, State of Santa Catarina, Southern Brazil. *Mar. Turt. Newslett.* **2012**, *133*, 23.
87. Goonetilleke, A.; Yigitcanlar, T.; Ayoko, G.A.; Egodawatta, P. *Sustainable Urban Water Environment: Climate, Pollution and Adaptation*; Edward Elgar: Cheltenham, UK, 2014; ISBN 1781004641.
88. Afonso, O.; Monteiro, S.; Thompson, M. A growth model for the quadruple helix. *J. Bus. Econ. Manag.* **2012**, *13*, 849–865. [CrossRef]
89. Yigitcanlar, T.; Bulu, M. Dubaization of Istanbul: Insights from the knowledge-based urban development journey of an emerging local economy. *Environ. Plan. A* **2015**, *47*, 89–107. [CrossRef]
90. Yigitcanlar, T.; Dur, F. Making space and place for knowledge communities: Lessons for Australian practice. *Aust. J. Reg. Stud.* **2013**, *19*, 36–63.
91. Pancholi, S.; Yigitcanlar, T.; Guaralda, M. Place making facilitators of knowledge and innovation spaces: Insights from European best practices. *Int. J. Knowl.-Based Dev.* **2015**, *6*, 215–240. [CrossRef]
92. Zhang, Y.S.; Zou, S.M. The governance mechanisms in innovation ecosystem of hi-tech enterprises. *Stud. Sci. Sci.* **2010**, *5*, 020.
93. Yigitcanlar, T.; Inkinen, T. *Geographies of Disruption: Place Making for Innovation in the Age of Knowledge Economy*; Springer: New York, NY, USA, 2019; ISBN 9783030032067.
94. Yigitcanlar, T.; Kamruzzaman, M. Does smart city policy lead to sustainability of cities? *Land Use Policy* **2018**, *73*, 49–58. [CrossRef]
95. Dutta, S.; Lanvin, B.; Wunsch-Vincent, S. *The Global Innovation Index 2018: Energizing the World with Innovation*; Cornell University: Ithaca, NY, USA, 2018; ISBN 9791095870098.
96. Lara, A.P.; Da Costa, E.M.; Furlani, T.Z.; Yigitcanlar, T. Smartness that matters: Towards a comprehensive and human-centred characterisation of smart cities. *J. Open Innov. Technol. Mark. Complex.* **2016**, *2*, 8. [CrossRef]
97. Trindade, E.; Hinnig, M.; Costa, E.; Sabatini-Marques, J.; Bastos, R.; Yigitcanlar, T. Sustainable development of smart cities: A systematic review of the literature. *J. Open Innov. Technol. Mark. Complex.* **2017**, *3*, 11. [CrossRef]
98. Lara, A.P.; Marques, J.S.; Santos, N.; Costa, E.M. Projeto Florip@ 21: A construção de uma região inteligente na cidade de Florianópolis, Brasil. In Proceedings of the XV Latin Iberian-American Congress on Management of Technology, Porto, Portugal, 27–31 October 2013; pp. 1673–1691.
99. Lorenzetti, J.; de Lima Trindade, L.; Pires de Pires, D.E.; Souza Ramos, F.R. Tecnologia, inovação tecnológica e saúde: Uma reflexão necessária. *Texto & Contexto Enfermagem* **2012**, *21*, 432–439.

100. Martins, C. O Papel das Incubadoras de Empresas do Polo Tecnológico de Florianópolis No Desenvolvimento do Processo de Empreendedorismo Inovador. Unpublished Master Thesis, Universidade do Sul de Santa Catarina, Florianópolis, Brazil, 2013.
101. Sarquis, A.B.; Fiates, G.G.; Hahn, A.K.; Cavalcante, F.R. Empreendedorismo inovador no polo tecnológico de Florianópolis. *REEN* **2014**, *7*, 228–255. [CrossRef]
102. Gaspar, J.V.; Menegazzo, C.; Fiates, J.E.; Teixeira, C.S.; Gomes, L.S. A revitalização de espaços urbanos: O case do Centro Sapiens em Florianópolis. *Revista Livre de Sustentabilidade e Empreendedorismo* **2017**, *29*, 183–205.
103. Menezes, A.G.; Lezana, Á.G.; de Abreu-Ronconi, L.F.; de Oliveira-Menezes, E.C.; de Melo, É.N. A pesquisa-ação como estratégia de avaliação da inovação social: Estudo de uma entidade educacional do município de Florianópolis. *NAVUS Revista de Gestão e Tecnologia* **2016**, *6*, 93–105. [CrossRef]

© 2018 by the authors. Licensee MDPI, Basel, Switzerland. This article is an open access article distributed under the terms and conditions of the Creative Commons Attribution (CC BY) license (http://creativecommons.org/licenses/by/4.0/).

*Article*

# City Branding, Sustainable Urban Development and the Rentier State. How Do Qatar, Abu Dhabi and Dubai Present Themselves in the Age of Post Oil and Global Warming?

**Martin De Jong [1,2], Thomas Hoppe [3] and Negar Noori [1,]\***

[1] Erasmus School of Law and Rotterdam School of Management, Erasmus University Rotterdam, 3062 Rotterdam, The Netherlands; w.m.jong@law.eur.nl
[2] School of International Relations and Public Affairs, Fudan University, Shanghai 200433, China
[3] Faculty of Technology, Policy & Management, Delft University of Technology, 2628 CD Delft, The Netherlands; T.Hoppe@tudelft.nl
\* Correspondence: noori@law.eur.nl; Tel.: +31-010-408-1510

Received: 4 April 2019; Accepted: 27 April 2019; Published: 30 April 2019

**Abstract:** In the past three decades Qatar, Abu Dhabi and Dubai have realised a meteoric economic rise. Whereas the former two can be considered 'rentier states' heavily depending on oil (and gas) revenues, the latter only leans on oil for a mere 6% of its gross domestic product (GDP). Although the economic rise has brought considerable welfare, it has also led these emirates to attain the world's highest per capita carbon footprint. To address this problem Qatar, Abu Dhabi and Dubai seem to have formulated policies with regard to sustainable urbanisation and adopted strong branding strategies to promote them internally and externally. In this paper we examine which steps have been taken to substantiate their claims to sustainable urbanisation, in branding as well as in actions taken towards implementation. We find that all three have been very active in branding their sustainable urbanisation policies, through visions and policy frameworks as well as prestigious development projects, but that the former is substantially more impressive than the latter. Results also show there is a difference between Abu Dhabi and Qatar on the one hand, and Dubai on the other. Dubai has large number of small 'free economic zones', academic institutions for developing a knowledge economy, and smart and/or sustainable urban neighbourhoods, while Qatar and Abu Dhabi have a small number of very large ones. From the three, it is currently Dubai which has taken the lead in this development, largely completing its industrial transition with vast economic diversification and urban expansion. However, across the board this has had little effect on its ecological footprint.

**Keywords:** city branding; sustainable urban development; rentier state; Qatar; emirates

## 1. Introduction

In the past three decades, a number of tiny gulf-states have realized a meteoric economic rise that very few analysts would have deemed possible [1–3]. Qatar, Abu Dhabi and Dubai were all emirates with population numbers well below 100,000 inhabitants whose prosperity depended heavily on rapidly declining revenue from pearl trading. When the British gave up their military protection in 1971, their independence hinged on a combination of decentralised unification and diplomatic agility [4–7]. Whereas Abu Dhabi and Dubai established an initially inconvenient federal arrangement with five other smaller emirates to their North-East (Sharjah, Ajman, Ras al-Khaimah, Umm al-Quwain and Fujairah) named the United Arab Emirates (UAE), Qatar and its tiny neighbour Bahrain maintained their independence and manoeuvred between regional superpowers Iran and Saudi Arabia, as well as each other [8]. As amply noted in the academic literature, the evolution of Qatar, Abu Dhabi and Dubai into global economic hubs by the year 2020, wielding considerable political and economic influence was something virtually nobody had anticipated. A superficial reading of their fate might have led one to conclude that it was primarily the exploration of oil and gas that had boosted their luck, but a closer look reveals that this is only true for Qatar and Abu Dhabi, but not for Dubai which leans on oil for a mere 6% of its GDP [3,9]. In many other emirates in fact, existing reserves have been largely depleted or even barely any oil was found. But even where large reserves were found, smart long-term oriented use of the cash-flow derived from it was used to set up immense sovereign wealth funds with substantial economic impact, nationally and internationally [10].

In fact, a much more credible explanation for their success has been a combination of personal, visionary and determined autocratic leadership in absolute monarchies, prudent, daring and diversified investment policies and skillful branding activities based on the revenues earned through oil and gas [11]. It is typically these features that differentiate them from Kuwait, which acquired its independence from Britain at an earlier stage and was far more developed at that time. The latter of the three partial explanations, branding, took the shape of constructing modern infrastructure hubs such as hypermodern ports and airports, establishing world-class airlines, developing landmark office towers, museums and shrines, offering homes to the wealthy and famous on iconic areas obtained through land reclamation, staging stunning world events like large exhibitions and sports championships, and boosting a powerful tourist industry by offering luxury resorts that lean on sea, sun, sand, camels and an expanding assortment of sophisticated and extreme shopping malls, and entertainment parks [9,12–14].

Visionary leadership, prudent, diversified investment policies and skillful branding gave Qatar, Abu Dhabi and Dubai a substantial amount of soft power, and has led them to occupy top positions on the international rankings for GDP per capita [6,15–17]. Moreover, the populations of these three giants have expanded from a few dozen thousands in the 1970s to a few million each by 2020. The percentages of foreign workforce in their economies have risen from virtually none to over 80% in each. The figures in the other emirates are significantly lower, but there too GDP per capita, population and percentage of imported workers have witnessed a dramatic increase [15,18].

In spite of this enormous growth and the credit they take for these achievements, policy-makers and analysts in each of the three emirates have gradually come to realise that there are significant drawbacks to this success story. Domestic consumption has expanded rapidly to the detriment of people's health and the natural environment. Excessive wealth has detracted from their eagerness to work hard to succeed. Growing populations have made locals largely invisible in public spaces and increasingly worried about the control they have over their own societies. Enhanced global ambitions and visibility have made Qatar, Abu Dhabi and Dubai vulnerable to international criticism of the way gender equality and the labour conditions of poor workers are handled. Last but not least, shrinking energy reserves have all pushed them to reconsider their model of (economic) development [3–5,10,15,16,19]. Qatar, Abu Dhabi and Dubai have formulated policies that aim at preparing for a post-oil economy characterized by a diversification of their industrial activities and preservation of their natural environment. They have opened a number of free trade zones to facilitate foreign direct investment and international trading, introduced elements of common law to strengthen rule of law in business operations and welcomed leading universities and research centres in 'education cities' to promote the knowledge economy [2,7,20]. In addition, they have undertaken action to strengthen environmental protection, promote the development of renewable energy and in the long run reduce carbon emissions.

As could be expected, they have chosen to do so through a plethora of charming international branding initiatives of which Qatar's 'Lusail Smart City', Abu Dhabi's 'Masdar City' and Dubai's 'Media City' and 'Internet City' have perhaps received the greatest international media coverage [3,11,21], although signs reveal that stakeholder approval is mixed [22]. The attraction of investment start-ups meets difficulties, and population numbers are surprisingly low [23]. While official government sources tend to be confident that they can replicate their previous economic success in the environmental realm and keep their societies stable, many international (primarily Western) academics are surprised or dismayed at the low degree of domestic stakeholder involvement in rolling out these policies. Some have predicted in a variety of ways that the regimes behind these 'rentier states' will barely, or not at all prove capable of the necessary reforms, or are even at risk of collapse [3,4,10,15,18]. Neither of these two extremes have actually happened (yet). Energy consumption and carbon footprints are still among the highest in the world while the political and administrative leadership and elites are also still firmly in place.

In this contribution, we examine what effective policy initiatives loom behind the shiny branding practices regarding economic diversification and sustainable urbanization. In particularly we are interested in answering the question: Which steps have been taken to substantiate Qatar, Abu Dhabi and Dubai's claims to economic and ecological modernization? In answering this question, we will not be leaning explicitly on theories or methodologies as developed in the field of place branding, as we have done in previous work [24–26], but rather take publications in the field of political science and international relations related to the three emirates under study, more specifically the concept of the 'rentier state' and its impact on the branding and implementation of policies regarding sustainable urbanisation as the starting point of our analysis.

The next few sections will offer the following content. Section 2 will present a gist of the existing theories and insights on the topic, mostly taken from scholarly work in political science and international relations. This work has a strong focus on the concept of city branding, institutional arrangements, branding and policy-making in the emirates and has been produced by European, American and Iranian academics and analysts often having one leg in Qatar, Abu Dhabi or Dubai (Arab authors are conspicuously missing from the literature). We analyse the dominant argumentation in this literature and assess the validity of their occasional extreme optimism about enhanced political and economic empowerment of common citizens and otherwise disheartening pessimism when it comes to the social and environmental sustainability underlying this growth and the resilience of the regimes in place. In Section 3, we connect this body of theory with our research design and provide information on data collection and analysis. Section 4 presents an overview of the basic ideas and assumptions underlying the branding practices and policy initiatives on sustainable urbanization in the three emirates, including the national visions, plans, frameworks and online texts and videos in the three emirates. Section 5 demonstrates branding strategies and policy initiatives undertaken in Dubai, Abu Dhabi and Qatar to diversify their economies in free economic zones, boost their research and higher education systems in academic cities, and given impetus to sustainable urban development in smart and/or sustainable cities. In Section 6, we compare the three emirates in how they deal with branding and investing in sustainable urban development overall and discuss the implications of our findings for the future credibility and resilience of the 'rentier states'.

## 2. The Politics of Branding and Sustainable Urban Development in Qatar, Abu Dhabi and Dubai

### 2.1. The Concept of City Branding

By 'branding' we mean the practice of place, which can be seen as a long-term strategic activity [27,28] aimed at the positioning of cities, regions and countries amidst their neighbours and peers, which is closely connected to the concept of 'place brand'. The latter can be defined as "a network of associations in the place consumers' minds based on the visual, verbal, and behavioral expression of a place and its' stakeholders. These associations differ in their influence within the network, and in importance for the placement of consumers' attitudes and behaviours." [29]. In relation to the city brand, 'place branding' can be seen as a strategic activity with cities trying to garner positive associations in the consumer's mind (Ibid.). City branding can also be seen as a holistic activity with city marketing and city promotion as contributing elements [30]. Lu and De Jong (2017) view city branding strategy as the practice of conveying a brand or symbolic essence of a nation, region or city to target audiences for enhancing one's fame and reputation or otherwise obtaining strategic gain [31]. 'City branding' then pertains to a communication system that connects the overall image and identity of a city [32,33]. According to Baker (2007) city branding, the image and identity of a given city are essential in making it unique among various alternatives [34]. Dinnie (2010) even argues that public authorities see a direct relation between the image of their city and its attractiveness as a place to attend, to live in, to invest in, or to study [12]. Others argue that current city factors address the city brand and city branding activities themselves. For instance, Anholt (2016) distinguishes a set of factors (image, identity, places, the position of city, suitable location, capacity and vitality of people) that influence the brand a given city adopts [35]. Shirvani Dastgerdi and De Luca (2019) developed a research model that addresses this issue, coining economic performance, media and advertising, cultural activities, policy making and urban planning as factors influencing city image, and indirectly city branding [36]. For the latter, the authors discern objectives such as creating global uniqueness, meaningfulness and acceptability, justifiable identity, stakeholder satisfaction, attracting resources and increasing the general wellbeing of citizens.

In practice, establishing a place branding strategy is often driven top-down by public administrators and materialized in a narrow-scope set of communication tactics to appeal to external, resourceful stakeholders [37]. City branding only seldom deploys the involvement of local citizens and stakeholders [29]. In acknowledging this problem Kavaratzis and Kalandides (2015) propose a rethinking of place brands by (i) incorporating more geographical understanding into place branding, and (ii) outlining a process that allows place elements and place-based associations to combine and form a place brand [38]. In a similar vein, Pedeliento & Kavaratzis (2019) argue that developing place brands cannot be devised as a top-down agenda designed by public administrators in which strategic objectives are imposed on local stakeholders [39]. Rather, they argue that place branding strategy must be pursued with the involvement of stakeholders in setting strategic objectives, and by raising awareness that achieving these objectives is largely owed to their own practices. Pedeliento & Kavaratzis (2019) argue that city branding is presumably more trustworthy, legitimate and effective once geographical and situational aspects are sufficiently addressed and setting the branding strategy and its objectives is not merely a top down endeavour, but also includes involvement of local stakeholders and citizenry [39]. As we will see below, not all of these aspects of city branding practices as advocated in the literature can be deemed common in gulf states.

## 2.2. City Branding in Qatar, Abu Dhabi and Dubai

"Be the flame, not the moth. You are the strength, you are the power, you are the decider, you are the thinker, you are the motivation, you are the reason, you are the purpose. Never forget that you are a brand [40]."

On August 10th 2018, Sara al Madani, (serial entrepreneur, public speaker, Board Member at the Sharjah Chamber of Commerce and Industry, Board Member at the UAE SME enterprise council of the Ministry of Economy, Board Member at ShjSEEN, and investor) posted the above message on LinkedIn. It epitomises the enormous drive among elites in wealthier gulf-states for a strong public profile, in social media as elsewhere. It has made Qatar, Abu Dhabi and Dubai world famous in recent decades and it is also beginning to affect other emirates such as Sharjah. Much of the academic work on the policy and politics of branding in the various emirates and its local, national and global impact has however been written by academics and journalists from outside the Arab world. Many of them have worked at universities in Qatar, Abu Dhabi and Dubai and subsequently moved on to other parts of the world. They all stress how important branding and public diplomacy have been in attracting foreign direct investment, tourism, business opportunities, qualified expatriate employees and international goodwill and how skilfully they were used. They are unanimous in their acclaim of what these gulf-states have achieved in terms of economic growth and global fame, and highlight how the leadership in these countries has shown vision and courage to think long term, undertake occasionally risky development projects and utilized generous revenues from oil and gas for strengthening their financial profile, growing a variety of industrial sectors and their own attractiveness.

But this academic praise is not unqualified. While some see in the positioning of the wealthier emirates a tendency to overemphasize Bedouin features at the expense of Persian and other elements in branding, their canonised national history and culture [6,20], others point at the large number of remaining severe social and policy issues the young and wealthy states are still faced with [3–5,15,18]. A very prominent challenge is the environmental one. The native population resident in the emirates, and that in the UAE and Qatar in particular, are the most resource intense consumers in the world. Their high levels of income and plentiful availability of natural resources make it almost natural for them to use large amounts of (their own) energy and import massive amounts of other products not readily available in the desert. Moreover, ambitious development programmes in land reclamation, real estate, infrastructure development, tourism, entertainment and post-oil heavy manufacturing industries also drive up the use of both natural resources and carbon emissions.

Mari Luomi (2012) was the first to examine the proclaimed shift in environmental policies in the emirates in response to their negative public image in environmental affairs [10]. She compared the policy responses in Qatar and Abu Dhabi and found that both had been affected by the dominant global discourse on reducing carbon emissions and experienced serious difficulty transforming their production and consumption patterns. However, while Qatar's energy and environmental policies had largely remained stagnant, Abu Dhabi had seriously begun changing its global position by diversifying its production economy and making friends in international arenas with more progressive partners than just their neighbouring gulf states. It was also investing heavily in Masdar City, at that time known as the world's first zero-carbon city, and an extensive development programme for nuclear energy, both backed by Crown Prince Muhammad al Zayed's powerful Mubadala investment company and undertaken in collaboration with prestigious international partners. Both programmes were aimed at scaling down dependency on traditional energy sources over time, strengthening new innovative high-tech industries and reducing Abu Dhabi's environmental footprint. Although Luomi had not studied Dubai and other emirates, she had reason to believe that efforts to improve the situation there had not even started in earnest. Dubai's reliance on risky financial policies combined with massive land reclamation projects, real estate development and mass entertainment, and Kuwait's lack of policy ambition and initiative due to paralysing struggles between political factions made environmental policy ineffective in these two gulf-states. Later however, in "Building a Sand Castle", Federico Cugurullo (2013) also offered an extremely critical assessment of the business model underlying Masdar City as well as of its actual implementation [41].

If making a transition in production patterns when much of one's economy has been built on oil and gas is a major challenge, realising substantial changes in consumption patterns among local Emiratis and Qataris is at least as problematic. Christopher Davidson (2012) eloquently describes the institutional mechanisms preventing meaningful policy change, even more so than in the rest of the world, as being the consequence of living under a 'rentier state' [15]. Ruling monarchies in the emirates (and other gulf-states such as Saudi Arabia and Oman) have been able to control, accumulate and utilise the national revenue generated through the annual sales of their national oil and gas reserves by establishing massive Sovereign Wealth Funds, but also by distributing financial favours over large segments of the population to keep them satisfied. Inhabitants pay no income taxes; housing, education and healthcare are totally free of charge and energy is available at virtually no charge. These incentives combined with very high incomes and generous conditions for those unable or unwilling to work have led to lifestyles where prosperity and an all-protective state are taken for granted. In this manner, the ruling families and tribes of the respective oil monarchies have been able to acquire popular support and stave off possible resentment and discontent against their leadership [9,15]. However, the downside of these rentier state mechanisms is that many nationals have evolved into complacent citizens and consumers that take this public generosity for granted. Once energy reserves dry up, sales prices abroad go down and the national population rises too fast, the regime will no longer be able to provide these services to the same extent and may well rapidly lose credibility. According to Davidson, this menace will eventually phase out these oil monarchies and replace them with more socially and economically inclusive regimes. This process will hit the poorer and less well-endowed emirates, such as Ras al-Khaimah, Umm al-Quwain, Ajman, Fujairah and Sharjah first, then hit Dubai and finally also reach Abu Dhabi and Qatar, which enjoy the largest reserves by far, last. Fromherz (2017), on the other hand, has argued that political scientists often fail to grasp the cultural complexities of Qatari and Emirati societies and purport simplistic analyses [2]. This can obviously be taken as indirectly distancing himself from such far-reaching predictions.

As we know in 2019, no such turnaround in power has taken place in any of these emirates (yet), and this in spite of lower oil and gas prices and a financial crisis from which these states have rebounded (especially Dubai which was hardest hit). Although public discontent has been recorded in the poorer emirates, in the three under study here (Qatar, Abu Dhabi and Dubai), the regimes are still firmly in place and public uproar is virtually absent. In fact, they have even formulated ambitious policy visions, industrial plans and urban master-plans in which they explain how they will engage in ecological modernization through further diversifying and updating their industrial structures, beefing up their education levels, developing into knowledge societies and become leaders in green building and sustainable urban planning. Their well-branded policy documents and project plans seem to suggest that they are well underway to replicate their economic miracle with an environmental one. In the remainder of this contribution, we will examine what holds true of these promises, whether and how the strong brands these emirates represent are indeed perpetuated in recent years and what holds true of Davidson's bold predictions.

## 3. Research Design and Methodology

Although we know that regime change has not happened in any of the emirates and is not immanent in any of them either, this does not immediately invalidate the credibility of the mechanisms on which Davidson's predictions are based. If the theorem of the rentier state in the emirates holds, one can expect that the leadership in Qatar, Abu Dhabi and Dubai will avoid adopting tough policy measures in the sphere of consumption which would negatively affect the comfortable lives their citizens are accustomed to (charges for energy use, subsidies for housing and various other public services, tax levels and prices for imported products). Instead, they would show high levels of ambition when it comes to the transformation of domains where this does not directly hurt citizens and consumers, primarily in the world of industrial and knowledge production, even if such interventions weigh heavily on the public purse. In other words, consumption patterns are unlikely to change much as long as this remains financially feasible, but impressive investments will be made in the promotion of new industries which make a transition possible from the primary and secondary industrial sectors to a more service-oriented economy. This includes the rise of research and education, science and technology, information and communication technology (ICT) infrastructures and renewable energy. Moreover, the focus on the economic growth vehicles that benefit the emirates as a place of consumption for both residents and visitors is boosted through a strong focus on high-quality infrastructure development, real estate development, conspicuous entertainment and tourism. These investments will invariably be labelled 'smart', 'sustainable' and/or 'green' and strong efforts will be made globally to brand these to the world as industrial and urban greening for maintaining a good international reputation. However, the manner in which this responsible attitude towards post oil and global warming is shown to the outside world takes the shape of excellent scores on simple indicators, such as the fastest growth in renewables production in the world and the highest percentage of green buildings in the Middle East. That is, the focus is on catchy but disjointed key performance indicators instead of more holistic figures regarding the ecological footprint, where good performance is hard if not impossible to realise in a desert environment with demanding consumers.

### 3.1. Research Design and Case Selection

The research design of the present study entails a multi-case study of three emirates. We use a comparative method to analyse key commonalities and differences between these cases in the way they, as rentier states, handle city branding and policy-making regarding sustainable urbanisation. We use a case study research approach because we deem sustainable urban development in the emirates (and cities and regions more generally) a complex phenomenon, which does not lend itself to other research designs such as experiments, surveys or modelling. Moreover, we aim to study the relation between rentier states on the one hand and place branding strategies and policy initiatives on the other in detail, as a current complex phenomenon in its real-life context. We therefore use multiple sources of evidence to gain more reasonable understanding of the situation [42].

Among all emirates only Qatar, Abu Dhabi and Dubai were selected for case study analysis because of what they have achieved in terms of economic growth and global fame. All three have had strong leadership with clear vision and courage to think long term, willingness to undertake risky development projects and aptitude to utilise generous revenues from oil and gas for strengthening their financial position and international profile. In this sense they performed better than other emirates and can be seen as rather 'extreme cases' [43]. We explore and analyse these cases to find evidence on theoretical claims pertaining to rentier states and their drive toward sustainable urbanization.

*3.2. Key Concepts and Operationalization*

In this study the main concepts are threefold: (i) rentier state (as the independent variable), (ii) branding strategies, and (iii) policy initiatives (the last two both as dependent variables). At their turn, both dependent variables are examined in three different realms: those of (1) economic diversification in 'free economic zones', (2) research and higher education in 'education cities' or 'academic cities', and (3) urban expansion in residential areas such as 'sustainable cities'. We decided not to include transport and other infrastructure development, because these are complex to examine and not confined in designated spatial enclaves. Moreover, the different spatial structures of the three emirates with Abu Dhabi having an extremely large size, Qatar in the middle and Dubai being only a fraction of Abu Dhabi and respective urban densities which are approximately the reverse of this, made meaningful comparisons an impossible task. If anything, existing studies indicate that because of its urban structure Dubai's public transport services can be operated reasonably efficiently, while metro and tram facilities in Qatar and Abu Dubai are far harder to operate in cost-effective ways. Table 1 shows how the key concepts in the present study are defined and measured.

Table 1. Definition and measurement of key concepts in the present study.

| Key Concept | Definition | Indicator |
| --- | --- | --- |
| Rentier state | A state which derives all or a substantial portion of its national revenues from the rent of indigenous resources to external clients. Dependent upon it as a source of income, it may generate rents externally by manipulating the global political and economic environment. Such manipulation may include monopolies, trading restrictions, and the solicitation of subsidies or aid in exchange for political influence [44] | Presence of an autocratic regime with one or a small number of dominant families able to control and redistribute the revenues derived from natural resources. It can use them intelligently and flexibly to appease its citizens and (more selectively) immigrant groups to stay in place. Qatar and Abu Dhabi have high amounts of natural resources and therefore ample opportunity to appease, Dubai has limited amounts of them left. Qatar and Abu Dhabi are therefore full rentier states, Dubai is a semi-rentier state. It is now practically a rentier-service hub, as it is effectively the major port and business centre for most of the region's oil and gas rentier states. |

**Table 1.** *Cont.*

| Key Concept | Definition | Indicator |
|---|---|---|
| Place branding strategy | The practice of conveying a brand or symbolic essence of a nation, region or city to target audiences for enhancing one's fame and reputation or otherwise obtaining strategic gain [31] while acknowledging that geographical and situational aspects are sufficiently addressed and setting the branding strategy and its objectives is not merely a top down endeavour, but also includes involvement of local stakeholders and citizenry, and a process be outlined that allows place elements and place based associations to combine and form a place brand [39,40] | Presence of a well-branded policy vision of economic diversification (1a) Presence of a well-branded policy vision of higher education (1b) Presence of a well-branded policy vision of sustainable urban expansion (1c) Presence of a large and conspicuous investment programme in green-labelled new industries (2a) Presence of a large and conspicuous investment programme in higher education (2b) Presence of a large and conspicuous investment programme in green-labelled residential expansion projects (2c) |
| Policy initiatives | Specific actions undertaken in response to generally formulated policy goals and ambitions and as such the physical embodiment of the actual implementation of the brand | Number and size of spaces displaying the emergence of non-oil and gas industries (3a) Number and size of spaces displaying the emergence of a strong research and higher education system (3b) Number and size of spaces displaying residential area development with green features (3c) Emerging variety and characteristics of the non-oil and gas industries (4a) Emerging quantity and quality of a thriving research and higher education system (4b) Emerging green and liveable features in sustainable residential areas (4c) |

*3.3. Data Collection and Analysis*

In order to collect the relevant data, we have conducted desk research in studying policy documents such as national visions, industrial plans, education plans, urban master plans, infrastructure plans and other relevant publicly available government publications. Online accessibility and completeness of these documents in both English and Arabic was such in Dubai, Abu Dhabi and Qatar that they could be considered reliable information for addressing the branding issues and indicators in Table 1. Availability of such documents proved problematical in Kuwait, Bahrain, Sharjah, Ras al-Khaimah, Ajman, Umm al-Quwain and Fujairah and little actual activity could be found in these emirates, a good indication of that the fact that the three cases we chose were indeed the appropriate ones for our purposes. In our case, the English versions of these documents were consulted. The availability and contents of these visions, plans and action programmes online and in printed documents allowed us to fill in the information on the branding strategies in the three emirates under study (indicators 1a–c and 2a–c). More detailed desk study and online searches into project documentation, statistics, site visits to several urban development projects (in Abu Dhabi and Dubai) and interviews with thirteen policy makers and experts in the United Arab Emirates and Qatar were used to collect more information on the translation of the brands and vision into actual policy initiatives and actions. Semi-structured questionnaires were used for the interviews pertaining to indicators on the policy initiatives (see Table 1). For all site visits and interviews notes were taken and site reports, and interview transcripts were made. These were analysed and used for filling in the required information on the features and implementation of the investment and development programmes in Table 1 (indicators 3a–c and 4a–c).

Given the strained political relations between the UAE and Qatar, between which direct transport connections have been made impossible, we were unable to visit Qatar. Given the quality of the online documentation available on Qatar, filling out the list of indicators regarding branding practices (indicators 1a–c and 2a–c) and policy initiatives (indicators 3a–c and 4a–c) were unaffected by this hurdle. To cover the list of indicators on practical policy measures we questioned people in the UAE on the situation in Qatar, on which some of them provided valuable information. Additional interviews were held in Barcelona during a smart city conference and at Delft University of Technology to verify and complement this information.

## 4. Visions, Plans, Frameworks and Other Online Sources in Place Branding and Policy-Making

When it comes to place branding, the three gulf emirates under study here are second to none. All three have reached international fame and their profiles are in many ways similar when it comes to wealth, luxury, entertainment and openness, but vary somewhat in terms of their social and cultural image, with Qatar and Abu Dhabi being more religious and conservative, and Dubai as coming out more profane and liberal. In addition to that Qatar has ventured more into valiant international diplomacy, Abu Dhabi and especially Dubai have remained more aloof from international conflicts under the safe umbrella of the UAE federal government [2,3,6,18]. When it comes to the branding of their smart and sustainable urban development, all three emirates have also done a reasonably convincing job, given that on the Sustainability Index issued by large international engineering consultant Arcadis, Dubai ranked first, Abu Dhabi second and Doha (Qatar's capital city) third as sustainable cities in the Middle East region. Still, in the domain of managing their balance between economic growth and environmental preservation, they face a task that is potentially far more challenging than reporting their impressive demographic and economic growth figures.

All three have adopted comprehensive visions or strategies for the future direction of their societies alongside more tactical or sector-specific strategies. The visions carry the respective names of "Qatar National Vision 2030", "Abu Dhabi Policy Agenda 2030", and the "Dubai Plan 2021" [44–46]. All three are directly supported by their respective highest leadership and contain a number of key priority areas or pillars on which government policies should be based.

In the case of Qatar these are: (1) human development (primarily education); (2) social development (justice and preservation of moral standards); (3) economic development (competitive and diversified); and (4) environmental development (harmony between social, economic and environmental aspects).

In Abu Dhabi, the government identified nine pillars constituting the architecture of the Emirate's social, political and economic future: (1) a large empowered private sector; (2) a sustainable knowledge-based economy; (3) an optimal, transparent regulatory environment; (4) a continuation of strong and diverse international relationships; (5) the optimisation of the Emirate's resources; (6) premium education, healthcare and infrastructure assets; (7) complete international and domestic security; (8) maintaining Abu Dhabi's values, culture and heritage; and (9) a significant and ongoing contribution to the federation of the UAE. The Abu Dhabi government has committed itself to direct public policy to strengthen and develop these by focusing on four key priority areas, which are: (1) economic development; (2) social and human resources development; (3) infrastructure development and environmental sustainability; and (4) the optimisation of government operations.

By comparison, the "Dubai Plan 2021" is by far the most explicit and hands-on of the three, by distinguishing six themes: (1) The People: "A City of Happy, Creative & Empowered People"; (2) The Society: "An Inclusive & Cohesive Society"; (3) The Experience: "The Preferred Place to Live, Work & Visit"; (4) The Place: "A Smart & Sustainable City"; (5) The Economy: "A Pivotal Hub in the Global Economy"; and (6) The Government: "A Pioneering and Excellent Government". In many of these areas, it is explicitly formulated that Dubai aspires to be the world's number one in many or all of these subject areas. As can be seen in Figure 1, all six themes are specified in remarkably sophisticated and optimistic ways and displayed in a special graph. The aspects of smart and sustainable urbanisation are included in various themes, but particularly in 'The Place'.

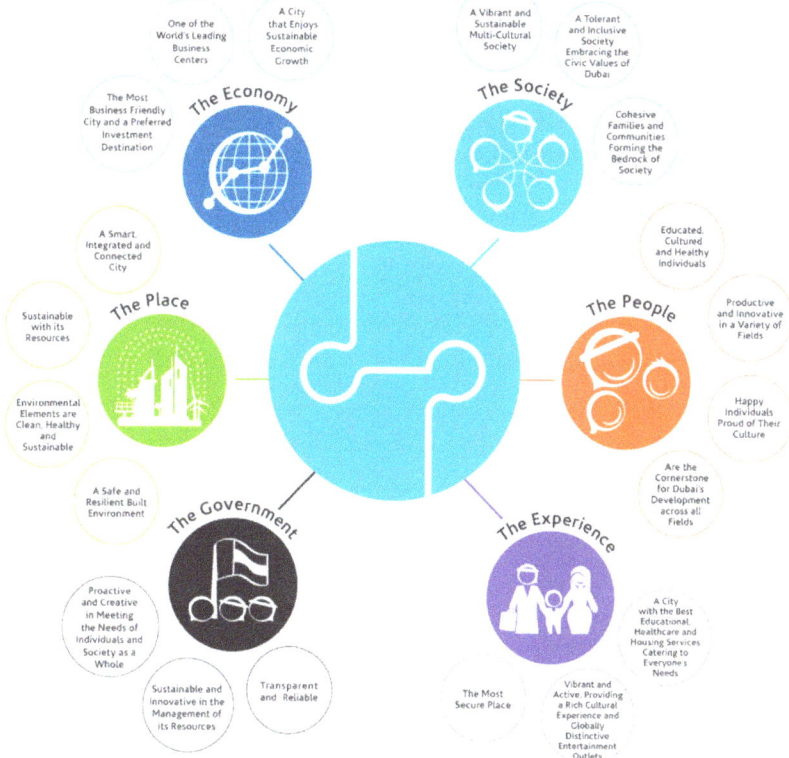

**Figure 1.** Schematic of the "Dubai Plan 2021".

Hanging below these comprehensive visions, in all three emirates we find more specific or tactical documents which specify these pillars, key areas or themes into investment programmes or actions for different policy areas. In Qatar, we find the 'Qatar National Development Strategy (2011–2016)' and 'Qatar National Development Framework (2032)', the 'Qatar National (urban) Master Plan', and Qatar's '2nd Human Development Report' [47,48]. Abu Dhabi has its 'Plan Abu Dhabi 2030/Urban Structure Framework Plan', 'Abu Dhabi Economic Vision 2030', and 'Abu Dhabi Education Reform: The Road to 2030'. Dubai has issued 'Dubai Industrial Strategy 2030', 'Dubai 2020 Urban Master Plan', and 'Sustainable Dubai 2016'. A cursory glance at all of these publicly available plans and programmes shows that Qatar and Abu Dhabi have relatively elaborate documents, which remain at comparatively strategic and visionary levels of abstraction, while Dubai's selection of government strategies has by far the most concrete, fleshed out, specific and action-oriented tone among the three emirates [48–54].

## 5. Branding and Implementing Sustainable Urban Development in Qatar, Abu Dhabi and Dubai

In their visions, policy plans and investment schemes, Qatar, Abu Dhabi, and Dubai pay ample attention to: (1) the promotion of economic diversification beyond the oil and gas industries alone, (2) the development of a mature innovation system by investing in research and education, and (3) smart and sustainable residential neighbourhood development. In the subsections below (Sections 5.1–5.3), we will address for each of these three types of developed spaces what the vision and plans say about the topic, what the focus of the investment efforts is, and what the number, size and important features of the newly developed sustainable spaces are.

## 5.1. Branding and Implementing Economic Diversification

In this paper, economic diversification implies the expansion of those economic sectors that go beyond exploiting natural resources alone, both by allowing the state-owned petroleum firms to develop related downstream economic activities, and the promotion of industries and services completely outside of the realm of oil and gas. In their visions and plans, Qatar, Abu Dhabi and Dubai all emphasize the importance of financial services, transport and communication, real estate, tourism and high tech. Nonetheless, the focus in Qatar is more on the promotion of light industry and knowledge exchange in areas where a variety of small and medium-sized enterprises are clustered, while Abu Dhabi tends to emphasize heavy manufacturing and export-oriented industries more. In the case of Dubai, little is said about petroleum related activities in general, which is understandable given the relatively small size of this sector as well as its rapidly declining role therein. All three emirates see the establishment of new, and the expansion of existing industrial and high-tech zones, as major contributors to industrial diversification and long-term global competitiveness. Some of these new industrial and high-tech zones are located in areas where natural resources are extracted (more on downstream processing and manufacturing), while others can be found nearby international ports and airports or elsewhere in town (more for high-tech industries, manufacturing, foreign export and various services). Most of these industrial zones have been designated free economic zones, where special favourable business conditions apply to attract foreign direct investment (FDI). Such conditions pertain to different types of fiscal advantages, such as complete 'tax holidays' up to 50 years, the absence of import duties, permission to establish 100% foreign owned firms, free repatriation of profits and/or capital, and simplified permitting procedures.

Table 2 shows considerable differences between the three emirates. While the types of favourable conditions granted are roughly the same, it appears that in contrast to Abu Dhabi and Qatar, Dubai began the practice of establishing free economic zones many years ago and now has many of them across its territory: 21 by the 1st of January 2019, to be exact. Within the UAE, Dubai stands apart as it has developed a reputation for international commerce long before its first free trade zone which was established in the 1980s [55]. Most are rather small in size but go by attractive names such as "Dubai Silicon Oasis", "Dubai Internet City", "Dubai Design District", "Gold and Diamond Park", and "Dubai Media City". Dubai uses sophisticated branding slogans. Abu Dhabi, on the other hand, merely enjoys the presence of four free economic zones, while Qatar only has one. As can be seen, these were established far later than those in Dubai. Nonetheless, they count far more square kilometres of developed economic zone area and can therefore contain a larger number of corporations. Qatar, in fact, only began this practice quite recently in response to the economic boycott of its Arab neighbours and has aimed to establish various spaces within its large "Education City" as free economic zones. In all three emirates though, the goal is to foster the emergence and evolution of technological clusters, such as in biotechnology, aerospace engineering, green technologies and services, ICT, health, and education. Apart from the Economic Free Zones mentioned above, all three emirates have developed a number of industrial zones and cities for their domestic manufacturing industries, which do not enjoy these special business conditions.

Table 2. Free economic zones and their main features in Qatar, Abu Dhabi and Dubai.

| Emirates | Name | The Year of Establishment | Profile/Brand | Area/Location | Investment/Source of Funding | Type of Industries/Institutions Are in |
|---|---|---|---|---|---|---|
| Dubai | Jebel Ali Free Zone | 1985 | A unique trade ecosystem that reduces cost, while enabling new opportunities for growth. | 57 sq. km/Jebel Ali area at the far western end of Dubai | governmental funding | Logistics, Warehousing, Economic Trade Zone, Real Estate, Property |
| | Dubai Airport Free Zone | 1996 | The fastest growing Zone in the world | 11 sq.km/Next to Dubai International Airport | governmental funding | Light manufacturing activities, Trading & General Trading, Services |
| | Dubai Internet City | 1999 | Innovation begins here | 139 sq. m/adjacent to Dubai Marina, Jumeirah Beach Residence | Dubai Holding subsidiary TECOM * Investments | Information and communications technology (ICT) companies |
| | Dubai Car & Automotive City Free Zone | 2000 | n/a | 743 sq. m/Ras Al Khor | | Automotive industry |
| | Dubai Media City | 2001 | The region's leading media hub | Next to the Palm Jumeirah | TECOM Group | Media industry and services |
| | Dubai Multi Commodities Center | 2002 | The world's #1 free zone | 2 sq.km/located in Jumeirah Lakes | DMCC is a government entity | Trading, Service, Industrial |
| | Dubai Healthcare City | 2002 | Your Health & Wellness Destination | 2.5 sq. km/Sheikh Zayed Road | governmental funding | Healthcare and clinical industry |
| | Dubai Techno Park | 2002 | n/a | 21.3 sq. km/Jebel Ali | governmental funding | High-profile companies specializing in technology, oil, gas and petrochemical industry and other industries |
| | Intl Media Production Zone | 2003 | n/a | 4 sq. km/Dubai International Financial Centre | TECOM Group | Media production Industry |
| | Dubai Silicon Oasis | 2004 | The integrated free zone technology park | 97740 sq. m/Middle of Dubai land | governmental funding | Services, Trading, Industrial: import raw material, manufacture, and process, assemble, package and export the finished product. |
| | Dubai Industrial City | 2004 | The leading manufacturing and logistics hub | 52 sq. km/Next to Jebel Ali International Airport | TECOM Group | Light and medium manufacturing |
| | Dubai Intl Financial Centre | 2004 | Gateway to Growth | 0.5 sq. km/Sheikh Zayed Road | governmental funding | Finance, Banking & Brokerage Services, Wealth Management, Reinsurance & Captive Insurance, |
| | Dubai Studio City | 2005 | Unleash your imagination | 2 sq. km/Sheikh Mohammad bin Zayed Road | TECOM Group | Film and broadcasting industry |
| | Dubai South | 2006 | The City of You—is an emerging master-planned city based on happiness of the individual. | 145 sq. km/around Al Maktoum International Airport | governmental and private funding | Light manufacturing activities, Logistics, Trading & General Trading, Educational and training, and educational consultancy services. |
| | Dubai Outsource Zone | 2007 | An outstanding business park dedicated to local and international outsourcing companies | Emirates Road | TECOM Group | Services: Business Process Outsourcing (BPO), HR Outsourcing, IT Outsourcing, back office and call center operations |
| | Gold and Diamond Park | 2011 | The Finest Creations, All Under One Roof | 47.5 sq. m/Sheikh Zayed Road | EMAAR group | Jewellery (gemstones, precious stones, gold, silver, platinum) trading, manufacturing, retails and services |
| | Dubai Design District(D3) | 2013 | A home for the region's creative thinkers | 130 sq. m/Next to Business Bay, Dubai Mall, and Burj Khalifa | governmental funding | Digital media, arts, design, and fashion |

Table 2. *Cont.*

| Emirates | Name | The Year of Establishment | Profile/Brand | Area/Location | Investment/Source of Funding | Type of Industries/Institutions Are in |
|---|---|---|---|---|---|---|
| | Khalifa Industrial Zone (KIZAD) | 2010 | The Integrated Trade, Logistics and Industrial Hub of Abu Dhabi | 410 sq. km/located almost equidistant between Abu Dhabi and Dubai | governmental funding | Trade & logistics, Manufacturing; aluminum, food & beverage, pharmaceutical Packaging |
| | Higher Corporation for Specialized Economic Zones | 2004 | A hub for a number of training | 14 sq. km/close to Musafah Sea Port, Abu Dhabi International Airport | governmental funding | Heavy to medium manufacturing, processing, and engineering activities |
| Abu Dhabi | Masdar City Free Zone | 2006 | An emerging global hub for clean technologies and renewable energy | 6 sq. km/6 kilometers away from the Abu Dhabi | Mubadala Development Company | Future green technology products |
| | Twofour54 | 2008 | One of the fastest growing media free zones in the region. | Proximity to Downtown Abu Dhabi | n/a | Media businesses |
| | Industrial City of Abu Dhabi (ICAD) | 2008 | n/a | 40 sq. m/located in the outskirts of Abu Dhabi city | n/a | Heavy-to-medium & Light-to-medium manufacturing, engineering and processing industries. |
| | Abu Dhabi Airport Free Zone | 2010 | A new global business address at the heart of Abu Dhabi Airports operations, will accelerate Abu Dhabi's economic diversification | 12 sq. km/near Abu Dhabi International Airport | governmental funding/owned subsidiary of Abu Dhabi Airports | Aerospace and Aviation industry, Airport &, Airline Services, Marketing and Events, Knowledge and Development |
| | Abu Dhabi Ports Company (ADPC) Free Zone | 2012 | Abu Dhabi Port Company (ADPC) is a good spot for industries and trading companies. | 2.7 sq. km/between Dubai and Abu Dhabi, in Taweelah | governmental funding | Trade, industrial production of goods, services like banking, management consultancy or other professional services |
| | Abu Dhabi Global Market Free Zone | 2013 | An ideal location for investors to set up a company in the financial sector. | 1.14 sq. km/Al Maryah Island | governmental funding in collaborating with the bigwigs (International funding) | Financial service industry |
| Qatar | Airport Free Zone—RAS BUFONTAS | 2018 | Technology, Trading and Logistics Hub | 3.96 sq. km/Adjacent to Hamad International Airport (HIA) | governmental funding/Qatar Free Zones Authority | Light Manufacturing, International Business Services, Aviation Sector, Emerging Technologies Logistics Hub |
| | Port Free Zone—UM ALHOUL | 2018 | Hub with Industrial Focus | 30.3 sq. km/Adjacent to the Mesaieed Industrial Zone & Hamad Port | governmental funding/Qatar Free Zones Authority | Maritime Industries, Heavy Manufacturing, Industrial Sectors Focus, Emerging Technologies, Logistics Hub |

To what extent have these policies indeed led to industrial diversification? Table 3 shows that Dubai now leans for over 90% of its GDP on economic activities unrelated to oil and gas, whereas Abu Dhabi and Qatar for which the industry breakdown is remarkably similar - still continue to lean on their natural oil and gas resources for roughly one-third of their GDP generation, with mining and quarrying still high on the list. If the aim is to move to more ecological modernisation, having these resource intensive and polluting industries in place might not be a good sign.

**Table 3.** Industrial diversification in Qatar, Abu Dhabi and Dubai.

| Dubai | | Abu Dhabi | | Qatar | |
|---|---|---|---|---|---|
| Economic Activity | GDP Share % (2017) | Economic Activity | GDP Share % (2016) | Economic Activity | GDP Share % (2016) |
| Wholesale and retail trade; repair of motor vehicles and motorcycles | 25.8 | Mining and quarrying (including crude oil and natural gas) | 35.9 | Mining and quarrying | 30.3 |
| Transportation and storage | 11.2 | Construction | 9.9 | Construction | 11.9 |
| Financial and insurance activities | 11 | Financial and insurance activities | 9.0 | Wholesale and retail trade; repair of motor vehicles and motorcycles | 10.0 |
| Manufacturing | 9 | Public administration and defence; compulsory social security | 7.3 | Financial and insurance activities | 9.6 |
| Real estate activities | 6.8 | Manufacturing | 6.5 | Manufacturing | 9.0 |
| Public administration and defense; compulsory social security | 6.8 | Wholesale and retail trade; repair of motor vehicles and motorcycles | 5.7 | Public administration and defense; compulsory social security | 8.7 |
| Construction | 6.5 | Real estate activities | 5.4 | Real estate activities | 7.7 |
| Professional, scientific and technical activities | 4 | Electricity, gas, and water supply; waste management activities | 4.1 | Financial intermediation services indirectly measured (FISIM) | 4.7 |
| Information and communication | 4 | Transportation and storage | 3.3 | Professional, scientific and technical activities; Administrative and support service activities | 3.7 |
| Accommodation and food service activities | 4 | Information and communication | 2.8 | Transportation and storage | 3.3 |
| Electricity, gas, steam and air conditioning supply | 3.4 | Professional, scientific and technical activities | 2.4 | Education | 2.1 |
| Administrative and support service activities | 3.3 | Administrative and support service activities | 1.6 | Human health and social work activities | 2.0 |
| Mining and quarrying | 1 | Education | 1.6 | Information and communication | 1.8 |
| Human health and social work activities | 1 | Human health and social work activities | 1.6 | Arts, entertainment and recreation; other service activities | 1.6 |
| Education | 0.7 | Accommodation and food service activities | 1.2 | Accommodation and food service activities | 1.2 |
| Other service activities | 0.5 | Agriculture, forestry and fishing | 0.7 | Electricity, gas, water supply, sewerage and waste management | 0.7 |
| Activities of households as employers; undifferentiated goods- and services-producing activities of households for own use | 0.5 | Activities of households as employers | 0.6 | Activities of households as employers; undifferentiated goods and services producing activities of households for own use | 0.7 |
| Arts, entertainment and recreation | 0.3 | Arts, recreation and other service activities | 0.3 | Import duties | 0.3 |
| Water supply; sewerage, waste management and remediation activities | 0.1 | | | Agriculture, forestry and fishing | 0.2 |
| Agriculture, forestry and fishing | 0.1 | | | | |
| Gross Domestic Product | 100 | | 100 | | 100 |

*5.2. Branding and Implementing the Development of an Innovation System*

The Qatar, Abu Dhabi and Dubai governments all consider high-level research a key asset to their long-term technological development and find a highly educated population with the skills required for a thriving economy in the 21st century vital to mobilise the rising numbers of local youth. Among the three emirates, Qatar is the most outspoken on this matter and even made human development one of its four key priority areas. Both Qatar and Abu Dhabi explicitly mention the relevance of social, moral and religious values in the education system, while tolerance and cosmopolitan attitudes come more to the fore in the case of Dubai's policy documents. In a similar vein as for economic diversification, special designated areas are sought that can function as receptacles for efforts to strengthen the national higher education system. The most famous one is 'Education City' in Qatar, in which elite international academic institutes were invited and on which $ US 2 billion was spent through the 'Qatar Foundation' (United Nations Global Compact, 2015). Established within this internationally oriented 'Education City' are campuses of Virginia Commonwealth University, Weill Cornell Medical College, Texas A&M University, Carnegie Mellon University, Georgetown University School of Foreign Service, Northwestern University and HEC Paris. Another large research and education programme, and one with a more national orientation, is 'Qatar Science and Technology Park', which received monetary resources from the Qatar National Research Fund. Qatar University, the main domestic university, is located there and still boasts higher student numbers than its international counterparts. In sum, Qatar is reported to spend around 9.3% of its total expenditure on education, which is clearly the highest in the Middle East/North Africa region [47]. Abu Dhabi has also found a sizeable number of international universities willing to set up subsidiaries within its borders, i.e., University of Strathclyde Business School, New York University, European International College, New York Institute of Technology, Paris Sorbonne University and Université Mohammed V-Agdal Abu Dhabi. The Abu Dhabi government targets its research funding mainly at the sectors microelectronics, health, culture and heritage and various forms of energy. The lion's share of its funds for research and educational development is earmarked to the Masdar Institute of Science and Technology (MIST), funded by Abu Dhabi's financial powerhouse, the Mubadala Investment Corporation. MIST is located within Masdar City, which historically was branded the first zero carbon town in the world. MIST, in which Massachusetts Institute of Technology plays an active role, targets the education and training of the future energy leaders in new specialized fields, it pushes R&D for technological innovation in sustainability, and supports the development of a knowledge-based economy [51]. Recent evidence (www.masdar.ac.ae) and insights gained from our own fieldwork (site visit and interviews) suggest that its expansion has stalled in recent years, and that it no longer exists as an independent institute and has been integrated with Khalifa University and become a part of the regular Abu Dhabi higher education system. In addition to MIST, Abu Dhabi also has a number of domestic universities (Abu Dhabi University, Khalifa University, Al Ain University of Science and Technology) and the federal Zayed University, which has campuses in both Abu Dhabi and in Dubai. Dubai's main higher education project is Dubai International Academic City, which has collected the highest number of international higher education institutes. However, as Table 4 shows, these academic institutes can all be found together in an area more or less the same size as Qatar's Education City while the funds in Dubai are mostly derived from private sector investments. Dubai's International Academic City lists academic institutes with lower positions on international rankings and is therefore far less significant as an attempt to grow a world-class high-tech innovation system than Qatar and Abu Dhabi.

**Table 4.** Academic institutions in Qatar, Abu Dhabi and Dubai.

| Emirates | Location | University | Level of Study | Established in Dubai | Number of Students |
|---|---|---|---|---|---|
| Dubai | Dubai International Academic City | University of Dubai | Bachelor, Master, PhD | 1997 | 768 |
| | | Zayed university * | Bachelor, Master | 1998 | 2114 |
| | | Higher Colleges of Technology * | Applied Diploma, Bachelor, Master | 1988 | 733 |
| | | The National Institute for Vocational Education | Diploma, Certificate | 2006 | 210 |
| | | Amity University Dubai | Bachelor, Master, PhD | 2011 | 1882 |
| | | Birla Institute of Technology and Science Pilani | Bachelor, Master, PhD | 2000 | 1603 |
| | | British University in Dubai | Bachelor, Master, PhD | 2004 | 1139 |
| | | Cambridge College International | n/a | 2007 | 69 |
| | | Curtin University | Undergraduate, Postgraduate coursework | 2017 | 10 |
| | | ESMOD French Fashion Institute | Bachelor, Certificate | 2006 | 103 |
| | | Heriot-Watt University | Bachelor, Master, PhD | 2005 | 3644 |
| | | Institute of Management Technology Dubai | Bachelor, Master | 2006 | 499 |
| | | MENA College of Management | Bachelor | 2014 | 292 |
| | | Manipal Academy of Higher Education | Bachelor, Master, PhD | 2003 | 2343 |
| | | Murdoch University | Diploma, Bachelor, Master | 2008 | 718 |
| | | S P Jain School of Global Management | Bachelor, Master, PhD | 2004 | 1628 |
| | | Shaheed Zulfikar Ali Bhutto Institute of Science and Technology | Bachelor | 2003 | 706 |
| | | University of Birmingham | Bachelor, Master | n/a | n/a |
| | | University of St. Joseph | Bachelor | 2008 | 69 |
| | Dubai Knowledge Village | Islamic Azad University | Bachelor, Master, PhD | 2004 | 423 |
| | | Michigan State University | Master | 2008 | n/a |
| | | Middlesex University | Bachelor, Master | 2005 | 3141 |
| | | SAE Institute | Bachelor | 2005 | 403 |
| | | The University of Manchester | Master | 2005 | 529 |
| | | University of Bradford | Master | 2009 | 132 |
| | | University of Exeter | Bachelor, Master, PhD | 2006 | 49 |
| | | University of Wollongong | Bachelor, Master, PhD | 1993 | 3905 |
| | Dubai Internet City | Emirates Aviation University | Diploma, Certificate, Bachelor, Master | 1991 | 1537 |
| | | Hult International Business School | Bachelor, Master | 2008 | 395 |
| | Dubai Media City | American University | Diploma, Certificate, Bachelor, Master | 1995 | 2297 |
| | Dubai Silicon Oasis | Rochester Institute of Technology | Bachelor, Master | 2008 | 891 |
| | Dubai South | University of South Wales | Bachelor, Master | 2017 | n/a |
| | Jumeirah Lake Towers | Moscow University for Industry and Finance | Bachelor, Master | 2013 | 95 |
| | | MODUL University | Bachelor, Master | 2016 | 255 |
| | Deira | London Business School | Master | 2006 | 208 |
| | Dubai International Financial Centre | CITY University of London | Bachelor, Master | 2007 | 240 |
| Abu Dhabi | City of Abu Dhabi | Khalifa University * | Bachelor, Master, PhD | 1989 | 1336 |
| | | Petroleum Institute | n/a | 2006 | 1654 |
| | | Abu Dhabi Polytechnic | Diploma, Certificate, Bachelor | 2010 | 642 |
| | | Emirates College for Advanced Education | Bachelor, Master, PhD | 1993 | 369 |
| | | University of Strathclyde | Bachelor, Master | 1995 | 201 |
| | | New York Institute of Technology | Bachelor, Master | 2005 | 163 |
| | | Sorbonne University | Bachelor, Master | 2006 | 630 |
| | | Mohammed V University | Bachelor, Master, PhD | 2009 | n/a |
| | | Abu Dhabi School of Management | Master | 2013 | 250 |
| | Khalifa city | Abu Dhabi University | Bachelor, Master, PhD | 2003 | 4374 |
| | Masdar City | Masdar Institute | n/a | 2007 | 417 |
| | Al Mafraq hospital | Fatma College of Health Sciences | Bachelor | 2006 | 612 |
| | Mohammed Bin Zayed City | Abu Dhabi Vocational Education and Training Institute | Diploma, Certificate | 2007 | 766 |
| | Saadiyat Island | New York University | Bachelor | 2010 | 618 |

Table 4. Cont.

| Emirates | Location | University | Level of Study | Established in Dubai | Number of Students |
|---|---|---|---|---|---|
| Qatar | City of Doha | Qatar University * | Bachelor, Master, PhD | 1973 | 14000 |
| | | American Education Center | Certificate | 2005 | 400 |
| | | Doha Institute for Graduate studies * | Master | 2011 | 350 |
| | | College of North Atlantic | Diploma, Bachelor | 2002 | 2000 |
| | Education City-Doha | Hamad Bin Khalifa University * | Master, PhD | 2010 | 6000 |
| | | Carnegie Mellon University | Bachelor | 2004 | 384 |
| | | Weill Cornell Medical College | Bachelor | 2001 | n/a |
| | | Northwestern University | Bachelor | 2008 | n/a |
| | | HEC Paris | Master, Certificate | 2010 | 4000 |
| | | Academic Bridge | Diploma, Certificate | 2001 | n/a |
| | | Georgetown university school of foreign service | Bachelor | 2005 | n/a |
| | | Texas A&M University | Bachelor, Master | 2003 | 635 |
| | | Virginia Commonwealth University | Bachelor, Master | 2010 | 339 |
| | Cultural Village | Doha Film Institute | Certificate | 2010 | n/a |

* These are public universities; all others are private

While the number of domestic and especially international universities in the three emirates is impressive, especially in view of their population size, it is less clear how comprehensive the range of the Bachelor, Master and PhD degrees is that these universities and research institutes offer and to what extent the campuses where they are located can be seen as promising building blocks for a thriving high-tech innovation system.

*5.3. Branding and Implementing the Smart and Sustainable Cities*

In light of rising population numbers and increased wealth, it goes without saying that all three emirates have experienced and are still experiencing rapid urban expansion. The question is to what extent smart and sustainable features highlighted in their plans are reflected in new urban development projects. In 2010, Abu Dhabi was the first among them to launch a green building code called 'Estidama'. It consists of environmental, economic, social and cultural aspects, and makes a distinction in five different types of buildings (office, retail, multi-residential, school and mixed uses). It awards 'Pearl credit points' for sustainable solutions (with a minimum score of 1 being legally compulsory) and is to be applied to all new buildings in Abu Dhabi. Qatar also adopted green building standards, entitled the 'Qatar Sustainability Assessment System' (later renamed to 'Global Sustainability Assessment System') which applies urban connectivity, site, energy, water, materials, indoor environment, cultural and economic values and management and operations as relevant themes. Government buildings have been required to adopt this assessment framework from 2016 on, but the enlargement of its wider application is yet to be realized [56]. Dubai currently has the most demanding regulatory system for green buildings in place: it has adopted compulsory green building regulations which apply to all types of building construction. These are to be used regardless of other rating systems and are not intended merely to substitute these. Here, distinction is made between villas, commercial buildings, public buildings and industrial buildings, and for each of these categories' application can occur to new constructions, additions/extensions/refurbishments and existing buildings. One could say that while Abu Dhabi came first among the three emirates, Dubai is now seen as the frontrunner in promoting green buildings. That said, it cannot be denied that in all three emirates standard buildings are still quite large and consume vast amounts of resources, water and energy, compared to international standards. Moreover, users and residents are gently encouraged to scale down their consumption rates rather than being legally obliged to do so: the carrot and the sermon are used instead of the stick.

As we can see in Table 5, when it comes to the developmental of new smart and sustainable residential areas, the emerging picture is much in line with what we saw for urban spaces promoting economic diversification and world-class innovation: Qatar and Abu Dhabi put most of their cards on a small number of very large project areas, while Dubai develops a higher number of smaller ones.

Abu Dhabi clearly boasts the globally best-known prestige smart city project: Masdar City. Likewise, in Qatar, Lusail Smart City is under development. Both mega urban development projects do not feature prominently in the official master plans, but their presence is very pronounced in online promotion videos and leaflets. Among the two, Masdar emerged earlier than Lusail; it was originally introduced as the first zero carbon city worldwide and later acquired a number of smart city features. Its general profile, however, still leans more on the creation of new green energy technologies and sustainable architectural features ('sustainable city') than on ICT and data processing. Lusail Smart City, on the other hand, is the essential interpretation of an Internet of Things style urban area where in a central command and operations centre incoming information taken from sensors all around the city is processed and acted upon. The smart city is part of the broader Lusail municipality expansion by the coast 23 kilometres north of Doha to accommodate the World Cup Football in 2022. Both Masdar and Lusail are generously funded by national investment corporations; subsidies and loans run in the billions of dollars. Disturbingly, it appears that progress for both of these mega-projects has slowed down and that ambition levels have been lowered, in spite of their impressive reputation among the uninitiated. In Masdar City, for instance, the personal rapid transit system never went beyond the testing stage, only the buildings constructed in the early stages are covered with solar panels and the exhibition hall shows few significant achievements in later years. In a similar vein, the optimism about Lusail Smart City has disappeared. It is in fact the much more small-scale and less actively branded Msheireb Downtown Doha (MDD) project that is gaining recognition. In MDD, data analytics is applied to a variety of sustainability aspects, such as the preservation of cultural heritage, transport systems and energy saving. Concerning the latter, data are collected and analysed on energy production and consumption, and use for awareness raising, enhancing energy efficiency and the smart metering of utilities. Notably, Dubai does not have prestigious smart or sustainable city projects boosted with similarly huge amount of investments, but it instead has a variety of more modest ones. Among them the Sustainable City, a mixed-use car-free township filled with 10,000 trees and organic farms with energy produced from solar parks, sustainable appliances installed in people's homes and various other features offering a low carbon lifestyle, is best known. Desert Rose City, with approximately 20,000 private residential units of the City for Emirati citizens and 10,000 for expatriates, was named after its shape. It runs on renewable energy produced in the city itself and is provided with a waste recycling facility. Dubai's South District, the location of the Expo 2020 site, has planned to have half of the electricity used in this event coming from renewable sources on the site and to reuse more than half of the material used in the construction of the site in future infrastructures. Finally, Smart Dubai is an emirate-wide data platform with a growing number of applications to the measurement and improvement of Dubai government services. Its main objective is to make inhabitants 'happy' through an improved quality of life. Smart Dubai is different from the others in the sense that it is not restricted to any specific urban space and therefore not an urban development project strictly speaking.

Table 5. Smart and sustainable cities and their key features in Qatar, Abu Dhabi and Dubai.

| | Name | Profile/Brand | Area/Location | Investment/Source of Funding | Key Features |
|---|---|---|---|---|---|
| Dubai | Dubai Smart City | The first happiest smart city in the world | Dubai land | Governmental | Renewable Energy<br>Electric vehicles (EVs)<br>Paperless government<br>Smart Health<br>Sustainability<br>Green building regulation |
| | Dubai Sustainable City | A sustainable lifestyle | 460000 sq. m/next to Dubai Studio City | Private funding (Diamond Developers) | Sustainable life style<br>Eco system services (eco system for birds, productive land with date palms, farms)<br>walkability<br>Innovation center<br>Green building regulation |
| | 'Desert Rose' Smart Sustainable City | A flower shaped environment-friendly city | 40 sq. km/a desert land at Dubai urban fringe | Dubai Municipality | Eco walk<br>Indicative accessibility (pedestrian and cyclists, light-rail, roads)<br>District Cooling<br>Vacuum Solid Waste Network<br>Multi-Utilities Tunnels Network<br>Electrical Network<br>Solar roofs & turbines |
| | Dubai South | The city of you- A city that defines itself by happiness of the individuals | 145 sq. km/next to Jebel Ali free zone | Governmental | The lieu of Expo 2020<br>Renewable energy<br>Self-sustained urban destination to empower businesses, families and individuals to grow and prosper |
| Abu Dhabi | Masdar City | A sustainable destination for residents and visitors to live, work, play and learn | 6 sq. km/beside Abu Dhabi International Airport | Mubadala Development Company | Clean Energy (Photovoltaic Power, Concentrated Solar Power, Wind, Waste-to-Energy, Energy Storage)<br>Sustainability<br>Eco-Villa prototype<br>Mobility (driverless Personal Rapid Transit)<br>Green building regulation |
| Qatar | Lusail | A city with a vision | 38 sq. km/located on the coast, about 23 km north of the city center of Doha | Qatari Diar Real Estate Investment Company | Eco Friendly Alternatives<br>Mobility (light rail transportation, Water transport system, Cycle and Pedestrian Ways System, Road hierarchy System)<br>Sustainable Infrastructures (District Cooling, Pneumatic Waste Collection, Sewage Treatment Plant)<br>Building rating (Gulf Sustainability Assessment System) |
| | MSHEIREB | Envisioning the city of the future | 764000 sq. m/Downtown Doha | Msheireb Properties (a subsidiary of Qatar Foundation) | Place Making and green building<br>Walkability, Mixed Uses<br>Authenticity<br>Sustainability<br>Mobility |

## 6. Results of the Analysis

Rentier states have generally been characterised as nations with governments that can rely on impressive amounts of revenue derived from their natural resources to keep their populations happy and themselves firmly in power. Some of these smaller rentier states have managed to convert their economic gain into reputation gain: they reached world fame at breakneck speed through impressive place branding activities. This why Qatar, Abu Dhabi and Dubai are now well-known around the world despite their small size and limited population size. In recent years, the era of post-oil and climate change has confronted them with a new and potentially even greater challenge: can they repeat the trick but then in terms of making an ecological miracle happen after having already created an economic one, and use their branding skills in a similar way to do this?

One of the areas where such a major environmental achievement is due is sustainable urbanisation. Major investments are made in developing new industrial and trade zones that go beyond oil and gas exploration and exploitation alone, in research and education cities where high-tech innovation is bred for future generations to lean on and in liveable and smart residential areas where residents may dwell in comfortable and ecologically friendly ways. In Sections 4 and 5, we have shown that the skillful approach Qatar, Abu Dhabi and Dubai developed in branding themselves as economically and financially successful places has indeed been replicated and possibly even further enhanced in coming across as being fully engaged in smart and sustainable urban development. This is visible both in their glossy and enthusing national visions and planning frameworks and in the urban development projects they engage in and promote globally through various channels.

The manner in which this was done, nevertheless, shows a remarkable difference between the larger and more conservative Qatar and Abu Dhabi on the one hand, and the more liberal Dubai on the other. Qatar and Abu Dhabi tend to initiate smaller numbers of far larger projects funded primarily by their national governments and/or large national investment corporations with dominant involvement of the national leadership. These mega-projects are widely touted as being cutting edge and have obtained global renown as being at the forefront of technological and urbanistic development. Education City in Qatar and Masdar City in Abu Dhabi are the most conspicuous examples of this. Abu Dhabi's Masdar City plays three roles at the same time: a free economic zone, home to a world-class research institute and a prestigious sustainable city. Qatar's Education was originally just set up for academic purposes, but recently pockets of it have been reserved as Free Trade Zones as well. This leads to conglomerate spaces where various functions are served at the same time. Conversely, Dubai has initiated a much larger number of relatively smaller urban development projects: it has more free economic zones, more foreign universities and more smart or sustainable urban neighbourhoods in its territory than the other two combined, although most of them do not exceed the size of just 1 square kilometer and they are all located at different sites. Much of the funding comes from private sources or is least mixed public-private in nature (although this certainly does not preclude Royal Family involvement). Dubai's branding capabilities are not inferior to those in Qatar and Abu Dhabi: its Media City, Internet City, Sustainable City and Smart Dubai have received or are beginning to receive wide international acclaim too. In other words, sustainable urbanisation in its various facets ranks highly on the policy agenda of all three emirates in the present study. All three have extensively engaged in place branding to boost their respective urban development initiatives derived from this policy agenda and all three have actually undertaken action to establish such smart and sustainable places. But there, the similarities end. An overview of key findings of our analysis is presented in Table 6.

Table 6. Summary table with key findings.

| Emirate | Rentier State | Active Place Branding | Free Economic Zones | Education/Academic Cities | Smart/Sustainable Residential Areas |
|---|---|---|---|---|---|
| Qatar | Yes (gas) | Yes | Small number of large ones | Small number of leading universities | One large project |
| Abu Dhabi | Yes (oil) | Yes | Small number of large ones | Small number of leading universities | One large project |
| Dubai | Not any more | Yes | Large number of small ones | Large number of middle-of-the-road universities | Several smaller projects |

Qatar and Abu Dhabi can both still rely on huge amounts of natural gas and oil, respectively, and may well continue to have sufficient reserves of these resources for decades to come. The post-oil policy is mostly a matter of preparing for climate change and a good reputation in handling international responsibilities. There is pressure to transform, but this pressure is primarily non-economic and therefore not directly felt by its governments and residents. In that sense, both emirates are still typical rentier states. Seen from that angle, it is unsurprising that their economic diversification is far from complete, that the intellectual output from their education and high tech areas (especially among Emirati and Qatari) is mediocre and that their smart and sustainable cities resemble modern versions of the 'technopoles' as described by Castells and Hall [57]. While Qatar has a stronger focus on promoting SMEs in light industrial sectors and Abu Dhabi on larger corporations in heavy manufacturing industry, in both cases their total economic weight is limited and the environmental gain of having them in place questionable. Moreover, their supply-driven approach has led to disappointing interest among private and international developers. Functionally they are less smart and sustainable than many would have wished. Their governments may well have the financial means to build facilities and places for the new age and change aspects of the production system, but they make this attempt in the absence of strict regulations which can be enforced. This becomes even more visible in the urging of their locals to conserve natural resources, like energy and water: this is all highly unpopular and mostly left to people's free will. Economic incentives are not administered, or rather they favour a lavish, rather than economical use of resources, and the political odds are likely to remain for some time to come. In this system, the branding of sustainable urbanisation is more sophisticated than the policy action itself can be, and being or having in place a rentier state lends much credibility to the argument that this context is notoriously difficult to change. Becoming smart and sustainable in action rather than words is tough if there is no physical or economic necessity to do so.

The story of Dubai is different. Its having depleted the lion's share of its natural resources has effectively made it an ex rentier state. Following the logic of Davidson's (2012) mechanisms for political stability and change, we might have predicted an imminent regime shift [15]. However, no such thing has occurred yet, and there are few if any signs of this being in the offing. What has in fact happened thus far is that the Dubai leadership encouraged a much more thorough economic transition by offering a great variety of differently tagged enclaved spaces all aimed at attracting different types of business sectors, educational audiences and residential target groups. Its urban expansion is far more pervasive and less selective than that of Qatar and Abu Dhabi and aims to offer different things to different corporations and inhabitants in different enclaves. A greater share of the investments is funded from private sources and there are many more free economic zones, academic institutes and smart and/or sustainable neighbourhoods, each of a smaller scale and less exclusive.

Given a higher level of dependence on private sector and international investors and settlers, the demand orientation in Dubai is more pronounced and the result is a better match with the needs of its intended clients. We should qualify this positive economic assessment to the extent that Dubai had to be saved from bankruptcy by Abu Dhabi in the aftermath of the financial crisis and therefore received external support. Its economic diversification has clearly been much more successful, but not all of it is top-notch. There is enormous variation in foreign academic institutes to choose from, but all of them are medium-range, private and expensive. And likewise, there are many profiles of differently styled residential areas and a small number of them reveal a serious attempt to create sustainable lifestyles. However, since many more others are built at the same time that offer impressive villas inviting high consumption levels without ecological features, the net environmental gain is limited. This practice in sustainable urbanisation allows for Dubai to appear as the number 1 on key performance indicators such as the fast rise in service oriented economic activity, number of foreign universities in its midst and presence of green buildings in its territory, while keeping its ecological footprint at unprecedented levels and almost on a par with Qatar and Abu Dhabi.

That said, Dubai is recognised by representatives from other emirates as being leader in taking steps to save energy and being serious about green building regulations. If anything dynamic towards sustainability will happen, it is most likely to be there. Transformative intentions in ex rentier states may indeed land on more fertile soil than in states still replete with gas and oil reserves. At the same time, it is tempting ultimately to draw the conclusion that from an environmental point of view, smart-sustainable urban forms in Qatar, Abu Dhabi and Dubai can be perceived as a 'branding hoax' [58]: their economic success is undeniable and the institutional mechanisms for growth and expansion operate as strongly as before, but the proclaimed ecological miracle at an aggregate level remains just a promise.

For their permanence, regimes in rentier states as well as ex rentier states seem to depend on continued economic success. Qatar, Abu Dhabi and Dubai have all three done a truly impressive job in redirecting financial revenues derived from their natural resources in sophisticated ways. All three have used advanced place branding strategies to achieve this, with Dubai having largely completed its industrial transition already since circumstances forced it to do so. The regime and leadership enjoy broad support and there are few indications that this will change any time soon. But successful recipes for dramatic industrial growth and urban expansion are unlikely bedfellows for reducing consumption rates of materials, water and energy, also when they go by the attractive name 'sustainable urbanisation'.

## 7. Conclusions and Discussion

This paper set out to answer two research questions:

1. Which steps have been taken to substantiate Qatar, Abu Dhabi and Dubai's claims to economic and ecological modernization?
2. Is there a difference in this between rentier states (Abu Dhabi and Qatar) and former rentier states (Dubai)?

Regarding the first question, steps undertaken to achieve economic and ecological modernization regimes seem to depend on continued economic success. Strategic policies in Qatar, Abu Dhabi and Dubai that aim at preparing for a post-oil economy are characterized by a diversification of their industrial activities and a preservation of their natural environment. All three have all three have indeed managed to redirect financial revenues and succeeded in in branding themselves as economically and financially successful places, for smart and sustainable urban development, as can be seen in their glossy and enthusing national visions and planning frameworks and the large-scale urban development projects they engage in.

Concerning the second question, we found a clear distinction between current rentier states Abu Dhabi and Qatar on the one hand, and semi-rentier state Dubai on the other. Whereas the former can still rely on the availability of oil and natural gas, the latter cannot and has been urged to cope with this new situation. However, it managed to do so through creating policies to diversify its economy and remain an attractive space in the region. It also led Dubai increasingly to differentiate itself economically from Abu Dhabi and Qatar, such as by going for quantity, variety and a demand orientation rather than prestige, selectiveness and a supply-driven approach. It now has a large number of relatively small free economic zones (instead a small number of large ones), a large number of average universities (instead of a few top universities), and a large number of sustainability projects (instead of one world famous sustainability project). Qatar and Abu Dhabi did the exact opposite.

All three emirates did their share of branding, but semi-rentier state Dubai truly excelled in this endeavour. It also managed to largely complete its industrial transition; circumstances forced it to do so. Their regime and leadership continue to enjoy broad support and there are few indications that this will change any time soon. This does not imply that successful recipes for dramatic industrial growth and urban expansion go hand in hand with lower rates of consumption. Sustainable urbanisation is clearly more an economic than an ecological success, and therefore it cannot be claimed that the environmental part of the branding promise is kept.

*7.1. Discussion*

Unlike worries or predictions made by authors like Davidson (2008, 2012), Coates Ulrichsen (2011, 2017), and Luomi (2012) no serious regime shift in any of the three emirates has been witnessed so far. Nor does radical political change seem nigh. As for Qatar and Abu Dhabi, one may claim that no such turbulence could have been expected, given the fact that their oil and gas reserves are vast enough to carry on with lucrative exploitation for many decades ahead. This situation continues to allow them to buy the support of their citizens through generous policies. In the case of Dubai, the Monarchy and elite families around it have shown strong economic leadership and orchestrated industrial, knowledge and industrial transformation without even coming close to being under threat. With regard to Luomi's earlier observations (2012) about a proclaimed policy shift in the emirates made in response to negative aspects in their global image in environmental affairs and the leading role Abu Dhabi was found to play in turning this around through active attempts towards an energy transformation, the present study shows that all three emirates are now engaged in formulating ambitious policies to foster sustainable urban development. However, the branding of these policies is decidedly more effective than their implementation. Moreover, while the excitement about Masdar City appears largely to have died down, it is Dubai that has taken thfe lead in many fields and referred to as a model to follow with its various themed cities, its relatively stricter building regulations and its Sustainable City which is planned to be replicated in the nearby emirate of Sharjah. Vis-à-vis the set of objectives city branding strategies can have according to Shirvani Dastgerdi and Di Luca (2019) we observed that the focus city branding strategies Qatar, Abu Dhabi and Dubai have is on creating global uniqueness, attracting resources, and arguably wellbeing of citizens (although about them, without them) but to a lesser extent on other types of objectives like meaningfulness and acceptability, justifiable identity, and involving stakeholders [36].

In sum, it is fair to state that a pro-claimed shift has been achieved and secured in policy making by reducing petroleum-related activities and establishing alternative industries (admittedly not all of them clean) by setting first steps in the development of a knowledge economy and by developing signature smart and sustainable residential areas. However, this has not (yet) really resulted in mitigating the environmentally harmful consequences of the extremely generous production and consumption patterns in either of the three emirates. Critics may well claim that the impactful branding of their environmental policies and landmark development projects should actually be seen as advanced forms of greenwashing. Nor do they reflect forms of active domestic stakeholder involvement that may trigger more thorough socio-cultural transformation. In that sense, the observations made in the present study are in line with some of the criticism vented by Cugurullo (2013, 2016) on the policy strategies looming behind Masdar City [41,59]. Similar remarks, mutatis mutandis, could also be raised with respect to Lusail City and other mega-projects in the emirates. Building an effective sustainable and/or smart city required a holistic approach across various technologies, policy areas and stakeholders which is often complicated to realise [60].

*7.2. Limitations of This Study and Suggestions for Further Research*

Limitations to the study mostly pertain to case selection and data collection. Regarding case selection only three particular cases in the Middle East were analysed. Although they are very significant examples of rentier states, the outcomes here beg the question whether using a wider set of cases for our analysis, and perhaps even cases of emirates that never were rentier states, would have led to other results. For instance, would the absence of sophisticated branding or convincing leadership have led to different outcomes? Assuming that valid data can be retrieved (which at the time of writing still seems doubtful here and there), taking the other emirates within the UAE, Bahrain, Kuwait and cities within other Middle Eastern states would offer insight in what the results of variations in the scores for various variables types of (ex) rentier states, (un)sophisticated branding strategies and (in)effective policy implementation, pro-active or reactive leadership) would lead to. Online data collection in Qatar, Abu Dhabi and Dubai proceeded smoother than it would have in many places elsewhere, but obtaining interviews was sometimes less straightforward. Eventually we were able to interview a number of civil servants, business employees and academic experts, but no interviews were conducted with either leaders, officials or citizens. This may have biased the findings in our work to a certain extent. Nonetheless, we feel that in this contribution, a bridge was built between concepts derived from political science on small Arab monarchies and policy practices in sustainable urbanisation. Few if any such efforts to collect empirical evidence of this connection have previously been made. Future studies could address shortcomings of the present study and build on its findings. They could also take up the challenge of more critically addressing and weighing the actual long-term environmental benefits of the developmental strategies chosen by the emirates. Moreover, they could examine the political and administrative processes underlying policy changes and the roles various stakeholders play in them.

**Author Contributions:** M.D.J. wrote most of the article and is the main author, data collection has been done by all, guided by N.N. and analyzed by all. Final editorial work has been done by T.H.

**Funding:** This research was funded by The Netherlands Organization for Scientific Research (NWO), with project number 467-14-153.

**Acknowledgments:** The authors are indebted to Zakaria el Khelloufi and Zakaria Mohamed for their support in the data collection. They are also grateful to Mustapha Aanzi and Tim Rogmans for introducing us to many of the respondents and to Christopher Davidson and Aziza Mayar for comments on an earlier version of this paper.

**Conflicts of Interest:** The authors declare no conflict of interest.

## References

1. Davidson, C. (Ed.) *Power and Politics in the Persian Gulf Monarchies*; Hurst & Company: London, UK, 2011.
2. Fromherz, A.J. *Qatar: A Modern History*; Georgetown University Press: Washington, DC, USA, 2017.
3. Coates Ulrichsen, K. *The United Arab Emirates: Power, Politics and Policymaking*; Routledge: London, UK, 2017.
4. Davidson, C. *Dubai: The Vulnerability of Success*; Hurst & Company: London, UK, 2008.
5. Davidson, C. *Abu Dhabi: Oil and Beyond*; Hurst & Company: London, UK, 2009.
6. Kamrava, M. *Qatar: Small State, Big Politics*; Cornell University Press: Ithaca, NY, USA, 2013.
7. Roberts, D.B. *Qatar: Securing the Global Ambitions of a City-State*; Hurst & Company: London, UK, 2017.
8. Commins, D. *The Gulf States: A Modern History*; I.B. Tauris: London, UK, 2012.
9. Syed, A. *Dubai: Guilded Cage*; Yale University Press: New Haven, CT, USA, 2010.
10. Luomi, M. *The Gulf Monarchies and Climate Change: Abu Dhabi and Qatar in an Era of Natural Unsustainability*; Hurst & Company: London, UK, 2012.
11. Tok, E.; Al Mohammad, F.; Al Merekhi, M. Crafting smart cities in the gulf region: A comparison of Masdar and Lusail. *Eur. Sci. J.* **2014**, *2*, 1857–1881.
12. Dinnie, K. (Ed.) *City Branding: Theory and Cases*; Palgrave Macmillan: Basingstoke, UK, 2011.
13. Govers, R.; Go, F. *Place Branding: Glocal, Virtual and Physical Identities, Constructed, Imagined and Experienced*; Palgrave Macmillan: Basingstoke, UK, 2009.
14. Westwood, S. Branding a 'new' destination: Abu Dhabi. In *Destination Brands: Managing Place Reputation*; Nigel, M., Pritchard, A., Pride, R., Eds.; Routledge: London, UK, 2011.
15. Davidson, C. *After the Sheikhs: The Coming Collapse of the Gulf Monarchies*; Hurst & Company: London, UK, 2012.
16. Coates Ulrichsen, K. *Qatar and the Arab Spring*; Hurst & Company: London, UK, 2014.
17. World Bank. United Arab Emirates. 2019. Available online: https://data.worldbank.org (accessed on 15 March 2019).

18. Coates Ulrichsen, K. *Insecure Gulf: The End of Certainty and the Transition to the Post-Oil Era*; Hurst & Company: London, UK, 2011.
19. Krane, J. *Dubaiu: The Story of the World's Fastest City*; Atlantic Books: London, UK, 2015.
20. Kazerouni, A. *Le Miroir des Cheikhs; Musee et Politique dans les Principautes du Golfe Persique*; Presses Universitaires de France: Paris, France, 2017.
21. Makadam, S.; Ramaswamy, R. Sustainable smart city: Masdar (UAE) (A City: Ecologically Balanced). *Indian J. Sci. Technol.* **2014**, 9. [CrossRef]
22. Al Naimi, A.; Karani, G.; Littlewood, J. Stakeholder Views on Land Reclamation and Marine Environment in Doha, Qatar. *J. Agric. Environ. Sci.* **2018**, *7*, 32–39.
23. Angelidou, M. Smart city planning and development shortcomings. *TeMA. J. Land Usemobil. Environ.* **2017**, *10*, 77–94.
24. Goess, S.; de Jong, M.; Meijers, E. City branding in polycentric urban regions: Identification, profiling and transformation in the Randstad and Rhine-Ruhr. *Eur. Plan. Stud.* **2016**, *24*, 2036–2056. [CrossRef]
25. De Jong, M.; Chen, Y.; Joss, S.; Lu, H.; Zhao, M.; Yang, Q.; Zhang, C. Explaining city branding practices in China's three mega-city regions: The role of ecological modernization. *J. Clean. Prod.* **2018**, *179*, 527–543. [CrossRef]
26. Noori, N.; de Jong, M. Towards Credible City Branding Practices: How Do Iran's Largest Cities Face Ecological Modernization? *Sustainability* **2018**, *10*, 1354. [CrossRef]
27. Baker, B. *Destination Branding for Small Cities*, 2nd ed.; Creative Leap Books: Portland, OR, USA, 2012.
28. Hankinson, G. Place branding research: A cross-disciplinary agenda and the views of practitioners. *Place Branding Public Dipl.* **2010**, *6*, 300–315. [CrossRef]
29. Zenker, S.; Braun, E.; Petersen, S. Branding the destination versus the place: The effects of brand complexity and identification for residents and visitors. *Tour. Manag.* **2017**, *58*, 15–27. [CrossRef]
30. Boisen, M.; Terlouw, K.; Groote, P.; Couwenberg, O. Reframing place promotion, place marketing, and place branding—Moving beyond conceptual confusion. *Cities* **2017**, *80*, 4–11. [CrossRef]
31. Lu, H.; de Jong, M.; Chen, Y. Economic city branding in China: The multi-level governance of municipal self-promotion in the Greater Pearl River Delta. *Sustainability* **2017**, *9*, 496. [CrossRef]
32. Kavaratzis, M. From city marketing to city branding: Towards a theoretical framework for developing city brands. *Place Branding* **2004**, *1*, 58–73. [CrossRef]
33. Kavaratzis, M.; Ashworth, G. City branding: An effective assertion of identity or a transitory marketing trick? *Ahmadreza Shirvani Dastgerdi Giuseppe De Luca Geographica Pannonica* **2005**, *96*, 506–514. [CrossRef]
34. Baker, B. *Destination Branding for Small Cities: The Essentials for Successful Place Branding*; Destination Branding Book: Portland, OR, USA, 2007.
35. Anholt, S. *Places: Identity, Image and Reputation*; Springer: Basingstoke, UK, 2016.
36. Shirvani-Dastgerdi, A.; De-Luca, G. Boosting city image for creation of a certain city brand. *Geogr. Pannonica* **2019**, *23*, 23–31. [CrossRef]
37. Hudson, S.; Cárdenas, D.; Meng, F.; Thal, K. Building a place brand from the bottom up: A case study from the United States. *J. Vacat. Mark.* **2017**, *23*, 365–377. [CrossRef]
38. Kavaratzis, M.; Kalandides, A. Rethinking the place brand: The interactive formation of place brands and the role of participatory place branding. *Environ. Plan. A* **2015**, *47*, 1368–1382. [CrossRef]
39. Pedeliento, G.; Kavaratzis, M. Bridging the gap between culture, identity and image: A structurationist conceptualization of place brands and place branding. *J. Prod. Brand Manag.* **2019**. [CrossRef]
40. Al Madani, S. LinkedIn Profile. Available online: https://ae.linkedin.com/in/dr-sara-al-madani-b4b84722 (accessed on 30 April 2019).
41. Cugurullo, F. Building a sand castle: An analysis of the genesis and development of Masdar City. *J. Urban Technol.* **2013**, *20*, 23–37. [CrossRef]
42. Yin, R. *Case Study Research: Design and Methods*; Sage Publications: Los Angeles, CA, USA, 2003.
43. Gerring, J. Qualitative Methods. *Annu. Rev. Political Sci.* **2017**, *20*, 15–36. [CrossRef]
44. Mahdavy, H. The Pattern and Problems of Economic Development in Rentier States: The Case of Iran. In *Studies in the Economic History of the Middle East*; Oxford University Press: Oxford, UK, 1970.
45. Government.ae. Plans and Initiatives for Sustainable Transportation. 1 July 2018. Available online: https://government.ae/en/information-and-services/education/importance-of-education-to-the-government (accessed on 30 February 2019).

46. *Qatar National Development Strategy 2011~2016*; Qatar General Secretariat for Development Planning: Doha, Qatar, 2016.
47. Qatar Higher Authorities. Qatar National Vision 2030. 2008, pp. 1–19. Available online: http://tinyurl.com/ha6fbgc (accessed on 30 February 2019).
48. Arcadis. Dubai, Abu Dhabi and Doha Are the Region's Most Sustainable Cities, Says New Index. 10 February 2015. Available online: https://www.arcadis.com/en/middle-east/news/latest-news/2015/2/dubaiabu-dhabi-and-doha-are-the-region-s-most-sustainable-citiessays-new-index (accessed on 15 April 2019).
49. DEWA. *Dubai Green Building Regulations & Specifications*; Government of Dubai: Dubai, UAE, 2013. Available online: https://www.dewa.gov.ae/en/consultants-and-contractors/policies-and-regulations/circulars-and-forms/green-building (accessed on 25 May 2018).
50. Dubai Government. Dubai Industrial Strategy 2030. 10 February 2015. Available online: https://www.dubaiplan2021.ae//wp-content/uploads/2016/06/Dubai-Industrial-Strategy-2030.pdf (accessed on 20 May 2018).
51. Government of United Arab Emirates. Annual Economic Report. 27 January 2017. Available online: http://www.economy.gov.ae/EconomicalReportsEn/MOEAnnualReport2017_English.pdf (accessed on 15 April 2019).
52. Government.ae. Dubai. 1 July 2018. Available online: https://government.ae/en/information-and-services/education/importance-of-education-to-the-government (accessed on 1 July 2018).
53. Government.ae. Dubai Clean Energy Strategy. 1 July 2018. Available online: https://government.ae/en/information-and-services/education/importance-of-education-to-the-government (accessed on 1 July 2018).
54. Qatar Education and Training. 4 May 2018; In www.export.gov. Available online: https://www.export.gov/article?id=Qatar-Education-and-Training (accessed on 04 May 2018).
55. Hazem Shayah, M.; Qifeng, Y. Development of Free Zones in United Arab Emirates. *Int. Rev. Res. Emerg. Mark. Glob. Econ.* **2015**, *2*, 286–294.
56. Qatar Green Building Council. QGBC & QNV 2030. 2018. Available online: https://qatargbc.org/aboutus/qgbc-qnv2030 (accessed on 15 May 2018).
57. Castells, M.; Hall, P. *Technopoles of the World: Making of 21st-Century Industrial Complexes*; Routledge: London, UK, 1994.
58. Yigitcanlar, T.; Lee, S.H. Korean ubiquitous eco-city: A smart-sustainable form or a branding hoax. *Technol. Forecast. Soc. Chang.* **2014**, *89*, 100–114. [CrossRef]
59. Cugurullo, F. Urban eco-modernisation and the policy context of new eco-city projects: Where Masdar City fails and why. *Urban Stud.* **2016**, *53*, 2417–2433. [CrossRef]
60. Yigitcanlar, T. Smart cities: An effective urban development and management model? *Aust. Plan.* **2015**, *52*, 27–34. [CrossRef]

© 2019 by the authors. Licensee MDPI, Basel, Switzerland. This article is an open access article distributed under the terms and conditions of the Creative Commons Attribution (CC BY) license (http://creativecommons.org/licenses/by/4.0/).

*Article*

# Smart Cities in Turkey: Approaches, Advances and Applications with Greater Consideration for Future Urban Transport Development

Can Bıyık

Department of Civil Engineering, Faculty of Engineering and Natural Sciences,
Ankara Yıldırım Beyazıt University, Ankara 06010, Turkey; cbiyik@ybu.edu.tr; Tel.: +90-312-906-22-53

Received: 26 April 2019; Accepted: 12 June 2019; Published: 17 June 2019

**Abstract:** The smart city transport concept is viewed as a future vision aiming to undertake investigations on the urban planning process and to construct policy-pathways for achieving future targets. Therefore, this paper sets out three visions for the year 2035 which bring about a radical change in the level of green transport systems (often called walking, cycling, and public transport) in Turkish urban areas. A participatory visioning technique was structured according to a three-stage technique: (i) Extensive online comprehensive survey, in which potential transport measures were researched for their relevance in promoting smart transport systems in future Turkish urban areas; (ii) semi-structured interviews, where transport strategy suggestions were developed in the context of the possible imaginary urban areas and their associated contextual description of the imaginary urban areas for each vision; (iii) participatory workshops, where an innovative method was developed to explore various creative future choices and alternatives. Overall, this paper indicates that the content of the future smart transport visions was reasonable, but such visions need a considerable degree of consensus and radical approaches for tackling them. The findings offer invaluable insights to researchers inquiring about the smart transport field, and policy-makers considering applying those into practice in their local urban areas.

**Keywords:** smart cities; mobility; visioning; policy

## 1. Introduction

In this century more than ever, cities need to construct smart transportation approaches, advances, and applications with much higher consideration for future urban development [1–4] since only then can they suggest new desirable urban environments and prioritize the aspects that are most critical and vital for their future [5–7].

Today, the smart transport concepts seem to pay more attention towards offering ecological [8–11] and economic development [12–15] and quality of living products using the abilities of innovative technologies [16–20]—perhaps as, in the close future, these are more gainful and remarkable tasks to deliver [21]. The concept of smart transport has been recognized by Smith et al. [22], where a critical level of participation among its experts and community can lead to approval of a desirable future place to make joint and shared means of action. According to Michaelson and Stacks [23], a wide variety of members from the public, practitioners, and scholars should be involved to draw an assessment of desirable future endpoints.

Amongst all smart transportation strategies, 'energy' is always paramount; that is why public and private companies play a chief role in future smart design [24–28]. On the other hand, smart transport is not only about energy and technology, but there are different combinations of applications, advances, and approaches that build the concept of future urban development.

Many different combinations of these applications exist in many worldwide cities—e.g., Adelaide, Amsterdam, Barcelona, Boston, Columbus, New York, San Francisco, Shanghai, Singapore, Tokyo, and Vienna—which goes together with these knowledge-based economic, ecological, and technological development efforts [29–32].

More recently, however, an increasing number of Turkish cities and metropolitan areas face a wide range of urban transport challenges from serious environmental illnesses, weak local governance, and a lack of efficient infrastructure [33,34]. Designing more efficient urban transport systems by using energy and technological solutions becomes more complex and uncertain when countless urban transport challenges emerge in growing Turkish cities due to a steady rise in urban population and in car-ownership [35,36]. Alternative visions can suggest aspirational ideas for the current transport challenges [37,38] and demonstrate a pathway of measures for future urban development [39,40]. However, until now there has not been academic research in Turkey that shows the role of vision assessments for smart transport urban development. It is obvious that if the reliability of smart future alternatives (and their implications of the pathways for the desired ones) are not explored and analyzed, then the existing urban transport challenges will remain in our urban environments. Therefore, the objective of this paper is to demonstrate the importance of vision developments and their justifications by a range of stakeholders involved in the implementation of future Turkish urban development. In addition, providing a great involvement process across the range of public, expert, and decision-makers is a primary target of urban policy development in this paper. Therefore, the intention was to gather a huge number of individuals from different geographic locations, different transport mode users, and diverse perspectives to design an efficient framework that integrates public engagement in future smart policy development.

Section 2 reviews the existing literature addressing smart transport approaches, advances, and applications. Section 3 details the theoretical framework for the research design, while Section 4 presents the main results are as follows: (a) public desires in scenario formation and (b) examples of changes to typical Turkish urban streets. Section 5 analyzes the reliability of the visions, according to the believability of key driver changes for alternative futures and the reactions of different transport mode users on the visions. Section 6 discusses how the targets of one desirable vision could be reached from the future to the present. Section 7 discusses and concludes the paper.

## 2. Smart Transportation Approaches, Advances and Applications

In the rapidly growing literature, there are countless smart transport definitions/descriptions—focusing on separate characteristics of key drivers or outcomes [41–45]. These are coined by researchers, practitioners, and government and international organizations and are generally vague or inchoate in conception [46]. However, there is not a commonly agreed explanation of smart transport cities [47]. Several various initiatives are now being implemented in different parts of the world. For example, 'ride sharing' is probably one of the most well-known implications. Private ride-sharing companies like Uber and Lyft have already introduced this mode of transportation, with more accessible services for using the high-occupancy vehicle lanes and providing more affordable costs for people [48]. Similarly, 'car-sharing applications' link drivers and passengers in real-time and offer alternative and easier transport journeys [49]. 'Smartphone mapping applications' show nearby cycling sharing systems and public cycling systems and increase convenience for users to see how many cycles are available at each station [50]. 'GPS-based route information on public transport' offers real-time GPS tracking from mobile devices thus optimizes public transportation journeys and thus increases the reliability of public transportation services [51]. The inclusion of 'traffic management' approaches provides great solutions to minimize traffic congestion [52]. 'Connected traffic signal' creates increased driving safety and fuel efficiency by re-arranging the current state of traffic lights and how they will change [53]. Another initiative is a sensor tracking system like a Radio-Frequency Identification (RFID) device, which shows parking availability within a paid meter on the street that can be visualized on a map [54].

The existing worldwide projects demonstrate that smart transportation city applications are frequently implemented in developed countries by various methods, but these interventions show prominent variances compared to different locations. For example, the United States Department of Transportation has submitted a grant of $40 million to provide support for the implementation of smart transport city ideas for their cities [55,56]. The Austin city council project includes an autonomous transport service provider from the airport to the nearby smart stations [57]. Columbus has decided to design towards expanding existing electric infrastructure and converting public vehicle fleets to electric and hybrid vehicles [58]. The Denver City Council has incorporated the data management ecosystem from several sources to provide a real-time picture of public and private transportation journeys and launched an electronic autonomous shuttle service along a one-mile route distance [59]. Detroit's smart city proposal includes new improved bus system [60] and a mobile application that maps out routes for users of public transport [61]. The New York City initiative includes a series of city service kiosks in the Link NYC network such as; free Wi-Fi, phone calls, device charging stations, local wayfinding, etc. [62]. In Pittsburgh, traffic light control systems are equipped to identify transit and freight vehicles and allow these vehicles to pass through the traffic lights quicker [63,64]. Portland has deployed building traffic sensors [65] and installed technology in fleet vehicles that can receive and transmit traffic condition data [66]. San Francisco uses wireless sensors to detect parking-space occupancy for a dynamic parking system that adjusts the cost of parking prices [67] and thereby reducing the time people spend looking for parking and avoiding distracted driving that disrupts pedestrians and cyclists [68].

The European Union (EU) has devised extensive use of transport city strategies to achieve a smart urban future for its metropolitan city-regions [69,70]. Amsterdam, for example, has improved smart street lighting which allows county municipalities to manage the brightness of streetlights [71]. These flexible traffic lights monitor dynamic traffic flow and provide information about existing travel patterns on a certain road to determine the best routes for different commuters [72]. Barcelona has made great strides in implementing information technology to reduce energy consumption and local emissions and in designing a new public bus network, utilizing primarily vertical, horizontal, and diagonal routes with several interchanges [73]. Bristol seeks to increase cycling mode share by 40% [74]. Kyiv Smart City initiative contains GPS trackers, installed on all public transportation fleets to decide favourable strategies for further infrastructural, technological and social development of the city in urban transport management [75]. The Kyiv Resident Card provides access to many local innovations and for all types of public transportation in the city [76]. The London Congestion Research Programme concluded that the city's economy would benefit from a congestion charge scheme by the restriction of private cars in central London [77] and by the introduction of extra on-street parking restrictions in the outer areas of the city [78]. In Vienna, all the subway lines, tram and bus routes are navigable through smartphone application updates for commuters [79].

In Asia, many Chinese and Indian cities launched smart cities missions to transform their cities into smarter and citizen-friendly local areas [80–82]. For, instance, in Indian cities, the smart systems assess real-time traffic information, particularly for emergency cases [83–87]. Shanghai's smart city mission focused on smart sensors in all buses and metro stations and the development of higher internet connection speeds have allowed increasing the productivity of the city [88,89]. Singapore has implemented several smart transportation initiatives, including an intelligent transport system, environmentally friendly transport, traffic management, smart airport initiatives, and driverless shuttle bus services [90,91]. For example, a shuttle bus service with 600 passengers per day was implemented by the Nanyang University of Technology in Singapore [92]. In Seoul, all the metro line systems are integrated with RFID and Near-Field Communication (NFC) for automatic ticket payment, helping customers to get their tickets with their smartphones [93,94]. Tokyo uses the 'Internet of Things' to design a safe and reliable rail network system and reduce the cost of maintenance closures [95]. In Australia, Adelaide's city council has evolved a more integrated network of villages in which the central business district is fed by outer suburbs [96,97].

Several African and South American cities have also started to work towards designing smart and modern transportation systems [98,99]. The Benin City transportation systems were upgraded with an advanced ITS system that assesses traffic flow in real time [100]. Medellín has implemented the sharing of transportation data between the different types of mass transit for providing a quick solution for unexpected shifts in the volume of passengers and massification of routes [101,102].

These are just some approaches, applications, and advances in how most developed cities are utilizing their innovative smart transportation strategies. It is obvious that the smart city innovations to date are evidently limited to developed regions and smart information knowledge from one city might not be properly transferrable to another city because of differences in human and technological frameworks, land-use, and transport patterns [10–31]. Considering these worldwide initiatives, Turkey is undoubtedly far behind in smart transportation city development [33–39]. There needs to be a lot of progress to create smart transportation initiatives in Turkish cities and it seems that the major problems faced by the cities in adoption of smart systems are lack of quality public transportation, road safety concerns, poor traffic management, lack of modal options and funding, and poor public transportation [103]. Although, a range of civil society activities was recently set up to energize planning, transport, and environmental authorities to develop strategies to promote smart systems, including The CityFix Turkiye, EkoIQ, Wrisehirler, and Embarqturkiye [104–110], there is still a small body of projects that are concerned with smart transport applications in Turkey [111]. In addition, a few good smart approaches, advances, and applications have been made in current practice, and these initiatives have been implemented by the central government, with public engagement being excluded from evidence, knowledge, debate, and the policy-making developments [112,113]. There is a relatively small body of literature that is concerned with the reliability of future transport visions in smart city development. Therefore, it is still not known whether the aspirational thinking of achieving desirable futures in Turkish urban areas could be workable. Thus, this paper seeks to add to the body of knowledge on the development of smart transport futures, thereby contributing to understanding the requirements to propose a radical change in future and imagination of archetypal areas inside a simulated Turkish transport environment. Together, this paper provides important insights on determining local policy actions and strategies to support policymakers in achieving their local objectives.

In addition, many international studies [114,115] have shown that smart transportation approaches, advances, and applications allude to deficits in a theoretical base and methodology. For example, much of the smart transport studies are limited to comprehensive public involvement. The previous studies on the ensuring of public involvement are not reliable [58–69]. Despite the emerging recognition that an organized involvement of stakeholders should become central to all the relevant features of smart transportation studies [70–84], limited attention has been paid to encouraging different stakeholders in all critical stages of smart city agendas [85–102]. Furthermore, visions are composed of several vital drivers and not all the key drivers are of similar desirability [115] and reflecting nuances of value-laden perspectives by separating different clusters of desirability makes it more distinctive to comprehend complex futures [116,117]. Just as two different urban areas do not have the same pathway for success, each area must construct their pathways to achieve the future targets based on a series of joint reflections, desires, and considerations [118–120]. The generalizability of much smart transport research on this issue is problematic. Although extensive research has been carried out on smart transport studies, the previous studies are not compatible with sustainability principles [121–124] and are not representing the interconnected mechanism of a desirable future through systemic relationship [71–78]. A more systematic approach would include main methodological principles of smart transport studies, in terms of providing public involvement in future urban transport development [125–129], developing systemic features of smart transport cities [130], establishing futures based on the concept of sustainability [114–126], as well as using 3D visualization techniques [131].

## 3. Methodology

The combination of several research methods in the paper was conducted due to the complexity of future transport policy development. Consideration of previous research and practical constraints [114–122] led to the development of a three-stage approach, with the engagement of the public, experts, and policy-makers a critical influence at all stages: (i) extensive online comprehensive survey; (ii) semi-structured interviews; and (iii) participatory workshops.

The visions were initially created through public outputs amongst the volunteers of the comprehensive online survey study and a series of extensive discussions, project meetings, and presentations with the members of the İstanbul Metropolitan Urban Design and City Planning Office. At the first round of these extensive meetings, Computer-Aided Design (CAD) and 3D modelling software were used to design archetypal components of an imaginary hypothetical Turkish urban area, but our survey findings from the online survey were also considered. Then, a visual description of each component for the current and future year was created along with the accompanying scenario narrative of the context in the next meeting. For each vision, visual images and scenario narratives were improved primarily based on the survey outcomes, with a very distinctive environmental, energy, urban change and transport mode-share context, though all with a radical restriction in private car use relative to current-time Turkish cities. Further meetings were organized to obtain feedback on the believability of our possible alternatives and their context. In the semi-structured interviews with the previous volunteer participants, the imaginary representations of each vision and their scenario assumptions were demonstrated to justify the reliability of alternative Turkish futures, according to imaginary descriptions and background assumptions of the visions and the travel behaviours of different mode users amongst the volunteer survey participants. The semi-structured interview was designed as a step to investigate what actions for the further stage might be required for smart development, rather than estimating real Turkish transportation future. In the last stage, a series of participatory workshops were organized to bring a group of experts and local and national policymakers (and a considerable number of the public from the previous stages) together to construct policy clusters and pathways for achieving the target of one specific vision in different geographical locations. The purpose of local workshops was to explore how one of the more desirable visions in this paper could be adapted to different local circumstances and how to create policy pathway measures across several Turkish cities from the present day to 2035.

To ensure that the above techniques could be carried out competently, the determination of stakeholders depended on the following criteria:

(i) "Public," including participants from different demographic and socio-economic groups and various mode choice transport users (pedestrians, cyclists, public transport users, drivers, and shop-owners). Recorded mailing lists from local associations were used to identify and select participants from this group in each case study city.
(ii) "Experts" with different experiences in the smart transport field, including urban planners, transport engineers, architects, civil engineers, academics, and civil society organizations. Recorded mailing lists were used from several Turkish professional companies, associations, and institutions to recognize participants.
(iii) "Policymakers," national and local transportation experts across five selected Turkish local municipalities (Ankara, Eskişehir, İzmir, İstanbul and Konya) to construct a timeline for the implementation of measures for their city.

*3.1. An Extensive Online Comprehensive Survey*

Setting up an efficient online email system for sourcing and recruiting research participants is the most appropriate way when the authors can quickly get access to lists of potential participants and the targeted participants in such lists already have an affiliation with an institution or sector [132]. E-mails were initially sent to only a portion of the people on the lists of potential participants. To get a higher participation response rate, the authors made several adjustments to the recruitment letters, then sent the revised letter to a different group of people on the long list of people to contact. The criteria for the urban location and stakeholder categories in which potential participants live were set out in the recruitment letter (see Section 3). A summary of the survey findings appears at this link: http://www.bisikletizm.com/bisikletli-ulasim-nasil-gelisebilir/ by clicking Anket Sonuçları.

A comprehensive research survey was conducted from March to June 2014 by using an online survey programme. The online survey link was sent to approximately 75,000 participants from the e-mail lists obtained from several Turkish public and private sectors. A total of 1135 people agreed to participate, and they were given a choice to provide their contact information for a possible follow-up interview and policy-development workshops. The main questions being addressed were how participants visualize their desirable futures. What are the expectations of participants regarding future smart transport visions? What are the key factors affecting the public choice of travel mode? What are the suitable policy measures to help achieve smart Turkish transport visions?

The socio-economic features of the participants are demonstrated in Table 1. Females accounted for 28.9% of participants, and males 71.1%. The age breakdown was: 18–35 years old (68%); 36–55 years (27.6%); and 55+ (3.7%). Less than a third (32%) indicated that they earn less than a €300 income per month, while 44% of the participants earn between €300 and €900. The remaining 24% earn over €1200 per month.

**Table 1.** Socio-economic characteristics of the participants.

| Factor | Subgroups | Number of Participants | Percentage |
| --- | --- | --- | --- |
| Gender | Female | 328 | 28.9 |
| | Male | 807 | 71.1 |
| Age | 18–25 | 390 | 34.4 |
| | 26–35 | 390 | 34.4 |
| | 36–45 | 194 | 17.1 |
| | 46–55 | 119 | 10.5 |
| | 56–65 | 42 | 3.7 |
| Income | No income | 203 | 17.9 |
| | Less than €300 | 160 | 14.1 |
| | €300–€600 | 240 | 21.1 |
| | €600–€900 | 260 | 22.9 |
| | €900–€1200 | 143 | 12.6 |
| | Over €1200 | 129 | 11.4 |

The survey text results were analyzed line by line and codes were assigned to the text. Then, the search for relations between conceptual survey texts and categories were examined. The goal was to understand possible drivers for smart transport developments in Turkey thoroughly. The critical changes in the vision development were categorized into four factors: (i) environmental solutions; (ii) technology; (iii) urban structure; and (iv) mode share. The scenario metaphors (see Section 4.1) and their visualizations (see Section 4.2) were mapped onto the possible Turkish urban areas designed, providing both a contextual description and associated generic representations of the vision storyline.

*3.2. Semi-Structured Interviews*

Semi-structured interviews offer a more open research process, where the interviewer has a series of general questions, as well as having some latitude to ask more detailed questions following up important issues. A total of 95 volunteer participants were asked to engage in the improvement and justification of the future visioning exercises through semi-structured interviews (see Table 2). Ninety-five in-depth interviews were conducted within the sample urban areas, and each semi-structured interview, based on a predefined guide, was designed to take about 15–25 min. The sample had a higher number of public participants (48), and the remaining participants were professionals (34) and decision-makers

(13) (Table 2). The interview work was completed in spring and summer of 2015. Each interview consisted of seven open-ended questions for three smart transport visions in Turkey by 2035, according to the following structure: (i) what are the views of the participants about the visions? (ii) Are these visions desirable? (iii) Do participants think the visions are consistent with their internal expectations? (iv) What differences would participants like to see regarding these systems? (v) What should central and local governments do? (vi) What are the requirements for such future changes? Moreover, (vii) What are the uncertainties regarding the visions?

Table 2. Description of participants in the semi-structured interviews.

|  | Ankara | Eskişehir | İstanbul | İzmir | Konya |
|---|---|---|---|---|---|
| Public | Two pedestrians Two drivers Three public transport users | Three drivers One public transport user One cyclist | Seven drivers Five public transport users Four pedestrians Two cyclists | Three cyclists Two public transport users Two pedestrians | Four drivers Three pedestrians Two public transport users Two cyclists |
| Experts | Three from university Two civil engineers Two urban planners | Three from civil society organizations Two from university One civil engineer | Four from civil society organization Three transportation engineers Two urban planners Two from university | Two urban planners Two traffic engineers Two from the private sector | Three transport planners One from university |
| Policy-makers | Two from the national government Two from local government | Two from local government | Two from local government One from the regional government | Two from local government | One from local government One from district municipality |

*3.3. Workshop*

The third stage of this study was operationalized through a series of five local policy construction workshops between January and April 2016. All policymakers involved in semi-structured interviews contributed to the local policy development workshop. The workshop size was between 25 and 30 people and this stage of the work mostly aimed to attract relatively senior participants from the local governments in each sample area and previous public and expert participants.

Each group were provided with samples of generic illustrations for specific visions in both a demonstration and in a hand-out. Further clarification was provided giving extended vision narratives for each vision (see Section 4.1). Questions to guide this clarification were asked as follows: what challenges exist for applying the strategies required for achieving the 2035 vision? For a specific vision, what policy measures will be implemented by 2020, 2025, and 2030? (for achieving the 2035 vision). The structuring of the qualitative data analysis process of the classification and the development of links between policy measures were simplified by using the NVivo (2.0) software programme.

## 4. Vision Development

The smart transport vision is future-oriented systems and offers urban mobility solutions for every individual [133–135]. The development of smart transport futures can be achieved through the Avoid-Shift-Improve (ASI) approach. These approaches are the most broadly adopted ways to deal with the challenges of existing urban transport systems [136,137]. Vision development is the result of a participatory process with a broad public, expert, and policy-maker involvement [23].

Figure 1 shows the most popular suggestions arising from 208 different responses from 1135 participants on designing future Turkish smart transport visions over the next two decades (see Appendix A). The most common factor is segregated cycle paths (72 participants), while the second highest measure was car speed reduction (53 participants). More cycling paths, restrictions for cars within the city centre, decreasing vehicle reduction, and the provision of pedestrian crossings are some of the other frequently described measures for the desirable future Turkish smart visions (Figure 1).

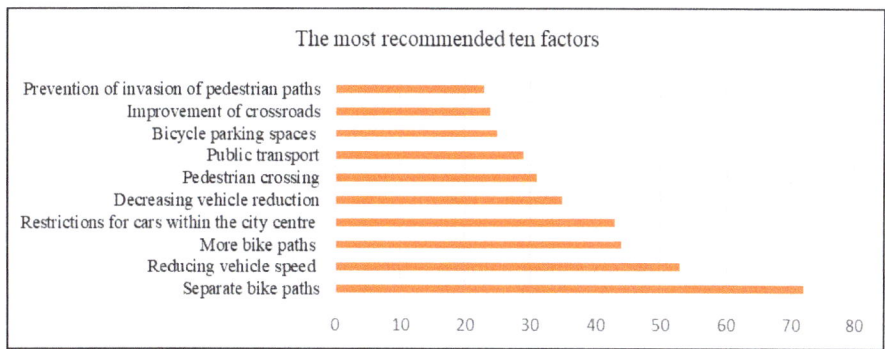

**Figure 1.** The most recommended 10 factors for Turkish transport visions.

These were initially analyzed by text line, and relevant quotations assigned to nine different conceptual codes (special groups; public awareness; incentives; smart transport strategies; urban features; walking; cycling; public transport systems; and preventing car use) (see Appendix A).

Each of the visions presents futures for parts of an ideal Turkish urban area where dependence on green transport systems has been increased and where provision for private cars has been substantially restricted. The future smart transport alternatives are distinctly different from the main perspectives, such as the approach to environmental solutions, technological innovations, urban structure changes, and mode-sharing arrangements (see Table 3). The visions were established based on participant recommendations from the comprehensive survey data, quotations of the volunteer participants from the open-ended questions in the survey work (see Section 3.1), and extensive discussion with the members of the İstanbul Metropolitan Urban Design and City Planning Office team through a series of committee meetings and presentations.

**Table 3.** Summary of 2035 visions for Turkish urban areas.

|  | Avoid Vision | Shift Vision | Improve Vision |
| --- | --- | --- | --- |
| Environmental solutions | - Reducing the need for car travel in urban areas | - Closing central parts of urban areas to the car | - Extensively expanded public transport systems for all commuters |
| Technological innovations | - Improvements to prevent possible traffic accidents among different users | - Implementing applications to promote walking and cycling | - Less traffic and emission-friendly public transport vehicles in traffic |
| Change in land use and urban form | - House prices in the city centre going up would make it harder to form compact cities | - More compact, mixed-use urban form<br>- The changes in urban form tend to happen slowly | - Rapid population growth is rapidly spreading to the outside of the urban area by forming new parks and forests. |
| Mode-sharing arrangements | - 40% walking; 5% cycling; 35% public transport; 20% car | - 45% walking; 10% cycling; 35% public transport; 10% car | - 40% walking; 5% cycling; 50% public transport; 5% car |

The scenario development meetings were carried out in a fast-moving brainstorm form, and the outcomes do not provide a straightforward input to tools for transportation planning such as mathematical modelling, travel behaviour changes or computational sciences. Integrated participant suggestions and participants quotations were elaborately structured by filling in the empty cell in the example table presented in Table 3. It was then thought most appropriate to shape the process of scenario development of the visions with the members of the urban design and planning office team who could be acquainted with some of the urban transport practices of the development proposed.

*4.1. Scenario Narratives*

4.1.1. Avoid Vision

People are being encouraged to meet their basic needs online to decrease the length of automobile travel in urban areas. Various conveniences and promotional coupons are being provided for online shopping and bill payments. Widespread use of digital technology would cause a significant decrease in transport demand. High energy prices are not effective in decreasing automobile dependency because the price of public transportation would also go up. Dedicated walking and cycling lanes would promote more children and young people to use non-motorized systems to go to school. In this vision, similar and moderate increases are foreseen in all three types of smart transport systems relative to the existing poor infrastructure systems.

Technological innovations would be most improved to decrease possible accidents and to develop a more environmentally friendly transportation system. Additionally, Intelligent Speed Adaptation (ISA) is installed in new vehicles. Digital technology is viewed as a meaningful solution to reduce road accidents. People would do major activities and pay their bills via state online programmes. Some meetings would be carried out from home by work platforms. Follow-up work systems are being monitored and reported more rigorously than before; however, there may be some implications of home working on active lifestyles.

There are no significant changes taking place in the physical structure of cities. House prices in the city centre going up would make it harder to form compact cities. Strategies to either provide good street lighting or physically separate cyclists from vehicle traffic would be expected to improve road safety significantly. Strengthening road infrastructure with a concern for safety and penalizing drivers for not giving priority to pedestrians.

Traffic and driver education programs become an integral part of compulsory activities and courses in Turkish primary and secondary schools, to encourage safe and responsible behaviour either as a driver, cyclist, or pedestrian. There would be more traffic signs visible around schools and shopping malls. Drivers will be prohibited from driving over 30 km/h on the busiest streets in the urban areas. Local administrations would receive funding for making cycling or walking transportation safer and more attractive by implementing calming traffic measures. The investments are highly associated with the automotive and technology sectors.

4.1.2. Shift Vision

Local administrations would encourage the public to use non-motorized transport modes mainly due to air pollution problems. With the development of newly pedestrianized locations, a decrease is expected in car dependency, and so a decline in air pollution emissions is expected. New settlement areas close to decrease car dependency and make a broader range of people easily able to use non-motorized transport modes. Public buses are cheaper and more comfortable than the current situation and enable different income groups to access town centres easily. Walking and cycling have significantly increased, and car use has dramatically reduced.

Technological applications help make walking and cycling more convenient. Weather reports, events, health measurement equipment, public transportation stops, and route information are easily accessible. Most of the people consider that digital technology will obliterate social interaction and they do not want to move to home working or internet shopping entirely. City centres would be reachable through small cars that operate on renewable energy systems.

Growing petrol prices would increase employment densities and so lead to denser urban areas, although the effects of increasing petrol prices on some features of urban form are hard to forecast. However, the changes in urban form tend to happen very slowly because local land use strategies constrain increases in urban density. Municipalities warn residents to park their automobiles in a way that would not block pavements and cycle lanes. High parking prices and cycling awareness events encourage people to utilize cycling. More extensive areas are assigned to cyclists and pedestrians. Pedestrianization projects and running parks are becoming more common to increase people's physical activity. Cycling and the integration of cycles with public transportation at transfer stations make cycling more appealing for different income groups.

Local administrations are encouraging cities to become more compact and multi-purpose. The pedestrianization of some locations in town centres offers significant advantages for the safe transport mode of cyclists and pedestrians. Denser urban areas would require automobiles to go slower, which in return helps to decrease possible accident risks in urban areas.

4.1.3. Improve Vision

Local administrations are trying to find efficient solutions for traffic jams and air pollution through a substantial reduction in car usage. Car drivers are encouraged to use workplace service buses or public transportation for their commutes. Car dependency shows the sharpest decline in this vision. Offering incentives to encourage the use of public transport options decreases the individual cost for such transport, and so encourages behavioural change. High energy demands would impose enormous hardships on private and public agencies in Turkey. The scarcity of energy resources will cause the price of fossil fuels to go up. Therefore, the government accepts the need to overcome the difficulty of procuring energy by investing in public transportation systems and by increasing awareness about sustainable energy. Supporting infrastructure developments have enabled public transport to become more convenient and people less dependent on car use compared to other visions.

A significant proportion of technological development consists of innovations regarding the improvement of fuel performance and economy. The new vehicles would operate on renewable energy that produces fewer emissions, or on electrical power. Technological developments will be limited, but city centres would be reachable through small cars that operate on renewable energy systems and would be integrated with cycles. Free-of-charge Wi-Fi systems spearhead public transportation systems, which are becoming a more practical transportation mode.

Local administrations are creating greenbelt areas to prevent city sprawl. The development of public transport would make new social and business locations in the outer parts of the city more appealing. The number of public transportation terminals would be increased in the outer parts of the city to increase accessibility. There would be a significant decrease in the number of severe accidents as drivers are encouraged to use public transportation services or non-motorized transport systems. Public transport drivers will be trained to be more aware of cyclists.

*4.2. Visualizations*

Three locations of the possible Turkish urban areas were designed as they were in 2015 in Figure 2. Specific archetypes used included a suburban area, an area close to a busy university campus, and part of the city centre. The residential area is a modern residential place to travel and live, but one where road parking is a crucial problem. The road is lacking the infrastructure to help pedestrians to cross the road safely and comfortably. There is a large taxi stand in the residential area, and unaffordable public transport links between the residential suburb and the outer locations of the city are the norm. The university campus is bound by a ring road, although beyond this, there was recent development such as scientific research and development centres and shopping areas. Pedestrian and bicycle access on the ring road are weak. The current roads for pedestrians and cyclists are narrow and uncomfortable. Illegal car parking along the road is common. A typical busy traffic corridor in the city centre has not successfully adapted to changing traffic circumstances over the years. The location is cluttered, and traffic congestion, noise, and local air pollution is the norm. The street lacks pedestrian infrastructure.

Figure 2. The current Turkish urban locations in 2015.

4.2.1. Avoid Vision

Avoid Vision decreases automobile dependency and proposes a more comfortable and safer active transportation system than the existing situation. Newly-arranged parking lanes prevent the pavement on the right side of the road being occupied by cars, enabling pedestrians to walk comfortably on the pavement. Measures to ease pedestrian use of the road have been implemented (Figure 3). In the new settlement areas, to reduce accident risks, suitable lighting systems improve the visibility of vulnerable road users. Properly placed barriers ensure safer cycling by separating the road from the motor vehicle. Cyclists can cross the road conveniently by using combined walking/cycling crossing places (see Figure 3). The fact that the cycle lane is to the left of the pedestrian walkway allows cyclists to travel faster. Some small-scale improvements have been made regarding the use of public transportation systems. There are reductions in available space for motor vehicles. Drivers are prohibited from driving over 30 km/h on the busiest streets, and greater enforcement is applied.

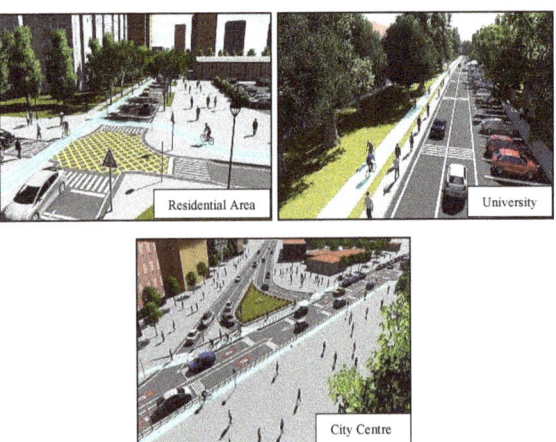

Figure 3. The Turkish locations as they might look in Avoid Vision.

### 4.2.2. Shift Vision

Figure 4 shows the same three locations as in Figure 3 and how they may look in 2035 under Shift Vision. The prevention of new settlement areas at the edges of the city is approached in a planned manner. The changes in urban form tend to happen very slowly because local land use strategies constrain increases in urban density. People tend to move through certain parts of the urban centres closer to workplaces. Shift Vision presents a broader and more socialized location for the users of non-motorized vehicles. There are numerous social facilities, such as cafes and art galleries, which enable people to socialize. Pedestrianization projects and open space are more familiar to encourage physical activity. A separate road at the far end of the broad pedestrian area was designed for cyclists. The pedestrianization of residential areas offers an opportunity for safe cycling. Public information spots are placed at the nearby university to raise awareness of cycling. Public buses are cheaper and more comfortable than the current situation and enable different income groups to access town centres more efficiently. Denser urban areas force slower driving, which in turn helps reduce accident risk and severity.

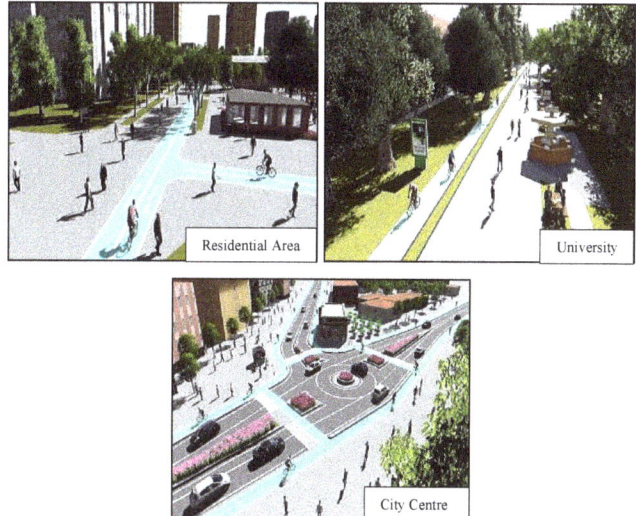

**Figure 4.** The Turkish locations as they might look in Shift Vision.

### 4.2.3. Improve Vision

Figure 5 shows the same three locations as in Figures 3 and 4 and how they may look in 2035 under Improve Vision. More affordable housing tends being encouraged in new developments with better transport connectivity and strong service provision, so it becomes increasingly possible for those on lower incomes and without an automobile to access jobs. Systems with pedestrian priority are designed, and roads for motorized vehicles are restricted. Road space is allocated to vulnerable roads users. There are more bicycle tracks for mainly recreational purposes in outer, natural areas of the city. Dedicated cycling lanes will pass behind bus stops, enabling cyclists to continue past a stationary bus, away from the traffic. Public transportation systems would become faster, more comfortable and will ensure that transportation to the city centre is less stressful. Public transport systems, school buses, and institutions' service buses will cover 50% of the traffic in urban areas. Car dependency has the sharpest decline of the visions, and non-motorized transport increases by nearly 10%.

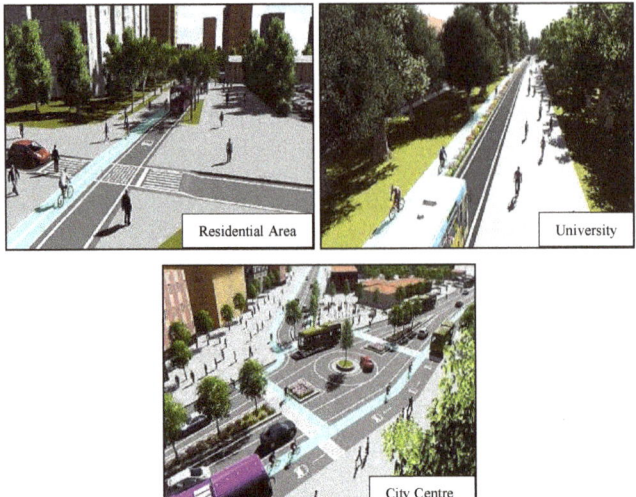

**Figure 5.** The Turkish locations as they might look in Improve Vision.

## 5. The Justification for the Visions

In general, participants stated that the core requirements for having active transport systems were explicitly considered in all three visions and all reduced the space available for cars to promote the development of smart transport.

As a public participant from İstanbul (Driver, male, aged 26–35) said *"we need to adopt these alternative transportation visions anyway. It looks more and more like we cannot live. If we do not emphasize walking, cycling and public transport more, İstanbul will become a giant car park. Small cities are experiencing similar concerns as well."*

*"A plan not having been made only for cyclists or only on the bus transport system. All transportation modes have been considered in a way to be balanced in different visions at the same time. One of the most neglected issues in our country is this. For example, if bicycle path is made, the place from where vehicles can go is not considered, or if one lane is removed, the places where cars will be parked are not taken into consideration".* (Expert, academy, female, aged 36–45, İstanbul)

Both experts and decision-makers asserted that the practices in Avoid Vision which improve the safety of people using different transport means would cause them to move to city centres rather than to suburban areas. In this vision, they said that it is not necessary for the car industry to invest more on safety in the presence of the improvements made to reduce accident risks, and there is no coordination between the municipalities and police units about speed control and fines due to a lack of inspection. Regarding Shift Vision, it was indicated that the variables created by the urban form could not be applied in Turkish cities because the growth of Turkish cities is dependent on unearned income in the construction industry. They advocated that radically reducing lanes is undefined as part of limiting the motorized vehicle traffic and expanding the areas for non-motorized vehicles will not increase socialization. They discussed the safety issue in Improve Vision, in which they stated that cycle accidents occurring in Turkish urban transportation are caused mostly by public buses, so bus drivers' awareness of cycle users must be raised, but if this is not achieved, this vision may pose high accident risks.

In general, public participants who prefer to walk explained that Shift Vision is the more desirable vision because wider spaces for their needs may be easily supplied, especially near the university campuses. Many cyclists thought the reason why Avoid Vision is an ideal transportation system for them is that traffic rules guiding the relationship between cycling and walking are shown in a clearer manner. A lot of public transportation users found Improve Vision as a preferred future in all locations compared to the other visions. This is because fast and comfortable public transport systems are crucial for people to go to work (or school) in the early hours of the day. Driver participants generally thought that future visions should not be compelling because the closure of some settlement areas to traffic would increase traffic congestion in other streets. Therefore, they found Avoid Vision is more reliable for them as private cars are restricted less in the settlement area. Most shop owner participants oppose any future visions where private car users could not pull up in front of their stores. That is because they think their customers are always vehicle-using customers, whereas public transportation users usually use the roads in transit, but some of them think a vision where the public transportation system is improved may cause more customers to visit their store in outer suburbs.

The general opinion of the participants is that Avoid Vision seems like a transitional approach, Shift Vision can only be implemented in limited locations, and Improve Vision appears to be the most complicated approach and solution. In the paper, although future smart transport systems were initially developed with much higher dependence on walking and cycling, the results show that the vision that developed the public transport system the most was seen to be most realistic by the participants.

### 5.1. Environmental Solutions

The environmental solution address issues such as air pollution, noise, and congestion. Generally, public participants think that the most critical change is reducing traffic jams through the development of existing smart transport systems. In addition, the expert participants think that "motor vehicle lane does not need to be closed off too much because the roads closed off somewhere create more traffic congestion in other locations" and "walking distances increase too much with the closing of the areas and will cause stressful situations rather than minimizing traffic jam."

> "In Shift Vision, instead of closing off the motor road and creating a social area, such areas could be formed in different areas where the road does not pass. There are adequate society areas at the university. Thus, I think if we narrow down the roads and close them to traffic this could then pose problems". (Driver, male, aged 18–25, Konya)

> "Even three minutes is necessary for the condition of students being late for class in the morning. Improve Vision can offer fast transportation, and everyone can drop off at his or her faculty". (Public transport user, female, aged 18–25, İstanbul)

The participants thought that, especially in the early hours of the day, there is a need for more efficient and fast public transportation systems within the campus, and therefore, they said even though Shift Vision presents a better campus environment, this vision creates a stressful situation by increasing walking distances, particularly for the students. One expert (Urban planner, female, aged 36–45, İstanbul) thinks dedicated road space for public transportation systems will permit traffic to flow faster; otherwise, the effects of public transport vehicles stopping in narrow lanes will cause traffic stoppages.

### 5.2. Technology

Technology has advantages and disadvantages in reducing travel demand and is an essential factor for business and smart life. In Turkish cities where home working may be appropriate, and to design smart technologies into working life, the spread of smart and digital innovations can lead to a decrease in daily travel trips. Several public and expert participants think that the introduction of digital technology into working life is awkward for the whole of Turkey; and said that it could cause declines in one-to-one people interactions.

> "You mentioned especially a scenario in which office works will be performed remotely. Is this an assumption? Is this an estimate? I wondered because of the subject in Turkey. It is likely for our cities,

*such as İstanbul and Ankara. Is it possible for the whole of Turkey?"*. (Driver, male, aged 26–35, İstanbul)

*"Life gets easier with increased information gathering opportunities by use of technology, online services cut down travels, but it would not be wrong to foresee a decrease in human and one-to-one interactions?"*. (Transport planner, male, aged 36–45, Konya)

One policy-maker considers the fact that the automotive sector does not need to make significant investments for safer urban transport environments in Avoid Vision. The policy-maker (Local, female, aged 46–55, Ankara) stated *"If speeding limits are decreased, the number of accidents will decrease. Death tolls in collision accidents will also decrease—no speeding. When there is no speeding, the driver can manage his safety. There are comfortable cars too. There is a new technology too. There are human-less drivers, sensitive pedestrian systems, but it seems like the automotive sector will not have to make these investments in this scenario."*

5.3. Urban Structure

Many of the experts interviewed in this study, tend to underestimate the importance of compact and high-density areas.

*"Taking measures for traffic safety particularly in the cities and settlements, which are dominated by motorized transportation, will make them more attractive and useful. The escape to suburban areas with dense traffic will be stopped"*. (Expert, transportation planner, male, aged 26–35, Konya)

*"The presence of dense areas is preferred. What we have is not compactness; it is an unplanned density. Bicycle transportation, to be improved has come down to such a compact area level that it is a problem in itself"*. (Expert, urban planner, female, aged 26–35, İzmir)

They highlighted that the Turkish economic strategy had driven urbanization for years and they said designing a compact urban area model is not possible with the current conditions unless the economic policies of the central government change.

*"Creating a compact city is possible with these improvements only for a very extended period since the growth of Turkish cities is dependent on unearned income"*. (Expert, academy, female, aged 36–45, Ankara)

*"Economy policies of the central government should be changed significantly. The economic strategy of Turkey has driven urban development and construction for years"*. (Expert, civil society organization, female, aged 26–35, Eskişehir)

Another public participant claimed that there is no need for vehicles inside the campus and therefore it should be a system that supports pedestrians and cyclists as in Shift Vision. *"Universities need to be made into more social areas. It could be easier to convert these places into a human-oriented urban environment, compared to the city centres"* (Public, pedestrian, female, aged 18–25, İstanbul).

Besides, several public participants said road spaces for cars had been limited too much in Shift Vision and that the need for cars is inevitable in some cases.

*"Sometimes there can be a situation of having to reach a place in the university; therefore, I may need the car. The bags I need to carry are heavy, and our campus is large, so carrying them can be tough for me. It would also not be possible for me to bring in from the university entrance; therefore, there should be lanes for the vehicles. If the drivers want to make an interim stop, a problem can arise. Still, it looks like having a two-lane road is essential. Otherwise, there would be transportation problems"*. (Driver, female, aged 36–45, İstanbul)

*"In a city like İstanbul, where 15 million people live, I think it is tough to apply simple solutions. I believe Shift Vision limits freedom of travel for motor vehicle users too much"*. (Driver, male, aged 56–65, İstanbul)

*"When coming to the university, I continuously must bring in and take away things. There could be a one-way road as in Improve Vision, and more attractive social areas could be created within the university. I believe Shift Vision limits freedom of travel for motor vehicle users too much"*. (Public transport user, male, aged 26–35, İzmir)

5.4. Mode Share

In general, participants think that Avoid Vision looks like a transitional vision in the short term, although they think this vision seems more reasonable since motor vehicles do not decrease as much as the other visions.

*"In Avoid Vision, automobile numbers do not decrease so much; this vision looks like a transition point. It can be a transition point for urban areas in Turkey as well"*. (Expert, academy, female, aged 26–35, Eskişehir)

*"Avoid Vision might be more realistic because it decreases automobile dependency less. There is a more consistent lane reduction in Avoid Vision, and there isn't a far-reaching reduction in the decreasing of traffic. It seems more reasonable since there is not as much lane reduction as Shift Vision and Improve Vision"*. (Expert, transportation engineer, male, aged 36–45, İstanbul)

The expert participants think that Shift Vision is not an alternative future that can be applied everywhere in different parts of typical Turkish urban areas, whereas it can be successfully implemented in certain parts of urban areas where motor vehicles are rarely restricted from entering the streets such as; narrow roads, campus areas, and historical places.

*"Some areas might be said that vehicles should get out, and just pedestrians and cyclists should be allowed. It could especially be historic urban centres. So, all three visions may have different application areas. For example, Shift Vision can be considered in some regions of the city where there are more bicycle and pedestrian transportation, and where some motor vehicles cannot enter some streets. It can be applied in city centres and university campuses, but it is not a vision that could be implemented to every location of Turkish urban areas"*. (Expert, urban planner, female, aged 26–35, Ankara)

Improve Vision prioritizes public transportation more, and that is why it seems more logical for the existing urban transport problems.

*"All three visions are meaningful but the vision that prioritizes public transportation includes the other visions more, and that is why it seems more logical. Especially three types of smart transportation futures are brought together. It shows it includes public transportation"*. (Expert, civil society organization, female, aged 36–45, Eskişehir)

*"Improve Vision seems an ideal vision since you suggest more complex transportation in the city centre as well. Improve Vision can also promote people to mass transport and can decrease the problems they live in daily transport. Otherwise, if we do not give more importance to public and active transport systems, İstanbul will be transformed into a big car park area"*. (Expert, academy, male, aged 36–45, İstanbul)

The participants emphasized that the economic strategy of Turkey is mostly dependent on the income-oriented building industry, so it is not convenient to design a compact urban model, as proposed in Shift Vision. Besides this, it was underlined that the spread of digital technology would lead to a decrease in communication between people in Avoid Vision. Additionally, they specified under the title of technology that producing smaller cars and lowering their carrying capacity in Shift Vision is in contrast with the environmental objectives and this approach can create disadvantages regarding fuel consumption, operating costs and conditions, traffic safety, and traffic jams.

## 5.5. The Response of Different Groups

### 5.5.1. Pedestrians

Participants preferring pedestrian transportation discussed that even though they liked the fact that in Avoid Vision, infrastructure systems were developed to support pedestrians' safety, this vision was more designed for motor vehicle users. It has been expressed that, rather than having car park arrangements on the main road, with each school having their car park area independent of the main road, wider pavements could be allocated for pedestrians.

> "In Avoid Vision, with making car-parking on the roadside, a less comfortable area was created for pedestrians. Instead of that, by setting up a car park arrangement within each school, wider pavements for the pedestrians should be created. For example, people want to walk when going to the cafeteria and in Avoid Vision, instead of the place allocated for the car park, widening the pavement could be more reasonable". (Id514, pedestrian, male, aged 36–45, İzmir)

Additionally, it was stated that the creation of a more people-oriented urban environment near the university would be easier compared to any location in the city centre. Many of the participants said universities have a young, student population, and traffic designs that are more social and people-oriented, as in Shift Vision, would be reasonable.

> "No need for vehicles inside the campus. One and two of every 100 people coming to the university provide for their transportation with their cars. In other areas, there should not be any need for a car. In the inner sections, the pavements need to be wide. University is especially a place where young people and studies are plenty; therefore, as well as walking will not pose a significant difficulty, it would also enable carrying out physical activities". (Id229, pedestrian, male, aged 18–25, Ankara)

### 5.5.2. Cyclists

Cyclists stated that in Avoid Vision, crossovers were shown more clearly and that in Shift Vision, conflicts between cyclists and pedestrians could happen. Another participant expressed that pedestrian and bicycle transportation could be more compatible. As both are slow transportation modes, it is better for them to go from the same place. In addition, the barriers to prevent automobiles violating the pavements, and the public transportation vehicles would not strike bicycles in the right and left turns.

> "In Avoid Vision, it is clearer on which side of the road bicycles and pedestrians can go across. In Shift Vision, it is not very clear, through where the bicycle path could go. For example, two different bicycle paths intersect at the midpoint of the road, and the bicycle path is provided over a single alternative track. It looks like at some points on the road; traffic confusion can arise between the bicycles and pedestrians. In Avoid Vision, providing a means for the bicycle and pedestrian ways to cross the road in a parallel way made it logical". (Id723, cyclist, male, aged 26–35, İstanbul)

### 5.5.3. Public Transport Users

Public transport users think Shift Vision limits freedom of travel for motor vehicle users too much. Improve Vision, on the other hand, both provide a faster transportation system by allocating separate roads for buses and offers a better urban area for bicycle and pedestrian transportation users as well.

> "Improve Vision has allocated single lanes and separate roads for public transportation; therefore, there also will be no chaos among the motor vehicles. For example, buses stopping at the bus stops to take in embarking passengers will not reduce the automobile's speed. At the same time, there will be less vehicle traffic in this area". (Id1011, public transport user, male, aged 26–35, Konya)

On the other hand, another participant maintained that because there is not much traffic at the university, a single-lane bus road could be adequate.

*"In Shift Vision, not having any vehicles would create problems. It does not seem possible that this could be implemented. As there will not be too much traffic in the university, having the single-lane road of Improve Vision could be adequate. In the existing campus, there needs to be a road line surrounding the university on the outside".* (Id553, public transport user, male, aged 36–45, Konya)

5.5.4. Drivers

Drivers generally consider residential areas to be transit regions for accessing main traffic roads and therefore, in the event of the road being closed off as in Shift Vision, traffic problems could arise in other streets. They believe that existing infrastructure systems are not adequate for helping people to use public transportation systems. The infrastructure system should be made adequately prevalent so that then the conditions can arise for people to use public transportation systems rather than their cars.

*"Many people have cars, and even though they know about traffic congestion, they do not want to use public transportation systems. That's because these systems cannot provide for comfortable transportation in the present situation. Instead of that, they prefer to travel with their cars even though they know of the traffic congestion. To the extent, public transportation systems are accessible, comfortable and cheap; people would quit using automobiles".* (Id193, driver, male, aged 36–45, Ankara)

Some car drivers said that the single-lane public transport road could create difficulties in its implementation because of minibuses and municipality buses race among themselves to take on passengers at the university. They also argued that Avoid Vision provides at the same time, adequate areas for the cyclists and pedestrians as well. Within the university, providing for transportation for the students with ring trips could not pose a problem, but for those who are coming to the university for the techno-city, like mentioned, or those coming to meet their various needs, Improve Vision could cause problems.

*"Sometimes there can be a situation of having to reach a place in the university; therefore, I may need the car. The bags I need to carry are heavy, and our campus is large, so carrying them can be tough for me".* (Id612, driver, female, aged 26–35, Ankara)

*"It would also not be possible for me to bring in from the university entrance; therefore, there should be lanes for the vehicles. If the drivers want to make an interim stop, a problem can arise. Still, it looks like having a two-lane road is essential. Otherwise, there would be transportation problems".* (Id452, driver, male, aged 36–45, İstanbul)

*"In Shift Vision, passages for vehicles have been limited too much. When coming to the city centre, I continuously must bring in and take away things".* (Id062, driver, female, aged 36–45, Konya)

5.5.5. Shop Owners

Shop owners expressed that narrowing the road in the city centre for car users would create great disadvantages for them. It has been observed that the participants, with their customers being car users, on the other hand, use city centre streets in transit and therefore they would prefer to Avoid Vision such a manner that car users could still pull by their shops.

*"The cars not being able to park means our business also being impacted to a significant level. That's because our customers are vehicle-using customers. Public transportation users use this road in transit. Our customers are car drivers, meaning not flowing customers".* (public transport users) (Id312, shop owner, male, aged 56–65, İstanbul)

*"If the road becomes a single-lane, it would create significant problems for the tradesmen. I do not think it's appropriate to reserve this much pavement for the pedestrians. That's because if vehicle traffic is not adequate, the vehicles could not park and if they want something from the tradesmen, they cannot buy it. The city centre roads still need to be two lanes. I mean it must be two-way and*

two-lanes. For this system to be realized, car parks need to be made underground, or pockets will be done here. If it is single-lane, when the cars stop, the cars behind will wait". (Id216, shop owner, male, aged 46–55, Ankara)

*"If automobile drivers cannot stop where they want to stop, it is definite that the businesses of the tradesmen will be very seriously impacted. Because of the renovation on this side, I have a daily €150 loss. No vehicle driver can pull up by the retail area".* (Id823, shop owner, male, aged 56–65, Eskişehir)

## 6. Policy Implications for Improve Vision

This section serves as a useful tool for helping future Turkish cities to understand how they might build their local policy pathways. As stated in Section 3.3, workshop participants were supplied with related resources (generic visualizations and the scenario statements for Improve Vision), in advance of each workshop, to discuss and create pathways with the context of their urban areas, as presented in Table 4.

We intended to identify distinctive policy measures that each sample needs to implement for achieving the target of one specific vision. Twenty-six participants were selected from across the selected urban samples: five from Ankara, Eskişehir, İzmir, and Konya, and six from İstanbul. It is interesting to mention that all selected urban areas in this study would have to apply different vital strategies to achieve their vision targets and that these local policy pathways could constitute exemplary approaches for many other Turkish cities. Meanwhile, it was noticed that the municipalities make their transport plans as part of their visions since the effect of the central authority on the local administrations is not strongly effective and satisfactory. It is somewhat surprising that the prominent smart transport practices of some Turkish urban areas are not applied, and the policy measures of each selected city can suggest important ideas for other cities.

### 6.1. Ankara

The first policy package for Ankara is to investigate new financial support for the construction of new metro lines (Keçiören and Airport) and increase the capacity of the metro and public transportation systems (especially in the direction of Çayyolu). As of the year 2020, it needs to be ensured that cyclists, especially in their commute to the university, can travel in a manner that is integrated with the metro systems. For public transportation systems in Ankara to become integrated with cycles, firstly, cycling sharing systems in universities need to be activated, and there needs to be cycling parking areas, especially in the metro stations by the universities. In 2025, there would be aims to increase the number of public transportation users with the improvement of price and physical integrations in all the public transportation systems through intelligent card systems. Then, the last phase was mapped to create a more comfortable pedestrianization area around some parts of the city centre by using high car parking charges.

### 6.2. Eskişehir

Eskişehir sets ambitious targets for the construction of new light rail infrastructure. In between 2015 and 2020, given significantly increasing numbers in public transportation systems, completion of the expansion of rail systems into three different areas (see Table 4) and the completion of the new cable car project should be planned. After the public transportation systems have a significant share in inner-city transport, on the roads where making cycle tracks were previously planned, these projects need to be put into operation. New cycle tracks need to be made into a continuous network, and after starting the operation of the new tramway line, bus services need to be brought to the neighbourhoods, where access to public transportation is still limited. Immediately after newly developed cycle networks, penalty proceedings would be applied for vehicle users occupying bicycle paths. As of the year 2025, in turn, to increase the efficiency of public transportation systems, the traffic control centre needs to be established. Between 2030 and 2035, a new park and green area projects should be started for increasing the square of green areas per person.

Table 4. Local policy pathways for Improve Vision applied in the selected Turkish urban areas.

| | Ankara | Eskişehir | İstanbul | İzmir | Konya |
|---|---|---|---|---|---|
| 2015–2020 | • Improvement of metro and mass transportation systems.<br>• Increasing the tramway's capacity and some navigations in M2Çayyolu metro.<br>• Finding additional resources for new metro investments.<br>• Start-up of Keçioren and airport metro lines. | • Start the construction of the new light rail transit line.<br>• The extension of tram lines into three separate regions (Yıldıztepe-Yenikent Çankaya, Camlıca-Batıkent and Emek-71 Evler) for increasing the capacity of tramway journeys.<br>• Activation of the ropeway transportation system. | • Improvement of underground and ground railway systems.<br>• Making private public transport lanes on some road corridors.<br>• To connect the new airport with the city centre by new rail systems.<br>• To create a High Occupancy Vehicle (HOV) lane on the two Bosporus bridges.<br>• To prevent the destruction of green areas because of the construction of the new airport and Bosporus bridge. | • Re-arrangement of recreational fields.<br>• Making bicycle roads in compliance with national standards.<br>• Enlargement of the bicycle sharing systems along the coast.<br>• Formation of vertical bicycle road connections from coastal areas to the city centre.<br>• The connection of the leading bike artery roads with each other. | • Providing the opening of new development in some areas (University and TOKİ lines).<br>• To create better public transportation facilities between Meram Medical Faculty and the Bus Terminal. |
| 2020–2025 | • Free car parks at metro stations.<br>• The realization of bicycle projects in universities.<br>• The presence of bike sharing systems on university campuses.<br>• Construction of bike parking in metro stations.<br>• Construction of new bicycle roads.<br>• Integrating the metro with bicycles. | • Determination of car park violation points in the city centre.<br>• Integration of bicycles into mass transportation.<br>• Penalizing for car users occupying bicycle routes.<br>• Formation of bicycle road networks.<br>• Increase the accessibility of public transport buses where passengers cannot easily use tramway systems. | • Construction of bicycle roads near the seaside.<br>• Integration of coastal mass transportation system with bikes.<br>• The performance of bicycle events and activities.<br>• Strengthening pedestrian infrastructures near the coastal sites<br>• The performance of pedestrianization works in historic areas. | • Integration of suburb and bus transportation with bicycles in all sub-provinces.<br>• Provide priority for bicycles in narrow streets.<br>• To apply high parking changes around the bus stops in the city centre. | • The implementation of the monorail project, which provides a great convenience for public transportation systems.<br>• Development of light railway systems and increasing the network length to 180 km.<br>• After the conversion of minibuses to buses, an electronic fare system should start in the city. |
| 2025–2030 | • Increasing car park changes.<br>• Improvement of smart card systems.<br>• Formation of new rights for minibus drivers.<br>• Development of the integration of all masses transportation systems. | • A review of public transport routes and lines according to the density of motor vehicle traffic.<br>• The design of smart stations for all bus and tramway stations.<br>• To convert 15% of the municipal fleet to electric vehicles. | • The creation of low emission zones<br>• To divide the city into different zones (high-density housing, commercial density, forestland, etc.) and evaluate each zone depending on their characteristics. | • Movement of the city centre's density to the Bayraklı region.<br>• The pedestrianization of Bayraklı, which would be the new town centre.<br>• The pedestrianization of the old Kadife Castle. | • Improvement of social life and public culture around part of the city centre.<br>• The minibuses will be removed from the city centre.<br>• The completion of road construction works for bicycle transportation. |
| 2030–2035 | • Designing the main centres of the city (like Ataturk Avenue, Kavaklıdere, Sıhhiye and Ulus) to provide priority to pedestrians. | • Activation of the pedestrianization works in some regions of the city.<br>• Increase green fields per individual. | • Implementation of traffic congestion changes on the Historical Peninsula. | • Construction of metro lines from the city centre to the north (Bergama) and to south (Ephesus) directions.<br>• Activation of the metro system towards the west (İzmir Institute of Technology) areas. | • Prohibition of car parking on certain roads to decrease vehicle use. |

## 6.3. İstanbul

In between 2015 and 2020, enhancing the connections of all transportation systems among themselves is the first stage for adopting Improve Vision in İstanbul. For example, the integration of cycle and sea transport and integrating the new airport with the rail systems of the city centre, are needed. Various ongoing underground and surface rail system projects should aim to be finished by the year 2020. For public transportation systems to be integrated with cycles in İstanbul, marine transport draws attention as a more feasible transportation mode in ensuring that integration. In the subsequent five years, activities directed at increasing the popularity of the cycling transport system in inner city transportation could be made. For example, to create different points of view regarding cycling transportation, different organizations and activities need to be activated at the same time. In addition, strengthening the connection of pedestrian transport with the coast and making cycle tracks, especially from these regions, are needed. After 2020, city zone applications must be arranged to create low-emission areas in the central parts of the city where motor vehicles are intense. By the year 2030, advances should be made which are directed at limiting car traffic in the Historic Peninsula. The aim of controlling traffic congestion in the Historical Peninsula in İstanbul would be reached by implementing optimum traffic congestion charges.

## 6.4. İzmir

It is initially advised that cycle tracks at the endpoints of the coastal road are not up to specific criteria and since there are dirt roads in these segments, cyclists cannot go at adequate speeds. Therefore, firstly the cycle tracks in these sections need to be widened according to the criteria before 2020. Then, on these expanded tracks, additional cycling sharing systems need to be placed, and in the next stage, cycling sharing systems need to be extended to the city's inner sections. Additionally, to enable integration in the last stage especially, as of the year 2020, for the cycle tracks on the coastal road, vertical cycle tracks need to be constructed towards the central parts of the city. By 2025, these vertically extending cycle tracks, in turn, should relate to the inner sections, and on the roads in between the bus and metro stops in these inner sections, car parameter fees should be kept high. To enable the integration between cycles and public transportation in İzmir efficiently and comprehensively, firstly the integration of public transportation systems in district centres shall be ensured. In the subsequent stage, public transportation and cycling transportation shall be integrated at locations in the city's inner sections. By 2030, for the realization of local policies directed at creating new social and business areas, some social and business activities in the Alsancak Region, where all activities of the city are collected, should be moved to the Bayraklı Region, which is situated in another central region of the city. Moreover, by the year 2030, the historic Kadife Castle locations would be transformed into a pedestrianized region. After 2030, it will be possible to start the construction of new railway systems since there is a vital tourism potential in the north (Bergama) and south (Ephesus) axes of İzmir.

## 6.5. Konya

In Konya, for Improve Vision to achieve its objective, in the first stage in the regions where the new university and the TOKİ residential projects are situated, new settlement areas need to be opened in the region. The public transportation facilities between the inner-city bus terminal and the Meram Medical School need to be developed. Additionally, in this time five-year period, in regions where bus operation costs show a deficit in inner city transport (new university and Toki residences), new areas open for settlement need to be created. The most important strategies that need to be realized by 2020 and 2025, on the other hand, are directed at increasing both the comfort and the capacity of public transportation systems along with the new routes and starting the construction stage of the new monorail project. In the steps following 2025, around 500 minibuses in the city centre will be removed from traffic. Then, with the removal of minibuses from the inner-city transfer, electronic fee systems in the public transportation systems will start to be used. In the determination of fees relating to the utilization of these systems, they should also be made to encourage the lower income segment to the public transportation systems. For public transportation to become modeless, perceived as a waste of time in the single centre city (journeys between different regions mostly run via the city centre route), developing social and business areas in some side-lined, and central parts of the city should be planned. After 2030, car parking prices in some busy streets of the city centre need to be high.

## 7. Discussion and Conclusions

This paper has outlined a participatory approach taken to designing future Turkish smart cities for the year 2035 which bring about a radical development in the level of green transport systems through the hypothetical Turkish urban areas.

The key message from our research is that switching much of the population to more active forms of transport for many journeys is entirely feasible, if such forms of transport are made accessible, comfortable, and can easily be integrated into the user's daily routine. Most people recognize the visions are reliable and solutions to the existing challenges are clear but robust to implement because they require relatively radical development not only in the habits in which people travel but also in the structure and organization of urban development.

Through this methodological perspective, it is useful to create remarkable and practical outcomes to enable comparison at the local level and provide enough knowledge to inform the local planning and development decision-making processes to construct smart transport cities, and at the same time provide generation of more informed regional and national policies and relevant actions in achieving a future urban development. For example, the key strategy for Ankara is to integrate cycling with metro and public transport systems, particularly at university stations, and to develop new smart card systems for all public transportation systems, with new rights and benefits for minibus drivers (see Table 4). On the other hand, the integration between the public transport systems or any smart card system has not yet been established in Ankara, whereas, in the other sample areas, the smart card systems and the public transport integration applications were being used much earlier for promoting the use of affordable public transport systems. Eskişehir was a single sample urban area that collaborates with the general security of the town for preventing car parks in cycle lanes. The general security and transportation departments of Eskişehir collaborated to prevent car parking in cycle lanes; however, similar applications are not implemented in other cities. This collaboration is essential for many Turkish cities where cyclists cannot use their routes due to the occupation of cycle lanes by cars (even for Konya, which has 240 km of cycling paths). The policy agenda of İstanbul includes new underground and ground railway systems and to implement a traffic congestion charge. İstanbul provides an essential message to the other major car-dependent cities that it is crucial to initially develop public transport systems for changing public behaviour from car journeys into public transport systems and then to improve non-motorized transport systems (see Table 4). İzmir was the only city that does not allow minibuses to enter the city centre as a transport mode, but their commercial use between definite terminals in the suburban areas and the counties was allowed. This practical application may offer an excellent point to other municipalities that have high traffic congestion due to the intense use of urban minibuses in other urban areas. The topography and urban structure of Konya offers a better urban environment for cyclists and pedestrians. However, cars occupy many cycle lanes, and current pedestrian planning projects were terminated due to public pressure. Making the city centre more attractive for the public needs may promote car drivers to use more smart transport systems because people in Konya must use some central parts of the city for their long transit journeys (see Table 4).

The research results corroborate the findings of a great deal of the previous worldwide city council projects [72–87], which devised extensive use of public transport strategies to achieve future urban development. In the context of providing public engagement to smart transport city development [58–69], the paper has highlighted a novel feature to the participatory approach, relating to the meaningful public involvement with a variety of participant groups throughout the construction of the local policy-pathways. While most of the smart transport studies do not involve public engagement during the future urban development [115–124], this paper reveals that the effectiveness of the local transport policy development can be notably increased by combining complementary participatory methods [125]. For example, the survey process represented a more closed process, where public participants have meaningful input into the imaginary future scenario developments.

However, the overall findings of the current study do not support a great deal of previous research [57–68], which has demonstrated that many advanced cities have tended to give greater priority to push factors such as parking management measures, building traffic sensors, traffic light control systems, implementing road prices, etc. On the other hand, to improve the attractiveness of desired future urban systems, different Turkish cities initially need to achieve their reduction in private car use through a greater proportion of pull measures (i.e., improvement of metro and mass transportation systems, integration of bicycles into mass transportation, development of light railway systems), rather than push measures. This approach may be somewhat related to fast-growing cities and towns [103]. Non-motorized improvements such as designing the main centres of Turkish cities to provide priority to pedestrians, development of the integration of all mass transportation systems, increasing green fields per individual, strengthening pedestrian infrastructures, and more, can be considered as other examples of pull measures in Turkey (see Table 4).

Finally, to generate smart transport city development, a wide-ranging approach needs to be synchronized among a variety of different public and private sectors to create an urban environment in which choosing to use green systems becomes noticeable. Such coordination may seem idealistic or utopian, but they are undoubtedly still some distance from the present Turkish government action plan on smart city transport development. Undoubtedly, such developments would involve a considerable degree of consensus that such a future is achievable, that the existing transportation problems are real, and that radical approaches to tackling them are essential.

**Funding:** This research received no external funding.

**Acknowledgments:** I would like to thank many anonymous participants, without whom the questionnaire surveys, interviews, and workshops may not have been carried out.

**Conflicts of Interest:** The author declares no conflict of interest.

## Appendix A

Table A1. Preventing car use.

| Key Factor | Response Number |
|---|---|
| Reducing private car usage | 5 |
| In the city centre | 2 |
| Reducing vehicle speed | 53 |
| A limited parking ban in the city centre | 3 |
| Resolving parking problems | 4 |
| Increasing petrol prices | 7 |
| A reduction in the number of cars parked | 2 |
| Complicating car purchase | 1 |
| Making harder to get a car license | 1 |
| Speed control | 5 |
| In residential areas | 4 |
| City centre | 1 |
| Annual quota system for vehicle usage | 1 |
| Decreasing vehicle reduction | 35 |
| Restrictions for cars within the city centre | 43 |
| Designing small cars | 3 |
| Deceleration of private cars in the crosswalk | 1 |

Table A2. Public transport systems.

| Key Factor | Response Number |
| --- | --- |
| Public transport | 25 |
| Better planned public transport system | 4 |
| Modern public transport systems | 2 |
| Increasing the number of public transport buses | 3 |
| Improving the quality of public transport services | 3 |
| Increasing public transport comfort | 1 |
| Improving public transport facilities | 1 |
| Intelligent public transport systems | 1 |
| Renewal of public transport vehicles | 2 |
| Upgrading public transport vehicles | 1 |
| Safer transport systems | 2 |
| More quiet transport systems | 2 |
| More enjoyable transport systems | 1 |
| Traffic management centre implements | 1 |
| Alternative systems | 2 |
| Intelligent road design | 3 |
| Smart design | 4 |
| Integrating cycling and public transportation | 9 |
| Technology advancements in public transport | 2 |
| Public transport users should respect each other | 1 |
| Reduced fare program | 2 |
| Minibuses | 1 |
| Tramway | 3 |
| Metro | 3 |
| Accelerating tram | 1 |
| Systematic road transportation systems | 1 |
| Increasing the frequency of time | 1 |
| Increasing the frequency of times during business hours | 2 |
| Public transport management service | 1 |
| Dissemination about public transport services | 2 |
| More accessible public transport systems | 1 |
| Increasing the share of renewable energy in public transport services | 1 |
| Route improvement project | 1 |

Table A3. Cycling.

| Key Factor | Response Number |
| --- | --- |
| Lowering the price of bicycles | 2 |
| Dissemination on cycling awareness | 13 |
| Safe bike paths | 19 |
| Aesthetic bike paths | 1 |
| Bicycle lifts | 1 |
| Electric bikes | 1 |
| Bike hire | 2 |
| Separate bike paths | 72 |
| Expansion of bike paths | 2 |
| Safety strips | 1 |
| More bike paths | 44 |
| Comfortable bike lanes | 3 |
| Safe bicycle parks | 1 |
| Security cameras near bike parking space | 3 |
| Bicycle parking spaces | 30 |
| Warning signs at the junction | 2 |
| Do not allow pedestrians to walk on bike paths | 4 |

Table A3. Cont.

| Key Factor | Response Number |
|---|---|
| Bicycle police | 1 |
| Signalized intersections | 2 |
| Traffic light priority for cyclists | 1 |
| Inserting helmet | 3 |
| Shower facilities | 2 |
| Improvement of bike paths | 4 |
| Creating complete bicycling networks | 6 |
| Better quality bike paths | 2 |

Table A4. Walking.

| Key Factor | Response Number |
|---|---|
| Increasing pedestrian paths | 4 |
| Safe pedestrian paths | 2 |
| Expansion of pedestrian paths | 15 |
| Giving priority to pedestrians | 2 |
| Seat benches | 3 |
| Reduction of the defect in the pedestrian path | 2 |
| Comfortable pedestrian paths | 1 |
| Better pedestrian paths | 3 |
| More comprehensive pedestrian paths | 2 |
| Regular pavement | 1 |
| Better pedestrian infrastructures in suburban areas | 4 |
| Less waiting times for pedestrians | 1 |
| Tree-lined pathways | 1 |
| Prevention of invasion of pedestrian paths | 23 |
| By motor vehicles | 19 |
| By electric poles | 1 |
| By cyclists | 4 |
| Improvement of crossroads | 25 |
| Suburban areas | 3 |
| More comfortable bike pathways for elderly | 3 |
| More comfortable bike pathways for disabled | 2 |
| Encouraging walking | 1 |
| Pedestrian crossing | 31 |
| Pedestrian signs | 2 |
| Designing more direct routes | 6 |
| Walking maps | 4 |
| Facilitating pedestrian access in hilly areas | 2 |
| Street lighting | 8 |
| Running parks | 1 |
| Editing underpasses and overpasses on the roads | 3 |
| For pedestrians | 1 |
| For cyclists | 2 |

Table A5. Urban features.

| Key Factor | Response Number |
|---|---|
| Increasing aesthetic | 16 |
| Park areas | 4 |
| Green areas | 5 |
| New modern squares | 4 |
| Visual beauty | 1 |
| More social and business places in residential areas | 3 |
| Urban design for family securities | 3 |
| More compact cities | 1 |
| Decreasing the population of major cities | 4 |
| By shifting into another city | 3 |
| Industries should be relocated outside of cities | 3 |
| Fixing distorted urban land | 2 |
| Improvement of urban environments | 2 |
| For better air quality | 1 |
| For decreasing noise pollution | 1 |
| City and regional planning for public | 1 |
| Ensuring security in the streets | 2 |
| Reduction of the population densities in major cities | 2 |
| Artistic places in major cities | 3 |

Table A6. Sustainable transport strategies.

| Key Factor | Response Number |
|---|---|
| Cyclists should have more rights | 3 |
| Protecting all rights of pedestrians | 2 |
| Preventing society from crazy young drivers | 1 |
| Penalty sanctions | 4 |
| Each transport mode users should comply with traffic rules | 3 |
| Improving the conditions of cyclists | 2 |
| Increasing the rules | 1 |
| Arrangements about passenger cars | 3 |
| Development of traffic laws | 2 |
| Development of local sustainable transport policies | 4 |
| Development of national sustainable transport policies | 6 |
| Ensure the observance of traffic signs | 1 |
| More comprehensive bicycle strategies | 2 |
| Developing bicycle culture in urban areas | 1 |
| Prevention of invasion of bike paths | 6 |
| By pedestrian | 1 |
| By motor vehicles | 5 |
| Tax incentives for cyclists | 1 |
| Cycling license law | 2 |
| The application of deterrent sanctions | 1 |
| Improvement of pedestrian rights | 1 |
| Making mandatory the use of pedestrian crossings | 3 |
| Penalizing car drivers who do not respect cyclists | 1 |
| Penalizing car drivers who do not give way to pedestrians | 4 |
| Campaigns and education | 6 |
| Different cultural campaigns based on sustainable transport | 2 |
| Giving bike education in kindergarten | 1 |
| Improving municipal management for cities to succeed | 1 |
| Shifting investments to small towns | 1 |
| People should live in or near areas where jobs are concentrated | 1 |

Table A7. Incentives.

| Key Factor | Response Number |
|---|---|
| Economic | 4 |
| For walking | 2 |
| For public transport | 4 |
| For decreasing passenger cars | 2 |
| For cycling | 1 |

Table A8. Public awareness.

| Key Factor | Response Number |
|---|---|
| Related to walking and cycling issues | 3 |
| Increasing the awareness of bicycle use | 2 |
| Awareness of pedestrians | 1 |
| Awareness of motor vehicles | 4 |
| Towards using walking and cycling for short trips | 3 |
| Towards sharing roads with cyclists | 2 |
| Towards respecting pedestrians in the pedestrian crossing | 1 |
| Expert and public events towards dissemination of walking and cycling as transport modes | 3 |
| Public spots for increasing the awareness of pedestrians | 2 |
| Dissemination of cycling | 4 |
| Organized cultural events for cycling | 6 |
| Extraction of traffic laws that increase people's consciousness | 1 |
| Increasing respect for pedestrians | 2 |
| Increasing respect for cyclists | 1 |
| Cyclists should have more rights | 6 |
| Protecting all rights of pedestrians | 1 |
| Organized cultural events for cycling | 5 |
| Cultural changes | 1 |
| People respect each other | 2 |
| Conscious and trained drivers | 1 |
| Prevention of unnecessary horn-blowing | 1 |
| Cultural innovation for a sustainable future | 3 |
| Public awareness | 1 |
| Training of public transport drivers | 4 |
| Solving social dimension problems | 6 |
| Giving importance to education | 2 |
| Health campaign for people using private vehicles | 1 |
| Training of people | 1 |
| Regular training | 1 |
| Granting of traffic education in schools | 1 |
| The public spot that expresses walking is good for heart health | 1 |
| Raise awareness about sustainable energy trends | 2 |
| Public spotlight on carbon emissions | 1 |
| Public spotlight on obesity | 1 |
| Raise awareness about sustainable energy trends | 1 |
| Public spotlight on carbon emissions | 1 |
| Education | 1 |
| Preventing society from crazy young drivers | 1 |
| Improving people attitudes towards less polluting public vehicles | 1 |

Table A9. Special groups.

| Key Factor | Response Number |
|---|---|
| Convenience for families with babies | 2 |
| Providing safety for child and young cyclists | 3 |
| Encouraging low-income people to use cycle | 1 |
| Electric bikes for adults | 1 |
| Better systems for disabled people | 2 |
| Build a shelter for stray dogs | 1 |

## References

1. Bratzel, S. Conditions of success in sustainable urban transport policy change in relatively successful European cities. *Transp. Rev.* **1999**, *19*, 177–190. [CrossRef]
2. Poister, T.H.; Streib, G. Elements of strategic planning and management in municipal government: Status after two decades. *Public Adm. Rev.* **2005**, *65*, 45–56. [CrossRef]
3. de Roo, G. Integrating city planning and environmental improvement: Practicable strategies for sustainable urban development. *Routledge* **2017**, *2*, 47–340.
4. Ding, G.K. Sustainable construction—The role of environmental assessment tools. *J. Environ. Manag.* **2008**, *86*, 451–464. [CrossRef] [PubMed]
5. Lele, S.M. Sustainable development: A critical review. *World Dev.* **1991**, *19*, 607–621. [CrossRef]
6. Sennett, R. Urban disorder today. *Br. J. Sociol.* **2009**, *60*, 57–58. [CrossRef]
7. Dahly, D.L.; Adair, L.S. Quantifying the urban environment: A scale measure of urbanicity outperforms the urban-rural dichotomy. *Soc. Sci. Med.* **2007**, *64*, 1407–1419. [CrossRef] [PubMed]
8. Zhang, X.; Hes, D.; Wu, Y.; Hafkamp, W.; Lu, W.; Bayulken, B.; Schnitzer, H.; Li, F. Catalyzing sustainable urban transformations towards smarter, healthier cities through urban ecological infrastructure, regenerative development, eco-towns and regional prosperity. *J. Clean. Prod.* **2016**, *122*, 4. [CrossRef]
9. Geldenhuys, H.J.; Brent, A.C.; de Kock, I.H. Literature review for infrastructure transition management towards Smart Sustainable Cities. In Proceedings of the IEEE International Systems Engineering Symposium, Rome, Italy, 1–3 October 2018.
10. Trindade, E.P.; Hinnig, M.P.F.; Moreira da Costa, E.; Marques, J.; Bastos, R.; Yigitcanlar, T. Sustainable development of smart cities: A systematic review of the literature. *J. Open Innov. Technol. Mark. Complex.* **2017**, *3*, 11. [CrossRef]
11. Chang, D.L.; Sabatini-Marques, J.; Da Costa, E.M.; Selig, P.M.; Yigitcanlar, T. Knowledge-based, smart and sustainable cities: A provocation for a conceptual framework. *J. Open Innov. Technol. Mark. Complex.* **2018**, *4*, 5. [CrossRef]
12. Khatoun, R.; Zeadally, S. Smart cities: Concepts, architectures, research opportunities. *Commun. ACM* **2016**, *8*, 46–57. [CrossRef]
13. Komarevtseva, O.O. Smart city technologies: New barriers to investment or a method for solving the economic problems of municipalities? *R-Economy* **2017**, *3*, 32–39. [CrossRef]
14. Höjer, M.; Wangel, J. Smart sustainable cities: Definition and challenges. In *ICT Innovations for Sustainability*; Springer: Berlin/Heidelberg, Germany, 2015; Volume 310, pp. 333–349.
15. Eremia, M.; Toma, L.; Sanduleac, M. The smart city concept in the 21st century. *Procedia Eng.* **2017**, *181*, 12–19. [CrossRef]
16. Letaifa, S.B. How to strategize smart cities: Revealing the SMART model. *J. Bus. Res.* **2015**, *68*, 1414–1419. [CrossRef]
17. Djahel, S.; Doolan, R.; Muntean, G.M.; Murphy, J. A communications-oriented perspective on traffic management systems for smart cities: Challenges and innovative approaches. *IEEE Commun. Surv. Tutor.* **2015**, *17*, 125–151. [CrossRef]
18. Ahvenniemi, H.; Huovila, A.; Pinto-Seppä, I.; Airaksinen, M. What are the differences between sustainable and smart cities? *Cities* **2017**, *60*, 234–245. [CrossRef]
19. Yigitcanlar, T. Smart cities: An effective urban development and management model? *Aust. Plan.* **2015**, *52*, 27–34. [CrossRef]
20. Yigitcanlar, T. Smart cities in the making. *Intern. J. Knowl. Based Dev.* **2017**, *8*, 201–205.

21. de Oliveira, M.J.; Homrich, A.S.; de Mello, R.; Carvalho, M.M. Applying backcasting and system dynamics towards sustainable development: The housing planning case for low-income citizens in Brazil. *J. Clean. Prod.* **2018**, *193*, 97–114.
22. Smith, A.; Stirling, A.; Berkhout, F. The governance of sustainable socio-technical transitions. *Res. Policy* **2005**, *34*, 1491–1510. [CrossRef]
23. Michaelson, D.; Stacks, D.W. Standardization in public relations measurement and evaluation. *Pub. Relat. J.* **2011**, *5*, 1–22.
24. Calvillo, C.F.; Sánchez-Miralles, A.; Villar, J. Energy management and planning in smart cities. *Renew. Sustain. Energy Rev.* **2016**, *55*, 273–287. [CrossRef]
25. Ejaz, W.; Naeem, M.; Shahid, A.; Anpalagan, A.; Jo, M. Efficient energy management for the internet of things in smart cities. *IEEE Commun. Mag.* **2017**, *55*, 84–91. [CrossRef]
26. Mosannenzadeh, F.; Bisello, A.; Vaccaro, R.; D'Alonzo, V.; Hunter, G.W.; Vettorato, D. Smart energy city development: A story told by urban planners. *Cities* **2017**, *64*, 54–65. [CrossRef]
27. Zygiaris, S. Smart city reference model: Assisting planners to conceptualize the building of smart city innovation ecosystems. *J. Knowl. Econ.* **2013**, *4*, 217–231. [CrossRef]
28. Kitchin, R. The real-time city? Big data and smart urbanism. *GeoJournal* **2014**, *79*, 1–14. [CrossRef]
29. Yigitcanlar, T.; Kamruzzaman, M.; Buys, L.; Ioppolo, G.; Sabatini-Marques, J.; da Costa, E.M.; Yun, J.J. Understanding 'smart cities': Intertwining development drivers with desired outcomes in a multidimensional framework. *Cities* **2018**, *81*, 145–160. [CrossRef]
30. Yigitcanlar, T.; Kamruzzaman, M. Does smart city policy lead to the sustainability of cities? *Land Use Policy* **2018**, *73*, 49–58. [CrossRef]
31. Yigitcanlar, T.; Foth, M.; Kamruzzaman, M. Towards post-anthropocentric cities: Reconceptualizing smart cities to evade urban ecocide. *J. Urban Technol.* **2019**, *26*, 147–152. [CrossRef]
32. Curtis, C.; Holling, C. Just how (Travel) Smart are Universities when it comes to implementing sustainable travel. *World Transp. Policy Pract.* **2004**, *10*, 22–33.
33. Yousefi-Sahzabi, A.; Unlu-Yucesoy, E.; Sasaki, K.; Yuosefi, H.; Widiatmojo, A.; Sugai, Y. Turkish challenges for low-carbon society: Current status, government policies and social acceptance. *Renew. Sustain. Energy Rev.* **2017**, *68*, 596–608. [CrossRef]
34. Balcı, V.; Özbek, O.; Koçak, F.; Çeyiz, S. Determination of the constraints of bicycle use in urban life Kent yaşamında bisiklet kullanım engellerinin belirlenmesi. *J. Hum. Sci.* **2018**, *15*, 35–50. [CrossRef]
35. Uçar, A.; Şemşit, S.; Negiz, N. Avrupa Birliği Akilli Kent Uygulamalari ve Türkiye'deki Yansimalari. *Suleyman Demirel Univ. J. Fac. Econ. Adm. Sci.* **2017**, *22*, 1785–1798.
36. Ayataç, H. Kentsel Ulaşım Planlaması ve İstanbul. *İTÜ Vakfı Dergisi* **2016**, *71*, 31–35.
37. Akbulut, F. Kentsel Ulaşım Hizmetlerinin Planlanmasi ve Yönetiminde Sürdürülebilir Politika Önerileri. *Kastamonu Üniversitesi İktisadi ve İdari Bilimler Fakültesi Dergisi* **2016**, *11*, 336–355.
38. Balaban, O. The negative effects of the construction boom on urban planning and environment in Turkey: Unraveling the role of the public sector. *Habitat Int.* **2012**, *36*, 26–35. [CrossRef]
39. Berberoğlu, S.; Akın, A.; Clarke, K.C. Cellular automata modelling approaches to forecast urban growth for Adana, Turkey: A comparative approach. *Landsc. Urban Plan.* **2016**, *153*, 11–27. [CrossRef]
40. Alphan, H. Land-use change and urbanization of Adana, Turkey. *Land Degrad. Dev.* **2003**, *14*, 575–586. [CrossRef]
41. Papa, R.; Gargiulo, C.; Cristiano, M.; Di Francesco, I.; Tulisi, A. Less smart more city. *TeMA J. Land Use Mobil. Environ.* **2015**, *8*, 159–182.
42. Basu, I. Elite discourse coalitions and the governance of 'smart spaces': Politics, power and privilege in India's Smart Cities Mission. *Polit. Geogr.* **2019**, *68*, 77–85. [CrossRef]
43. Paschek, F. Urban Sustainability in Theory and Practice-Circles of Sustainability. *Town Plan. Rev.* **2015**, *86*, 745.
44. Hollands, R.G. Will the real smart city please stand up? Intelligent, progressive or entrepreneurial? *City* **2008**, *12*, 303–320. [CrossRef]
45. Caragliu, A.; Del Bo, C.; Nijkamp, P. Smart cities in Europe. *J. Urban Technol.* **2011**, *18*, 65–82. [CrossRef]
46. Brenner, N.; Schmid, C. The 'urban age' in question. *Int. J. Urban Reg. Rese.* **2014**, *38*, 731–755. [CrossRef]
47. Gladwin, T.N.; Kennelly, J.J.; Krause, T.S. Shifting paradigms for sustainable development: Implications for management theory and research. *Acad. Manag. Rev.* **1995**, *20*, 874–907. [CrossRef]

48. Jin, S.T.; Kong, H.; Wu, R.; Sui, D.Z. Ridesourcing, the sharing economy, and the future of cities. *Cities* **2018**, *76*, 96–104. [CrossRef]
49. Shaheen, S.; Chan, N. Mobility and the sharing economy: Potential to facilitate the first-and-last-mile public transit connections. *Built Environ.* **2016**, *42*, 573–588. [CrossRef]
50. Romanillos, G.; Zaltz Austwick, M.; Ettema, D.; De Kruijf, J. Big data and cycling. *Transp. Rev.* **2016**, *36*, 114–133. [CrossRef]
51. Mintsis, G.; Basbas, S.; Papaioannou, P.; Taxiltaris, C.; Tziavos, I.N. Applications of GPS technology in the land transportation system. *Eur. J. Oper. Res.* **2004**, *152*, 399–409. [CrossRef]
52. Dresner, K.; Stone, P. Multiagent traffic management: A reservation-based intersection control mechanism. In Proceedings of the Third International Joint Conference on Autonomous Agents and Multiagent Systems, New York, NY, USA, 19–23 July 2004; Volume 2, pp. 530–537.
53. Goodall, N.J.; Smith, B.L.; Park, B. Traffic signal control with connected vehicles. *Transp. Res. Rec.* **2013**, *2381*, 65–72. [CrossRef]
54. Rahman, M.S.; Park, Y.; Kim, K.D. Relative location estimation of vehicles in the parking management system. In Proceedings of the 11th International Conference on Advanced Communication Technology, Gangwon-Do, Korea, 15–18 February 2009; Volume 1, pp. 729–732.
55. Geller, A.L. Smart growth: A prescription for liveable cities. *Am. J. Public Health* **2003**, *93*, 1410–1415. [CrossRef] [PubMed]
56. Hall, R.E.; Bowerman, B.; Braverman, J.; Taylor, J.; Todosow, H.; Von Wimmersperg, U. The vision of a smart city. *Brookhav. Natl. Lab.* **2000**, *5*, 41.
57. Segal, M.; Kockelman, K.M. Design and implementation of a shared autonomous vehicle system in Austin, Texas. In Proceedings of the Transp. Research Board 95th Annual Meeting, Washington, DC, USA, 10–14 February 2016; Volume 16, p. 1837.
58. Barkenbus, J.N. Eco-driving: An overlooked climate change initiative. *Energy Policy* **2010**, *38*, 762–769. [CrossRef]
59. Mogk, J.E.; Wiatkowski, S.; Weindorf, M.J. Promoting urban agriculture as alternative land use for vacant properties in the city of Detroit: Benefits, problems and proposals for a regulatory framework for successful land use integration. *Wayne L. Rev.* **2010**, *56*, 1521.
60. Barbeau, S.J.; Borning, A.; Watkins, K. OneBusAway multi-region–rapidly expanding mobile transit apps to new cities. *J. Public Transp.* **2014**, *17*, 3. [CrossRef]
61. Sinky, H.; Khalfi, B.; Hamdaoui, B.; Rayes, A. Responsive content-centric delivery in large urban communication networks: A LinkNYC use-case. *IEEE Trans. Wirel. Commun.* **2017**, *17*, 688–1699. [CrossRef]
62. Möller, D.P.; Fidencio, A.X.; Cota, E.; Jehle, I.A.; Vakilzadian, H. Cyber-physical smart traffic light system. In Proceedings of the IEEE International Conference on Electro/Information Technology (EIT), DeKalb, IL, USA, 21–23 May 2015.
63. Park, K.; Willinger, W. *Self-Similar Network Traffic and Performance Evaluation*; Wiley-Interscience: Hoboken, NJ, USA, 2000; Volume 4, pp. 21–52.
64. Leontiadis, I.; Marfia, G.; Mack, D.; Pau, G.; Mascolo, C.; Gerla, M. On the effectiveness of an opportunistic traffic management system for vehicular networks. *IEEE Trans. Intell. Transp. Syst.* **2011**, *12*, 1537–1548. [CrossRef]
65. Shaheen, S.A.; Mallery, M.A.; Kingsley, K.J. Personal vehicle sharing services in North America. *Res. Transp. Bus. Manag.* **2012**, *3*, 71–81. [CrossRef]
66. Mathur, S.; Jin, T.; Kasturirangan, N.; Chandrasekaran, J.; Xue, W.; Gruteser, M.; Trappe, W. Parknet: Drive-by sensing of road-side parking statistics. In Proceedings of the 8th International Conference on Mobile Systems, Applications, and Services, San Francisco, CA, USA, 15–18 June 2010.
67. Hayashi, H.; Inomata, R.; Fujishiro, R.; Ouchi, Y.; Suzuki, K.; Nanami, T. Development of pre-crash safety system with pedestrian collision avoidance assist. In Proceedings of the 23rd International Technical Conference on the Enhanced Safety of Vehicles, Seoul, Korea, 27–30 May 2013.
68. Hull, A. Policy integration: What will it take to achieve more sustainable transport solutions in cities? *Transp. Policy* **2008**, *15*, 94–103. [CrossRef]
69. Papa, R.; Gargiulo, C.; Galderisi, A. Towards an urban planners' perspective on Smart City. *TeMA J. Land Use Mobil. Environ.* **2013**, *6*, 5–17.

70. Gharaibeh, A.; Salahuddin, M.A.; Hussini, S.J.; Khreishah, A.; Khalil, I.; Guizani, M.; Al-Fuqaha, A. Smart cities: A survey on data management, security, and enabling technologies. *IEEE Commun. Surv. Tutor.* **2017**, *19*, 2456–2501. [CrossRef]
71. Shahzad, G.; Yang, H.; Ahmad, A.W.; Lee, C. Energy-efficient intelligent street lighting system using traffic-adaptive control. *IEEE Sens. J.* **2016**, *16*, 5397–5405. [CrossRef]
72. Roca-Riu, M.; Estrada, M.; Trapote, C. The design of interurban bus networks in city centres. *Transp. Res. Part A Policy Pract.* **2012**, *46*, 1153–1165. [CrossRef]
73. Goodman, A. Walking, cycling and driving to work in the English and Welsh 2011 census: Trends, socio-economic patterning and relevance to travel behaviour in general. *PLoS ONE* **2013**, *8*, e71790. [CrossRef] [PubMed]
74. Pozdniakova, A. Digitalization process in Ukraine as a prerequisite for the smart city concept development. *Balt. J. Econ. Stud.* **2017**, *3*, 14–19. [CrossRef]
75. Hurkovskyy, V.I.; Mezentsev, A.V. International experience of applying of the electronic identification of citizens as a technological basis of electronic petitions: Organizational and legal aspects. *Public Manag.* **2017**, *1*, 63–73.
76. Leape, J. The London congestion charge. *J. Econ. Perspect.* **2006**, *20*, 157–176. [CrossRef]
77. Marsden, G. The evidence bases for parking policies—A review. *Transp. Policy* **2006**, *13*, 447–457. [CrossRef]
78. Alpopi, C.; Silvestru, R. Urban development towards the smart city-a case study. *Adm. Manag. Public* **2016**, *27*, 107.
79. Datta, A. New urban utopias of postcolonial India: 'Entrepreneurial urbanization in Dholera smart city, Gujarat. *Dialogues Hum. Geogr.* **2015**, *5*, 3–22. [CrossRef]
80. Joss, S.; Cook, M.; Dayot, Y. Smart cities: Towards a new citizenship regime? A discourse analysis of the British smart city standard. *J. Urban Technol.* **2017**, *24*, 29–49. [CrossRef]
81. Kumar, T.V.; Dahiya, B. *Smart Economy in Smart Cities*; Springer: Singapore, 2017; Volume 11, pp. 3–76.
82. Sundar, R.; Hebbar, S.; Golla, V. Implementing intelligent traffic control system for congestion control, ambulance clearance, and stolen vehicle detection. *IEEE Sens. J.* **2014**, *15*, 1109–1113. [CrossRef]
83. Thakur, T.T.; Naik, A.; Vatari, S.; Gogate, M. Real-time traffic management using the Internet of Things. In Proceedings of the 2016 International Conference on Communication and Signal Processing (ICCSP), Madras, India, 6–8 April 2016.
84. Radhakrishnan, P.; Mathew, T.V. Passenger car units and saturation flow models for highly heterogeneous traffic at urban signalised intersections. *Transportmetrica* **2011**, *7*, 141–162. [CrossRef]
85. Marisamynathan, S.; Vedagiri, P. Modeling pedestrian delay at signalized intersection crosswalks under mixed traffic condition. *Procedia-Soc. Behav. Sci.* **2013**, *104*, 708–717. [CrossRef]
86. Sharma, A.; Vanajakshi, L.; Rao, N. Effect of phase countdown timers on queue discharge characteristics under heterogeneous traffic conditions. *Transp. Res. Rec.* **2009**, *2130*, 93–100. [CrossRef]
87. Cervero, R.; Day, J. Residential relocation and commuting behaviour in Shanghai, China: The case for transit-oriented development. *US Berkeley Cent. Future Urban Transp.* **2008**, *2*, 4–16.
88. Wang, Y.; de Almeida Correia, G.H.; de Romph, E.; Timmermans, H.J.P. Using metro smart card data to model location choice of after-work activities: An application to Shanghai. *J. Transp. Geogr.* **2017**, *63*, 40–47. [CrossRef]
89. Mega, V.P. Transport for Sustainable Cities. In *Sustainable Cities for the Third Millennium: The Odyssey of Urban Excellence*; Springer: Berlin/Heidelberg, Germany, 2010; Volume 11, pp. 61–74.
90. Debnath, A.K.; Chin, H.C.; Haque, M.M.; Yuen, B. A methodological framework for benchmarking smart transport cities. *Cities* **2014**, *37*, 47–56. [CrossRef]
91. Hoe, S.L. Defining a smart nation: The case of Singapore. *J. Inf. Commun. Eth. Soc.* **2016**, *14*, 323–333. [CrossRef]
92. Liu, W.; Zhao, C.; Zhong, W.; Zhou, Z.; Zhao, F.; Li, X.; Fu, J.; Kwak, K. The GPRS mobile payment system based on RFID. In Proceedings of the International Conference on Communication Technology, Istanbul, Turkey, 11–15 June 2006; Volume 4, pp. 1–4.
93. Pelletier, M.P.; Trépanier, M.; Morency, C. Smart card data use in public transit: A literature review. *Transp. Res. Part C Emerg. Technol.* **2011**, *19*, 557–568. [CrossRef]
94. Kröger, W. Critical infrastructures at risk: A need for a new conceptual approach and extended analytical tools. *Reliab. Eng. Syst. Saf.* **2008**, *93*, 1781–1787. [CrossRef]

95. O'HARE, D. A history of visions and plans for the transformation of a coastal tourism city into a knowledge city: Australia's Gold Coast. In Proceedings of the International Planning History Society Proceedings, New Delhi, India, 11–14 December 2016.
96. Tian, T. Bowden main park in Adelaide, Australia. *Landsc. Archit. Front.* **2017**, *5*, 86–95.
97. Wood, A. The politics of policy circulation: Unpacking the relationship between South African and South American cities in the adoption of bus rapid transit. *Antipode* **2015**, *47*, 1062–1079. [CrossRef]
98. Haarstad, H. Who is driving the 'smart city'agenda? Assessing smartness as a governance strategy for cities in Europe. In Services and the green economy. *Palgrave Macmillan* **2016**, *8*, 199–218.
99. Slavova, M.; Okwechime, E. African smart cities strategies for Agenda 2063. *Afr. J. Manag.* **2016**, *2*, 210–229. [CrossRef]
100. Boko-haya, D.D.; Li, Y.D.; Yao, C.R.; Gu, Y.; Qiang, B.; Xiang, Q.Q. Development of a conceptual model for overcoming the challenges of road and bridge infrastructure development: Towards innovative solutions in the Benin Republic. *Int. J. Eng. Res. Afr.* **2016**, *26*, 161–175. [CrossRef]
101. Bejarano, M.; Ceballos, L.M.; Maya, J. A user-centred assessment of a new bicycle sharing system in Medellin. *Transp. Res. Part F Traffic Psychol. Behav.* **2017**, *44*, 145–158. [CrossRef]
102. Martínez-Jaramillo, J.E.; Arango-Aramburo, S.; Álvarez-Uribe, K.C.; Jaramillo-Álvarez, P. Assessing the impacts of transport policies through energy system simulation: The case of the Medellin Metropolitan Area, Colombia. *Energy Policy* **2017**, *101*, 101–108. [CrossRef]
103. Stead, D.; Pojani, D. The urban transport crisis in emerging economies: A comparative overview. In *The Urban Transp. Crisis in Emerging Economies*; Springer: Berlin/Heidelberg, Germany, 2018; pp. 283–295.
104. Aydemir, P.K.; Yilmazsoy, B.K.; Akyüz, B.; Akdemir, Ç. Kentsel Ulaşımda Yaya Öncelikli Planlama/Tasarım ve Transit Odaklı Gelişimin Metropol Kentlerdeki Deneyimi, İstanbul Örneği. *Kent Akademisi* **2018**, *11*, 523–544.
105. Bilbil, E.T. The operationalizing aspects of smart cities: The case of Turkey's smart strategies. *J. Knowl. Econ.* **2017**, *8*, 1032–1048. [CrossRef]
106. Bulu, M. Measuring competitiveness of cities: Turkish experience. *Int. J. Knowl. Based Dev.* **2011**, *2*, 267–281. [CrossRef]
107. Gonel, F.; Akinci, A. How does ICT-use improve the environment? The case of Turkey. *World J. Sci. Technol. Sustain. Dev.* **2018**, *15*, 2–12. [CrossRef]
108. Kuşçu, S. Avrupa Birliği Ulaştırma Politikası ve Türkiye'ye Yansıması. *Gazi Akademik Bakış.* **2011**, *9*, 77–92.
109. Yüksel, A.N.; Sener, E. The reflections of digitalization at the organizational level: Industry 4.0. in Turkey. *J. Bus. Econ. Finance.* **2017**, *6*, 291–300. [CrossRef]
110. Yavuz, M.C.; Cavusoglu, M.; Corbaci, A. Reinventing tourism cities: Examining technologies, applications and city branding in leading smart cities. *J. Glob. Bus. Insights* **2018**, *3*, 5. [CrossRef]
111. Gazibara, I.; Goodman, J.; Madden, P. Megacities on the Move. Available online: http://forumforthefuture.org/sites/default/files/project/downloads/megacitiesfullreport.pdf (accessed on 9 April 2019).
112. Tuğaç, Ç. Türkiye İçin İklim Değişikliğine Dayanıklı Kentsel Planlama Modeli Önerisi: Eko-Kompakt Kentler. *Atatürk Üniversitesi İktisadi ve İdari Bilimler Dergisi* **2018**, *32*, 1047–1068.
113. Toprak, D. Sürdürülebilir kalkınma çevresinde çevre politikaları ve mali araçlar. *Süleyman Demirel Üniversitesi Sosyal Bilimler Enstitüsü Dergisi* **2006**, *2*, 146–169.
114. Scott, A.J.; Shorten, J.; Owen, R.; Owen, I. What kind of countryside do the public want: Community visions from Wales UK? *GeoJournal* **2011**, *76*, 417–436. [CrossRef]
115. Iwaniec, D.; Wiek, A. Advancing sustainability visioning practice in planning—The general plan update in Phoenix, Arizona. *Plan. Pract. Res.* **2014**, *29*, 543–568. [CrossRef]
116. Nevens, F.; Frantzeskaki, N.; Gorissen, L.; Loorbach, D. Urban Transition Labs: Co-creating transformative action for sustainable cities. *J. Clean. Prod.* **2013**, *50*, 111–122. [CrossRef]
117. Komninos, N.; Pallot, M.; Schaffers, H. Special issue on smart cities and the future internet in Europe. *J. Knowl. Econ.* **2013**, *4*, 119–134. [CrossRef]
118. Haasnoot, M.; Kwakkel, J.H.; Walker, W.E.; ter Maat, J. Dynamic adaptive policy pathways: A method for crafting robust decisions for a deeply uncertain world. *Glob. Environ. Chang.* **2013**, *23*, 485–498. [CrossRef]
119. Molotch, H. The city as a growth machine: Toward a political economy of the place. *Am. J. Social.* **1976**, *82*, 309–332. [CrossRef]
120. Fouracre, P.R.; Sohail, M.; Cavill, S. A participatory approach to urban transport planning in developing countries. *Transp. Plan. Technol.* **2006**, *4*, 313–330. [CrossRef]

121. Song, H.; Srinivasan, R.; Sookoor, T.; Jeschke, S. *Smart Cities: Foundations, Principles, and Applications*; John Wiley & Sons: Hoboken, NJ, USA, 2017.
122. Grant, J.L. Theory and practice in planning the suburbs: Challenges to implementing new urbanism, smart growth, and sustainability principles. *Plan. Theory Pract.* **2009**, *10*, 11–33. [CrossRef]
123. Monfaredzadeh, T.; Krueger, R. Investigating social factors of sustainability in a smart city. *Procedia Eng.* **2015**, *118*, 1112–1118. [CrossRef]
124. Tregoning, H.; Agyeman, J.; Shenot, C.; Sprawl. Smart growth and sustainability. *Local Environ.* **2002**, *7*, 341–347. [CrossRef]
125. Soria-Lara, J.A.; Banister, D. Dynamic participation processes for policy packaging in transport backcasting studies. *Transp. Policy* **2017**, *58*, 19–30. [CrossRef]
126. Yigitcanlar, T.; Velibeyoglu, K. Knowledge-based urban development: The local economic development path of Brisbane, Australia. *Local Econ.* **2008**, *23*, 195–207. [CrossRef]
127. Yigitcanlar, T.; Dur, F. Developing a sustainability assessment model: The sustainable infrastructure, land-use, environment and transport model. *Sustainability* **2010**, *2*, 321–340. [CrossRef]
128. Banister, D. Sustainable urban development and transport-a Eurovision for 2020. *Transp. Rev.* **2000**, *20*, 113–130. [CrossRef]
129. Meyer, M.D. Transport planning for urban areas: A retrospective look and future prospects. *J. Adv. Transp.* **2000**, *34*, 143–171. [CrossRef]
130. Papa, R.; Galderisi, A.; Majello, V.; Cristina, M.; Saretta, E. Smart and resilient cities. A systemic approach for developing cross-sectoral strategies in the face of climate change. *TeMA J. Land Use Mobil. Environ.* **2015**, *8*, 19–49.
131. Tiwari, A.; Jain, K. GIS Steering smart future for smart Indian cities. *Int. J. Sci. Res. Publ.* **2014**, *8*, 442–446.
132. Orosz, G.; Dombi, E.; Tóth-Király, I.; Roland-Lévy, C. The less is more: The 17-item Zimbardo time perspective inventory. *Curr. Psychol.* **2017**, *36*, 39–47. [CrossRef]
133. Banister, D.; Hickman, R. Transport futures: Thinking the unthinkable. *Transp. Policy* **2013**, *29*, 283–293. [CrossRef]
134. Rotmans, J.; Kemp, R.; Van Asselt, M. More evolution than revolution: Transition management in public policy. *Foresight* **2001**, *3*, 15–31. [CrossRef]
135. Zimmermann, M.; Darkow, I.L.; Heiko, A. Integrating Delphi and participatory backcasting in pursuit of trustworthiness—The case of electric mobility in Germany. *Technol. Forecast. Soc. Chang.* **2012**, *79*, 1605–1621. [CrossRef]
136. Bakker, S.; Kees, M.; van Bert, W. Stakeholders Interests, Expectations, and Strategies regarding the Development and Implementation of Electric Vehicles: The Case of the Netherlands. *Transp. Res. Part A Policy Pract.* **2014**, *66*, 52–64. [CrossRef]
137. Fulton, L.; Wright, L. Climate Change Mitigation and Transport in Developing Nations. *Transp. Rev.* **2013**, *25*, 691–717.

© 2019 by the author. Licensee MDPI, Basel, Switzerland. This article is an open access article distributed under the terms and conditions of the Creative Commons Attribution (CC BY) license (http://creativecommons.org/licenses/by/4.0/).

Article

# Determining Favourable and Unfavourable Thermal Areas in Seoul Using In-Situ Measurements: A Preliminary Step towards Developing a Smart City

**You Jin Kwon [1], Dong Kun Lee [2,*] and Kiseung Lee [3]**

[1] Interdisciplinary Program in Landscape Architecture, Seoul National University, Seoul 08826, Korea; eugene.kwon@snu.ac.kr or eugeneugene.kwon@gmail.com
[2] Department of Landscape Architecture and Rural System Engineering, Seoul National University, Seoul 08826, Korea
[3] Graduate School of Public Health, Seoul National University, Seoul 08826, Korea; sslks93@snu.ac.kr
* Correspondence: dklee7@snu.ac.kr; Tel.: +82-2-880-4885

Received: 15 May 2019; Accepted: 13 June 2019; Published: 17 June 2019

**Abstract:** Urban heat island effects (UHIE) are becoming increasingly widespread, thus, there is an urgent need to address thermal comfort, which significantly influences the daily lives of people. In this study, a means of improving the thermal environment by spatial analysis of heat was implemented to ensure basic thermal comfort in future smart cities. Using Seoul as the study site, the relationship between sensible heat and land cover type was used to identify heat islands in this city. Thereafter, k-means clustering was employed to extract unfavourable and favourable thermal areas. High sensible heat indicates locations where environmental heat needs to be mitigated. Sensible heat distribution data were used for spatial typification to formulate an effective land cover factor to mitigate the UHIE. In-situ net radiation data measured at six sites were utilised to confirm the spatial typification of the thermal environment. It was found that expanding the green space by 1% reduces the sensible heat by 4.9 W/m$^2$. Further, the building coverage ratio and green coverage influence the sensible heat in compact residential areas. The study results can be used to establish spatial planning standards to improve the thermal environments of sustainable cities.

**Keywords:** energy budget; land cover ratio; sensible heat flux; heat mitigation; thermal environment improvement; sustainability; in-situ validation; spatial typification by heat flux

## 1. Introduction

### 1.1. Urban Heat and Spatial Typification for Sustainability

The urban heat island effect (UHIE) depends on the land cover, roughness of the urban surface, presence of buildings, and albedo. The land cover type significantly affects the thermal environment, which determines the local climate zone (LCZ) [1,2] and is affected by thermal radiation in urban canyons. The UHIE is considered a function of time, weather conditions, and structural characteristics in practical applications such as road engineering, climatology, phenology, energy conservation, and weather forecasting [3,4]. Oke et al. [5] noted that although studies on the UHIE have explained this phenomenon, they are still lacking with respect to its causes, suggesting that the situation is well described but not well understood. Because the UHIE is influenced by various spatial factors, its assessment via a spatial multi-scale approach is required.

Thus far, the causes of heat generation with respect to space have been studied macroscopically to understand the UHIE [6]. From this perspective, the UHIE is caused by the temperature difference between urban and suburban areas and is usually studied on a regional scale. However, thermal comfort at the micro-scale (that is, at the building scale) is of immediate concern to the residents of urban areas. Therefore, research on heat generation and its mitigation is increasing in importance. Micro-scale studies [7] need to be approached from the meso-scale [8,9] and multi-scale [10], and the irradiation of land must be accounted for [11].

Moreover, increases in population and population density cause the thermal environment to deteriorate [12,13]. As rapid urbanization and the corresponding increase in global warming are becoming worldwide concerns [14], every country is aiming to develop sustainable cities to promote heat island reduction. Energy sustainability is being emphasized to achieve this objective [15]. Amongst anthropogenic factors, the use of air conditioners, especially during summers [16], has led to further deterioration of the urban thermal environment [17]. This deterioration is characteristic of especially large cities [18]. Moreover, smart solutions are applied mainly in medium and large cities. Smart cities are being increasingly developed and are among the means of attaining energy sustainability and improving the thermal environments of urban areas [19]. Therefore, a practical means of implementing methods of enhancing sustainability is necessary. However, the development of smart cities also involves ideological issues that favour business-led technological solutions [20] without practical means. The development and evolution of the smart city concept are more complex than they are described. For instance, a report was published on an integrated conceptual model for Vienna, which states that the framework of a smart city must compensate for insufficient environmental sustainability [21]. The implementation of the smart city framework should also take natural resources into consideration [22]. Yigitcanlar emphasises that the effects of smart cities, such as $CO_2$ reduction, are insufficient for attaining sustainability [23]. To improve quality of life, the comfort provided by the thermal environment is important [24,25]. Detailed spatial analysis of the thermal environment to ensure this comfort merits further investigation, as a preliminary step for thermal environment improvement with the objective of developing smart cities in the future.

*1.2. Heat Energy and Spatial Typification*

Concrete spatial analysis of the thermal environment is a current priority. As stated by Howell, 'Energy radiates from one object to another under all conditions and at all times' [26]. Since heat energy is intangible in tangible space [27], one spatial solution for improving the thermal environment is the typification of the latter. In meteorology, sensible heat flux, or sensible heat, is defined as the conductive heat flux from the land surface to the atmosphere and is evident in the transportation of energy being an important factor in the energy budget of the surface of the earth [28]. Sensible heat flux is a less popular quantity for defining the thermal environment than temperature, but it is effective for presenting the physical characteristics in space or performing quantitative estimations, such as heat transfer prediction [29,30].

In this research, we analysed the UHIE in terms of sensible heat flux instead of temperature over a wide area, considering factors such as the presence, size, and density of buildings as well as the presence of greenery in the relevant area. However, the relationships between the causes of thermal radiation generation and the effects of the spatial factors that are the fundamental causes of the UHIE are difficult to comprehend. Moreover, the thermal approach is challenging to apply to planning or designing in a practical geographical range. However, the urban surface temperature and ambient temperature are the results of heat exchange. By accurately diagnosing the urban land cover elements and analysing the heat distributed in an urban space, it is possible to understand the characteristics of the space to improve the urban thermal environment. In particular, to reduce or adapt the UHIE, it is necessary to understand this phenomenon itself and to examine the relationships between the physical elements of the urban space and the UHIE as well as the factors affecting the sensible heat flux.

*1.3. Studies Using the Existing Spatial Approaches*

The increase in summer temperatures affects individuals vulnerable to heat [31,32]. In this context, many studies have been conducted with the objective of mitigating the UHIE. These include studies on the effects of green spaces on temperature reduction [33–36] (shadows of trees [37–39], grass in parking spaces [40,41], and rooftop greenery [42–44]), changes in the surface temperature according to the materials of buildings [45,46], temperature reduction due to the optimal spatial arrangement of buildings (ventilation path, road width, and pith of building) [47,48], land use, land cover change, and urban ecological service function [49,50].

Further, researchers have attempted to identify heat-vulnerable areas by investigating the distributions of social indicators and heat [51]. Efforts have also been made to identify the causes of high temperatures in urban areas in summer and the consequent formation of urban heat islands [52,53]. However, because of the characteristics of temperature, it is difficult to identify physical, spatial features. For example, the surface of a coating or building is composed of various materials but the amount of radiant heat incident on the surface and reflected off it varies with time according to its composition ratio [54]. The physical method of measuring radiant heat is a quantitative approach [26]. Radiant heat has been measured using the physical approach on both a global scale [55] and an urban scale (meso-scale) from an atmospheric or meteorological perspective.

There have also been micro-scale studies on indoor temperature reduction in summer from a green architectural perspective [56]. Further, global-scale studies have been conducted using physical models of heat generation or transfer [57,58]. A study on climate change involving satellite remote sensing captured the long-term trends of many meteorological variables [59] between heat and space that digitise the characteristics of the latter, and remote sensing studies have been conducted to explore the relationship between surface temperature and land cover factors [60,61]. This approach is not applicable on a small scale, that is, considering the microclimate; thus there is little difference in identifying linkages with the spatial elements that form cities on land. In addition, several computational fluid dynamics studies have been conducted, such as those involving ENVI-met or RayMan [62–64]. However, there are limits on the edges of the domains used to simulate urban spaces including all relevant urban characteristics.

### 1.4. Typification Method

Several researchers have attempted to classify the energy budgets of areas based on their spatial characteristics; for instance, Köppen established a climate classification system [65]. Regional zones can be spatially characterised depending on their temperatures (high or low). Based on the climate classification system developed by Köppen, intangible climatic elements have been used in other systems to identify climate zones [65], ecosystem maps [66], agricultural land [67], and hydrological areas [68]. In this study, thermal spatial typification was employed to determine climate zones and map the heat, based on which another method was derived. Several means of classifying thermal environment into climate zones exist; in this study, a statistical approach called k-means clustering analysis [69] was chosen. This approach is a representative non-hierarchical clustering method and can quickly group large amounts of data, such as weather data. Further, in the case of choosing the initial configuration for the relative distinction of the characteristics of isothermal heat, k-means clustering is advantageous compared to hierarchical analysis [70]. ased on the k-means clustering results, we define an 'unfavourable thermal area' (UTA) as a region in which heat is concentrated; these regions should be prioritised for heat reduction. Conversely, a region with a relatively low heat concentration is defined as a 'favourable thermal area' (FTA).

### 1.5. Objectives and Application

In the aforementioned micro-scale studies, the UHIE has been explored in terms of temperature changes with respect to the presence of buildings, as well as their size and density, and the presence of surrounding green spaces. The urban surface temperature and atmospheric temperature are the results of accumulating heat. As mentioned by Ma [71], differences in albedo due to land use and land cover type cause spatial changes in surface heat flux [71]. By accurately determining urban land cover elements and analysing the heat distributed in an urban space, it is possible to improve the thermal environment of that space. In this study, based on an existing empirical model, we classified the estimated sensible heat distribution. Thereafter, using the classification results, in-situ measurements of the net radiation and sensible heat were performed to confirm the classification accuracy (as explained in Section 2).

The purpose of this study was to identify the distribution of sensible heat, which affects thermal comfort, through spatial typification with the objective of improving the thermal environment. The heat typification is based on five elements: the buildings, green spaces, water spaces, roads, and impervious surfaces in areas inhabited by urban residents. Further, the study results provide useful tools for sustainable urban development, including (1) the quantitative validation of two types of thermal areas through measurement at the micro-scale for proper spatial typification, (2) the suggestion of urban land cover ratio criteria for urban planning and design, and (3) an evaluation of the contribution of green spaces to improving thermal comfort [72].

### 1.6. Research Process

Our study was focused on the validation of spatial typification for thermal environment improvement through in-situ measurement. Firstly, we reviewed the current research related to the questions addressed in and purposes of our study (Figure 1).

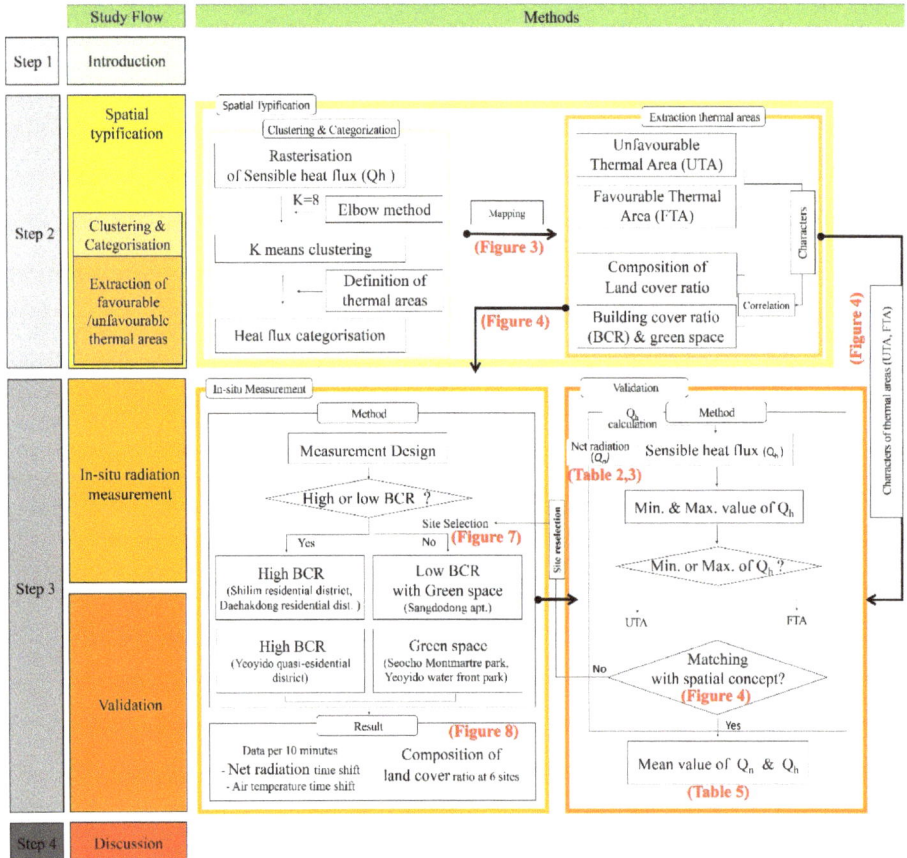

**Figure 1.** Research flow (see Figure A1 in Appendix B for details).

Secondly, we analysed the thermal environment of Seoul with reference to the method employed in our previous work [73]. In this step, we rasterised the sensible heat flux based on existing data, considering the energy balance, net radiation, latent heat, sensible heat, storage heat, and artificial heat. Then, the heat flux was clustered through the k-means approach and the results were categorised. Next, in the 'thermal area extraction' stage, the UTA and FTA were extracted through the mapping. In addition, the land cover ratios of the thermal areas were derived from the daytime shift of the sensible heat flux and local land cover scale data. To select the study sites for the in-situ measurements, correlation analysis of the land cover ratio and sensible heat flux was conducted. The building cover ratio (BCR) and green spaces were found to be two significant factors affecting the changes in and distribution of the sensible heat flux (Appendix C).

Thirdly, we performed validation using field measurements. The two factors derived from the previous step become the criteria for selecting six study sites for in-situ measurement. We measured the sensible heat flux of the spatially typified thermal areas, i.e., the UTAs and FTAs, at the study sites. From the in-situ measurements, we collected net radiation and air temperature data. We also analysed the land cover ratios at each site. In the validation stage, we analysed the calculated values of the sensible heat flux based on the net radiation measured at the six sites corresponding to the two types of thermal areas. Furthermore, we validated the typification method by matching the sites with high sensible heat fluxes according to the in-situ measurements to the areas identified as having high sensible heat fluxes (UTAs) according to the spatial typification. If they matched, the results were analysed, and if not, the study site was re-searched.

We further considered how to create an implemental tool, a spatial typification system, and planning standards for thermal environment improvement. We also suggest that this approach has an indirect effect on sustainable energy use.

## 2. Extraction of Favourable and Unfavourable Areas

Intangible heat is dynamic in nature. However, in an environment experiencing continuous heat inflow, the heat stagnates and accumulates, causing a considerable increase in temperature [74] and making the region/space heat-intensive. Mitigating this excessive heat improves the thermal comfort of the residents [75]. Therefore, to reduce heat enhancement, we attempted to elucidate the causes of spatial concentration of heat by identifying the common spatial characteristics of areas with high heat. Thus, in this section, we describe the characteristics of spaces where heat is concentrated.

### 2.1. Spatiotemporal Scope

The city of Seoul is an interesting space because its thermal comfort related problems can be solved in a sustainable manner using the various surface characteristics of its compact urban area [76]. For validation, we conducted measurements at Gwanak, Mapo, and Seocho, which are residential areas in Seoul with high sensible heat concentrations. Temporal data for the summer of 2015, based on the automatic weather station (AWS) data format, time, and spatial resolution, were provided by the Seoul Metropolitan Government Meteorological Agency and SKTech X, a private company. The measurements for validation of the typification were conducted in 2018.

### 2.2. Input Data

The hourly heat flux was calculated using the air temperature, atmospheric pressure, and relative humidity collected from 287 stations in total, including 38 meteorological stations in Seoul and 249 SKTech X meteorological stations. We typified the heat flux and land cover data (Figure 2) on a day with clear weather, low cloud cover, and peak air temperature in August (Table 1). We also created a 100 m × 100 m grid to map the thermal environment and thermal distribution data.

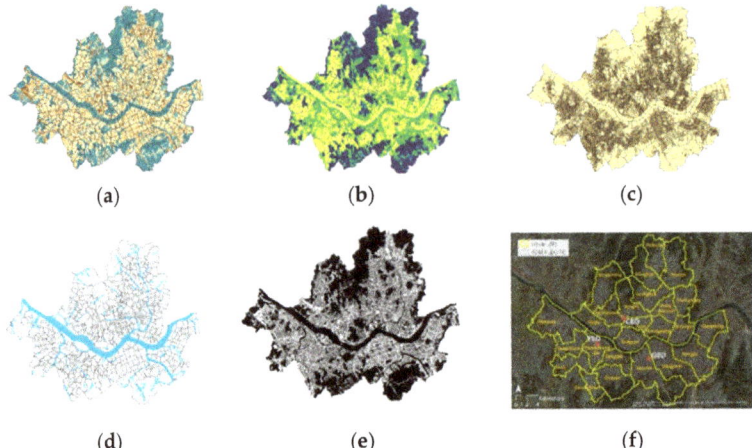

**Figure 2.** Input data: (**a**) impervious surfaces, (**b**) green spaces, (**c**) buildings, (**d**) water and wetlands, (**e**) roads, and (**f**) administrative map.

**Table 1.** Metrology and spatial attributes (land cover factors): data and sources.

| Classification | Input Data | Source |
|---|---|---|
| Meteorological data for heat flux distribution | -Air temperature, relative humidity, cloud cover, saturated water vapour pressure | -Korea Meteorological Administration (38 stations) -SKTech X (249 stations) |
| Extraction of FTAs and UTAs | -Latent heat model, sensible model, storage heat model | Holtslag, 1983 Loridan, Grimmond 2011 |
| Spatial attributes | -Subdivided land cover map (green spaces, wetlands, impervious surfaces) -.shp file of Seoul administrative district–building .shp file -.shp file depicting the widths of roads -.shp file depicting the width of roads | -Ministry of Environment, -Statistical Geographic Information Service (SGIS), -Seoul Information Communication Plaza |

## 2.3. Methodology

In k-means clustering, centroid data are used as prototypes and are divided into clusters composed of the data closest to the prototype. This process is based on the theory established by Tobler stating that ' ... near things are more related than distant things ... ' [77]. K-means clustering is characterised by rapid grouping of vast amounts of data and selection of initial configurations; further, it is suitable for thermal analysis, which is an isometric rather than hierarchical approach [78]. Therefore, thermal distribution typification through k-means clustering allows the group to be expressed as a heat value while classifying heat-concentrated spaces based on the land cover ratio, which is a spatial factor [73].

To reflect only the land cover characteristics for the urbanised parts of the city, we initiated the cluster analysis based on the heat values, excluding the main mountains and large hydrological systems around the major ecological axis that considerably influence the latent heat flux ($Q_e$) and sensible heat flux ($Q_h$).

The k-means clustering was performed on sensible heat flux data distributed every 100 m × 100 m in Seoul to derive the UTAs. The number of divided groups was derived from the natural break function of the sensible heat flux. The mean values of the sensible heat clusters were compared with those obtained from research on spatial typification based on heat flux [73] to distinguish the values as unfavourable and favourable.

## 2.4. Results of K-Means Clustering

The thermal areas that were grouped through k-means clustering are referred to as 'heat-concentrated regions'. Among these, the UTAs were identified as areas with sensible heat fluxes higher than the mean value in Seoul; similarly, the values of low sensible heat-concentrated regions, called FTAs, are lower than the mean value of the sensible heat flux in Seoul. In Figure 3, the reddish areas correspond to UTAs and the blue areas represent FTAs.

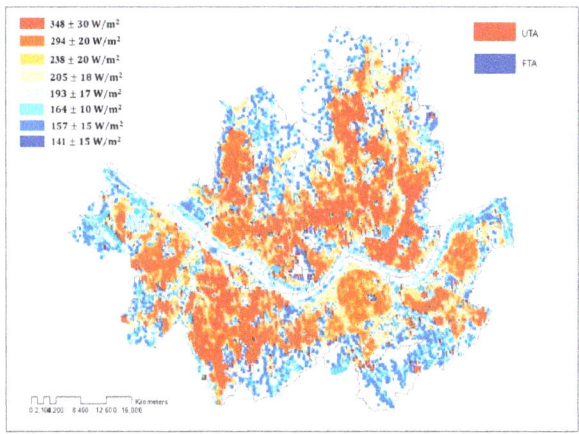

**Figure 3.** UTAs and FTAs.

We analysed the building composition and BCR in the UTAs and FTAs corresponding to residential areas. The UTAs are characterised by high building density with no green cover, while the FTAs are characterised by low building density and low population density of greenery or water (Figure 4). The verification of these areas is discussed in detail in Section 3.

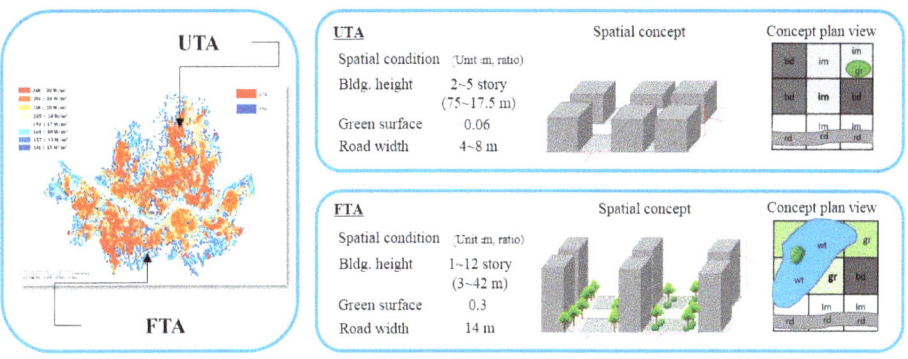

**Figure 4.** Concepts of UTAs and FTAs.

## 3. Sensible Heat Flux Calculations

Since there is no dedicated equipment for measuring sensible heat flux, we estimated the heat flux using the existing models for calculating the energy budget by defining the relationships among the net radiation and the four heat fluxes. The energy budget is composed of four elements, namely, the anthropogenic heat flux ($Q_F$), sensible heat flux ($Q_h$), latent heat flux ($Q_e$), and storage heat flux ($Q_s$),. The horizontal heat flux due to wind-borne transportation ($\Delta Q_A$) was included in the energy budget at a value of 0, because of the warm and mild wind conditions on the measurement day. The sensible heat flux was extracted from these four heat fluxes. The model for calculating the sensible heat flux using the net radiation ($Q_n$) is as follows [79–82]:

$$Q_n + Q_F = Q_h + Q_e + Q_s + \Delta Q_A. \tag{1}$$

The net radiation was derived from the urban energy balance in (1), which was derived by Offerle [79]. All of the heat fluxes have units of W/m$^2$:

$$\Delta Q_S = \sum_{i=1}^{l} f_i \times (a_{1i} Q_n + a_{2i} \Delta Q_n + a_{3i}) \tag{2}$$

$f_i$: land cover ratio (unit: ratio)
$i$: green cover, water cover, impervious land, building cover, and road cover
$\frac{\gamma}{s} = -0.00003178 \times temp^3 + 0.03 \times temp^2 - 0.092 \times temp + 1.463$

The storage heat flux in (2) was derived from an equation considering the land cover ratio and empirical coefficients. $\gamma$ is a psychrometric constant, and s is the slope of the curve of saturation vapour pressure versus temperature:

$$Q_h = \left[ \frac{\{(1-\alpha) + \left(\frac{\gamma}{s}\right)\}}{\{1 + \left(\frac{\gamma}{s}\right)\}} \right] \times (Q_n - \Delta Q_S) + 20 \tag{3}$$

$$Q_e = \left[ \frac{\alpha}{\{1 + \left(\frac{\gamma}{s}\right)\}} \right] \times (Q_n - \Delta Q_S) + 20 \tag{4}$$

$\alpha$ : an empircal parameter related to the moisture status of the surface

Both the sensible heat flux (3) and latent heat flux (4) were estimated using the model developed by Holtsalg [79]. In these equations, 20 (W/m$^2$) is an empirical constant. Both $\alpha$ and 20 (W/m$^2$) were determined based on the Penman Monteith approach [82]:

$$Q_F = 6.8 \, (T_C - T_d) + 12 \text{ (for } T_d \leq T_C) \tag{5}$$

where $T_C$: Daily maximum temperature (unit: K) and $T_d$: Daily mean temperature (unit: K)

The anthropogenic heat flux (5) was estimated using a model considering temperature. The advection ($\Delta Q_A$) was negligible at the six investigated sites on the day of measurement because we chose a day with a wind speed of approximately 0 m/s.

The land cover ratio ($f_i$) was calculated using the relative area occupied by each type of land cover within a 100 m × 100 m grid (Tables 2 and 3).

Table 2. Empirical land cover coefficients.

| Land Cover Coefficient | $a_1$ (Ratio) | $a_2$ (h) | $a_3$ (W/m$^2$) |
|---|---|---|---|
| Green | 0.34 | 0.31 | −31 |
| Building | 0.07 | 0.06 | −5 |
| Impervious | 0.83 | 0.4 | −54.2 |
| Water | 0.5 | 0.21 | −39.1 |
| Road | 0.61 | 0.41 | −27.7 |

Source: Grimmond and Oke [83], Roberts and Oke [84].

Table 3. Anthropogenic heat flux at the neighbourhood scale.

| Neighbourhood | LCZ | Anthropogenic Heat Flux (W/m$^2$) | Site |
|---|---|---|---|
| Large dense, city centre | 1,2 | 100–1600 * | 4 |
| Medium dense, city centre | 3 | 30–100 * | 6 |
| Low dense, open, low-rise | 6 | 5–50 * | 2 |
| Open high-rise | 4 | 26–80 ** | 1 |
| Green (Low-planted), Water | D,G | - | 3,5 |

Source: * Oke, Mills, Christen and Voogt [85]; ** Pigeon, G., et al. [81].

## 4. In-Situ Validation

In this section, we present the results of our examination of whether the UTA and FTA identification method can be applied to actual sites, considering the land cover rate (in particular, the BCR). The degree of sensible heat flux loss derived using the urban canyon structure can also be determined. Therefore, an empirical study was conducted to verify the accuracy of the classification method by dividing the study area based on sensible heat, which was done to develop methods that can help improve the thermal environment.

In the previous section, we presented the characterization of the disadvantageous thermal zones using low-rise buildings in residential areas with high BCR. We considered such residential areas and focused on heat transfer to estimate the UTAs and FTAs. Because no instrument has been developed to measure sensible heat, we estimated the sensible heat using a physical model based on the net radiant heat data measured by a sensor. Therefore, to validate the results of the in-situ measurements aimed at controlling the outdoor conditions during measurement, the following in-situ measurements were conducted:

a   BCR and net radiation time shift data
b   Time shifts with variations in BCR
c   Mixed-deployment observation of the overall effects of land coverage
d   Various urban climate data for calculating the sensible heat flux

Observations were made to obtain net radiation data to calculate the sensible heat flux under different land cover compositions, especially BCRs. In addition, we collected air temperature, wind speed, humidity, albedo, and sky view factor (SVF) data for sensible heat flux estimation (Equation (1), Section 4).

*4.1. In-Situ Measurement Process (Appendix B)*

4.1.1. Considerations for Measurement Design

According to the BCR values and measurement techniques reported previously [86], we designed a net radiation measurement technique. The observation sites were required to be in residential areas with mild wind speed, as the wind speed is often used as an independent variable when measuring heat flux. Wind speed causes surface turbulence; in addition, heat flux transfer can be determined by airflow. Furthermore, the heat flux incident on a space could be controlled by varying the relative humidity. When both the wind speed and humidity were controlled, the effects of other parameters on the energy budget could be determined. Because these two variables cannot be controlled easily, the measurement data were obtained by varying the BCR in areas under similar microclimate conditions.

4.1.2. Observation Site Selection

This study was focused on validating FTAs based on the influence of the BCR on the net radiation and sensible heat flux of a micro-scale urban area, by performing in-situ measurements in Seoul. Thus, we identified the study areas that were extracted as FTAs and UTAs and had continuous, similar densities and shapes of buildings in the neighbourhood. Moreover, we sought locations with high sensible heat flux ($Q_h$), UFA, and the place with low $Q_h$, which represented FTA. Therefore, areas with different land coverage included in the UTAs and FTAs in Seoul were selected as sites to measure the net radiation. Most residential and commercial complexes in Seoul have high BCRs. Among the residential complexes in Seoul, the six sites we chose have plane topography; further, permission was obtained from the resident committees prior to performing the measurements.

### 4.1.3. Measurements and Data Collection

The observations were conducted from May to June 2018. This date range was selected considering the availability of the location for measurement and the presence of favourable weather conditions, because cloudy weather is often observed from 9:00 to 19:00 in Seoul. The types of equipment used to obtain the measurements were as follows. The net radiation was collected using a CNR4 (Campbell Scientific, Inc., Logan, UT, USA; Figure 5, Table A1 in Appendix A) with two-directional sensors, where the upper sensor was used to collect the net radiation and the data were collected in a logger (CR1000, Campbell Scientific, Inc.). A thermometer, an anemometer, and loggers were used to measure and collect the air temperature and wind data (Figure 5). A camera with a fish-eye lens was used to determine the SVF. The measurements were performed when the mean wind speed was 0 m/s and temperature was 18–28 °C. The humidity was 55% (measured 7–10 m from the 12-story building). The data were collected from 9:00 to 18:00 in 1 min intervals. The times of sunrise and sunset in Seoul from May to June are 7:00 and 18:30, respectively. All instruments were deployed within 1 m from each building, and each instrument was elevated 1.2 m above the ground.

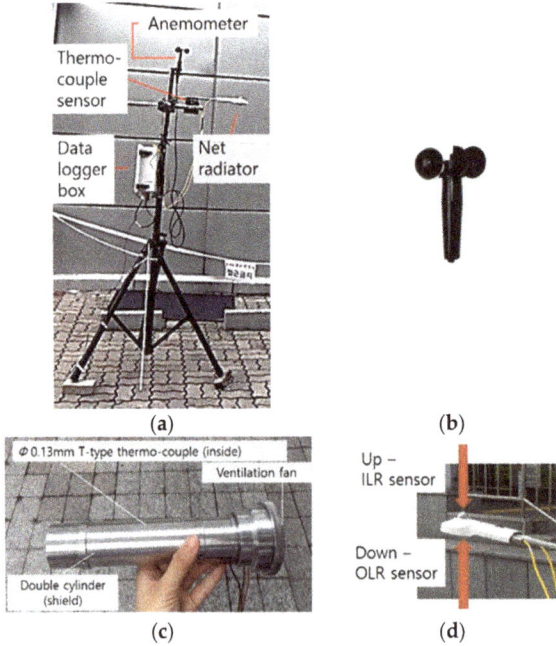

**Figure 5.** Observation instruments: (**a**) mobile net radiometer instrument kit, (**b**) anemometer, (**c**) thermometer (revised from Park, [87]), and (**d**) bi-directional radiation sensors (Kwon and Lee, [86]).

### 4.1.4. Context of the Study Sites

For the six sites mentioned earlier, net radiation measurements were performed to calculate the energy budget and air temperature, albedo, wind speed, and relative humidity data were collected on warm, sunny days.

According to the objective of obtaining in-situ measurements, residential areas in Seoul were divided into two categories: low density and high density (Figure 6). Most apartments are exposed concrete slab or brick structures and face south. The distances between the buildings and the instrument deployment locations were 3–36 m. The SVF of the complex for in-situ observation is 0.46–0.90, and the building heights are in the range of 7–40 m. Two open spaces exist: one offers a plain view with green land coverage and the other is a riparian area (Sites 3 and 5, Figure 7).

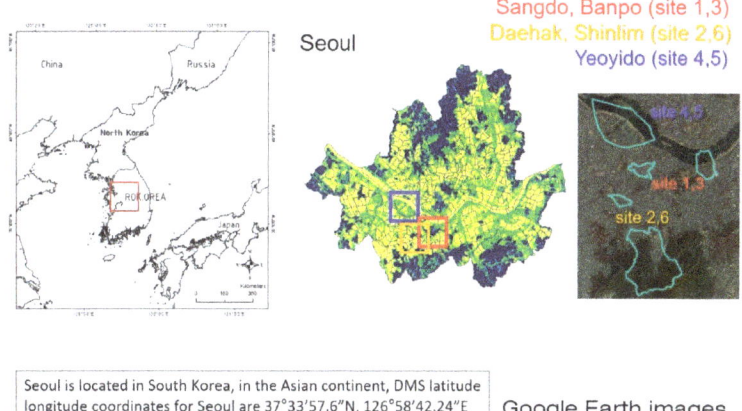

**Figure 6.** Sites: Sangdo (1), Daehak (2), Montmartre Park (3, Banpo), Yeoyido (4), Riverside Park (5), Shilim (6).

**Figure 7.** Street views of the study sites.

## 4.2. Limitations of the in-situ Measurement

The measurements are intended to show the influence of the BCR on the net radiation and sensible heat flux in the spaces around different densities of buildings [88]. However, because the study sites were residential areas, the residents did not readily allow us to obtain measurements because they think that vulnerability to heat directly impacts land and housing prices. Thus, the study was limited by the restricted number of observation locations and hours. Moreover, it was difficult to obtain reliable data owing to the frequent movement of the floating population in the study areas. Finally, the CNR4 (a net radiometer) is expensive and limited in number; thus, it was impossible to examine the multiple sites simultaneously.

To answer the research question posed at the beginning of the study, we performed validation by obtaining the sensible heat flux and comparing the results with those reported previously [73]. The sensible heat was estimated based on the in-situ measurements of the net radiation. The equations related to the estimation are addressed in Section 3. The residential areas of Sites 2, 4, and 6 have high densities (Table 4), and Sites 2 and 6 are low-rise residential areas. However, Site 4 is also a high-density building area but with high-rise quasi-residential units (Table 2). The building areas of Sites 1, 3, and 5 are low-density areas, and Site 1 is high-rise residential area including green land coverage. Site 3 is a neighbourhood park, and Site 5 is a riverside park located in a residential district. To collect net radiation and land cover data from the various locations, we deployed a mobile net radiometer instrument kit at six measuring stations (Sites 1–6; Figure 7, Table 2).

**Table 4.** Two types of sites (O: pertinent; X: NA).

| Type | Site 1 | Site 2 | Site 3 | Site 4 | Site 5 | Site 6 |
|---|---|---|---|---|---|---|
| High density | X | O | X | O | X | O |
| Low density | O | X | O | X | O | X |

The four land cover ratios of Site 1 are evenly distributed, without water (Figure 8). Because Seoul has a high density of buildings, the building ratio is the highest among the five land cover ratios at Site 6 (0.73), Site 2 (0.60), Site 4 (0.52), and Site 1 (0.35), in sequence. As a neighbourhood park, Site 3 has the highest green cover ratio. Lastly, the water ratio is the highest at Site 5, the riverside.

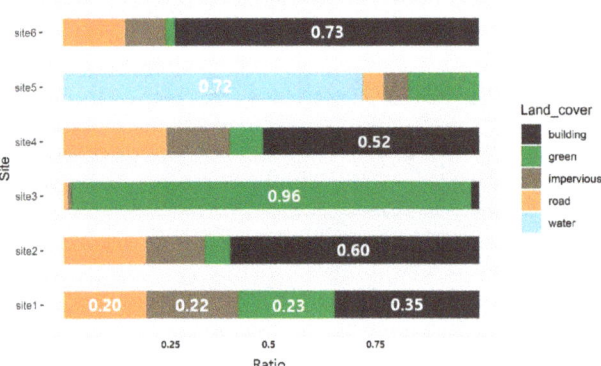

**Figure 8.** Land cover ratios at six sites.

## 5. Validation Results: Comparison of Sensible Heat Fluxes from the Six Sites

The sensible heat fluxes at the six sites were compared, indicating that FTAs and UTAs are more likely to be validated in an urban landscape under the same microclimatic conditions. Therefore, we examined the changes in $Q_h$ by classifying the measurement sites into high and low density areas using the spatial characteristics of the FTAs and UTAs in Seoul, Korea.

Site 1 is a low-density residential area with a neighbourhood garden: the Butterfly Garden. The maximum heat flux there is 357.97 W/m², and the minimum is 52.19 W/m². The average sensible heat flux at Site 1 is 269.13 W/m² (Figure 9a, Table 5). Site 2 is a high-density residential area composed of low-height buildings. The maximum heat flux there is 524.60 W/m², and the minimum is 65.02 W/m². The average sensible heat flux at Site 2 is 389.05 W/m² (Figure 9b, Table 5).

Site 3 is a neighbourhood park near a residential area. The maximum heat flux there is 123.12 W/m², and the minimum is −7.69 W/m². The average sensible heat flux at Site 3 is 83.39 W/m² (Figure 9c, Table 5). Site 4 is a high-density, quasi-residential area composed of high-rise buildings. The maximum heat flux there is 616.72 W/m², and the minimum is 165.50 W/m². The average sensible heat flux at Site 4 is 379.01 W/m² (Figure 9d, Table 5). Site 5 is a riverside park along the Han River, which is the main waterway in Seoul. The maximum heat flux there is 432.61 W/m², and the minimum is 32.61 W/m². The average sensible heat flux at Site 5 is 311.19 W/m² (Figure 9e, Table 5). Site 6 is a high-density residential area without green space. The maximum heat flux there is 595.07 W/m², and the minimum is 170.67 W/m². The average sensible heat flux at Site 6 is 444.06 W/m² (Figure 9f, Table 5).

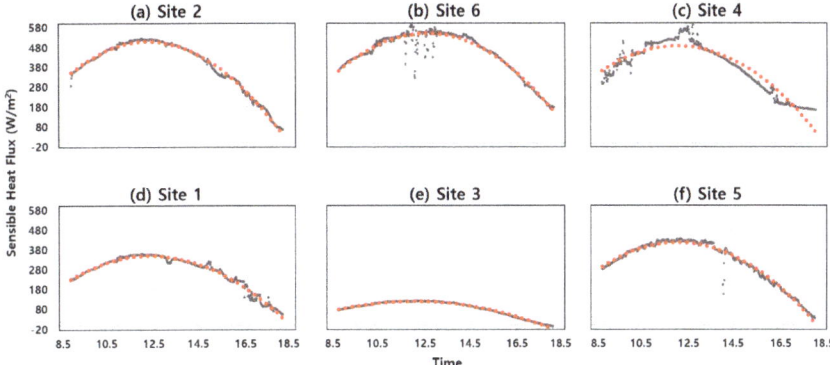

**Figure 9.** Sensible heat flux versus time plots of the six sites. (**a**) Site 2: Daehak; (**b**) Site 6: Shilim; (**c**) Site 4: Yeoyi quasi-residential; (**d**) Site 1: Sangdo; (**e**) Site 3: Montmartre Park; (**f**) Site 5: Hangang riverside.

**Table 5.** Mean values of net radiation and sensible heat flux (W/m²).

| Site | Site 1 | | Site 2 | | Site 3 | | Site 4 | | Site 5 | | Site 6 | |
|---|---|---|---|---|---|---|---|---|---|---|---|---|
| Time | $Q_n$ | $Q_h$ | $Q_n$ | $Q_h$ | $Q_n$ | $Q_h$ | $Q_n$ | $Q_h$ | $Q_n$ | $Q_h$ | $Q_n$ | $Q_h$ |
| Peak ± 0.5 (h) * | 540 | 278 | 586 | 398 | 351 | 85 | 548 | 376 | 456 | 315 | 672 | 466 |
| After sunset ** | 168 | 89 | 184 | 124 | 52 | 6 | 256 | 174 | 96 | 85 | 323 | 222 |

* average values over an hour extending 30 min before and after the peak time; ** average values over an hour after sunset.

Besides the high-density residential areas, Site 6 (Shilim) exhibits the highest values of both $Q_n$ (672 W/m²) and $Q_h$ (466 W/m²) (please see Table 5). Site 6, which is a high-density residential area with low-rise buildings and minimal green cover, is the most suitable place to apply thermal environment improvement strategies in Seoul. The greatest temporal change in $Q_n$ (456 W/m²) is evident at Site 5, a riverside area, but the most rapid temporal transition in $Q_h$ is at Site 2, a dense residential area (398 − 124 = 274 W/m²). At Site 3, a green space, on the other hand, the heat flux changes moderately from the peak until sunset (85 − 6 = 79 W/m²).

The sensible heat values of the six sites exhibit parabolic trends (red dotted line: parabolic fit) similar to those of the net radiation, but with time shifts (Figure 9). Each measured site exhibits the highest $Q_h$ value between 12:00 and 13:00 but has a different slope. The sites corresponding to high-density residential areas (Sites 2, 4, and 6) display higher slopes, and the heat values decrease rapidly between 100 W/m² and 200 W/m² after the peak points. However, at 16:00, the sensible heat slope for Site 3 (a neighbourhood park) decreases more gradually than those of the other sites. The $Q_h$ range of the high-density residential areas (50–600 W/m$^2$) is higher than those of the non-high-density residential areas, including the low-density residential areas, green spaces, and waterfronts (−20 to 400 W/m$^2$) (Figure 9). It is assumed that the wider space of the urban canyon receives active natural convection that transfers heat flux [89] from the buildings, ground, or surrounding environment into the air [90].

## 6. Discussion

Because of the rapid densification of cities and climate change, which are contrary to sustainability developments, summer heatwaves occur, and many means of reducing these heatwaves have been explored. Most previous research has been focused on blocking solar radiation to improve thermal comfort. However, in this study, we investigated whether the FTAs and UTAs determined spatially as presented in Section 2 correspond to those at the actual sites chosen. Therefore, our research was mainly focused on finding the appropriate thermal environmental conditions for residents by validating the spatial typification after determining the FTAs in summer.

We investigated the net radiation values of the sites representing FTAs and UTAs, with additional emphasis on the temporal shifts in the sensible heat flux; further, the heat fluxes of these different thermal environments were compared from the perspective of FTAs and UTAs, considering the BCR.

### 6.1. Spatiotemporal Shift in Sensible Heat Flux

We determined the effects of spatial and environmental factors on $Q_h$ in Seoul by measuring the net radiation and calculating $Q_h$ in residential areas located in urban canyons and green spaces. The temporal shifts in $Q_h$ in the urban canyons were determined for two types of areas: high-density residential areas (Sites 2, 4, and 6) and low-density areas (Sites 1, 3, and 5).

The study areas were enlarged to examine the correlation between the sensible heat fluxes measured in-situ and those based on two types of areas broadly classified as FTAs and UTAs. The high-density residential areas (Sites 2, 4, and 6; UTAs) have higher sensible heat flux ranges, while the low-density residential areas (Sites 1, 3, and 5; FTAs) have lower ranges, because natural convection enables sensible heat flux transfer [89]. Further, green spaces increase the latent heat, thereby decreasing the sensible heat. By designing wider spaces between buildings and providing green spaces within a residential complex, the sensible heat flux can be decreased; therefore, green spaces should be considered in the layouts of residential areas.

The measurement equipment used in this study was located in the urban canyons 1.2 m above the ground; thus, the sensor readings may not adequately reflect the land cover conditions beyond the nearest buildings. Because this research was conducted in a limited number of areas, the net radiation trends in more places should be investigated to obtain more accurate results.

Firstly, the extent of the effects of the five land cover types on the sensible heat was observed to follow the sequence green area > road > building > impervious surface > water surface (Appendix D). Secondly, we found that green areas impact the sensible heat the most. However, Site 6 has the highest sensible heat flux (444 W/m$^2$) with the lowest green surface ratio (0.02), and Site 3, with the highest green surface ratio (0.96), has the lowest sensible heat flux (82 W/m$^2$) (Table 6). Thirdly, the gradients of the sensible heat flux trends for each site were examined, which reflect the shift rate ($\Delta y/\Delta x$) of $Q_h$. The change rates were found to follow the order Site 2 > Site 6 > Site 4 > Site 5 > Site 1 > Site 3. This order corresponds to that of daytime $Q_h$, except for the order of Sites 6 and 2. The reason that Site 6 has the greatest slope is that the higher buildings there cause slower cooling than at the other sites (nighttime $Q_h$) [91]. Fourthly, the sensible heat trend line was examined, and it was found that the greatest negative slope corresponds to a rapid decrease in heat flux. According to the relative effects of the land cover types, the order of green surface ratio effects on sensible heat the most. Moreover, the green coverage ratio follows the order Site 3 > Site 1 > Site 5 > Site 4 > Site 2 > Site 6. Based on comparison with the results in, the green areas promote low sensible heat flux, except at Sites 2 and 6.

In other words, green surfaces have the greatest effects on restricting heat flux among the various land cover types. Lastly, regression analysis between the building, green area, and impervious coverage ratios indicated that the sensible heat flux decreases by 4.9 W/m² when the green coverage increases by 1%. Because of the low correlation (Appendix D), the riverside site, Site 5, was excluded from the regression analysis.

Table 6. Comprehensive results.

|  |  | Site 1 | Site 2 | Site 3 | Site 4 | Site 5 | Site 6 |
|---|---|---|---|---|---|---|---|
| $Q_h$ day (W/m²) | | 278 | 398 | 82 | 379 | 311 | 444 |
| $Q_h$ night (W/m²) | | 89 | 124 | 6 | 174 | 85 | 222 |
| $Q_n$ day (W/m²) | | 540 | 586 | 351 | 548 | 456 | 673 |
| $Q_n$ night (W/m²) | | 168 | 184 | 52 | 256 | 96 | 323 |
| Trend line | a | −10 | −15, | −4, | −13 | −12 | −14 |
| formula: y ($Q_h$) | b | 252 | 358 | 100 | 308 | 293 | 353 |
| = ax² + bx + c | c | −1213 | −1692 | −490 | −1382 | −1368 | −1696 |
| Green | | 0.23 | 0.06 | 0.96 | 0.08 | 0.17 | 0.02 |
| Building | | 0.35 | 0.6 | 0.02 | 0.52 | 0 | 0.73 |
| Impervious | | 0.22 | 0.14 | 0.01 | 0.15 | 0.06 | 0.1 |
| Road | | 0.2 | 0.2 | 0.01 | 0.25 | 0.05 | 0.15 |
| Water | | 0 | 0 | 0 | 0 | 0.72 | 0 |
| Location | | Sangdo | Daehak | Montmartre | Yeoyi | Riverside | Shilim |
| Aspect (land use) | | Resid. * high-rise | Resid. * low-rise | Park | Quasi_Resi. ** | Waterfront | Quasi_Resid. ** |

* Resid.: Residential district, ** Quasi Resid.: Quasi-residential district.

We found that the sites with high densities, low-rise residential buildings, and high percentages of impervious paved areas, Sites 2 and 6, have higher sensible heat fluxes than the site with high-rise residential buildings and green surfaces, Site 1. In these cases, the in-situ measurement results do correspond to the concept of UTA and FTA in residential areas (Figure 4, Table 6).

In this study, we found that focused sensible heat flux directly affects the urban thermal environment relative to the surface type and spatial conditions in an urban canyon [92]. The in-situ data agreed well with the thermal spatial typification results overall. Thus, net radiation measurement allows the proposal of new methods to locate areas in which the UHIE can be mitigated preferentially. Moreover, our study elucidates techniques for increasing energy efficiency by manipulating the spatial aspects and land cover types of overpopulated residential areas with high densities of buildings. Research on the temporal changes and energy budgets of low-density areas with green spaces will help urban planners develop microclimate-change adaptation methods.

We expect that these findings regarding the interactions among the green cover ratio, BCR, and sensible heat flux can contribute to rationalising landscape and architectural standards for the redevelopment of residential apartment complexes in megacities. However, to solve the heat problems in such apartment complexes, the role of green coverage, which reduces the sensible heat flux, is important. The development of high-density residential areas will require the establishment of guidelines for construction, considering the temporal changes in sensible heat in the canyon widths between buildings in residential areas. In addition, land use characteristics such as those of the terrestrial and surrounding environments affect the sensible heat flux variations and resulting trends.

### 6.2. Development of Thermally Sustainable Smart Cities

Criteria for ensuring energy sustainability [93] and micro-scale residential space comfort should be considered in detail for thermally sustainable smart city development projects. This approach will help promote the participation of residents [94] in developing smart, environmentally sustainable cities [21]. In particular, it is necessary to establish criteria to describe the thermal environment and spatial scale of the design of a city [95]. For this purpose, sustainable development perspectives are required and it is necessary to study these criteria to understand the limitations related to the use of space with awareness of the thermal vulnerability of the plan. In addition, the increase in the green coverage rate is due to the reduction in UHIs in the case of smart urban planning, and this increase in green space also contributes to the enhancement of thermal comfort and resident welfare. Sustainable urban development also requires small-scale investigation of the function of green space in rendering ecosystem benefits to society [96], as a substantial part of it.

For energy sustainability, it is necessary to remove energy imbalances to promote the equality [97] and comfort of residents as a precondition for the development of smart cities. Based on the finding of this study, green designs and technologies for smart cities should be examined and a green digital charter for energy conservation and typification should be implemented. In particular, an integrated conceptual model that incorporates considerations other than environmental sustainability should be established, because environmental sustainability alone is insufficient for the construction of a suitable smart city framework.

## 7. Conclusions

The objective of this research is to examine the effectiveness of a spatial typification method in improving environmental conditions for the sustainable development of smart cities. Spatial types were identified for thermal environmental improvement considering heat flux and land cover, which influence the sensible heat flux, using empirical formulas instead of approaches using temperature [1]. The UTAs identified using this method are prioritised for heat reduction. Then, the spatial typification was verified through in-situ measurement of the net radiation in UTAs and FTAs at six research sites in which such areas were derived by thermal spatial typification. The k-means clustering method was applied to classify the values of three kinds of heat flux: latent, sensible, and storage heat, and a type of unfavourable urban thermal environment was defined to identify measures that would increase thermal comfort. The characteristics of each thermal environmental type are based on the land cover type. The ratios of impervious surfaces, roads, and buildings in UTAs are higher than those of FTAs (relatively comfortable thermal environments). For thermal environment improvement, the following measures are proposed to urban planners and designers based on the results of this study: (a) green surfaces promote sensible heat flux mitigation, (b) typification of thermal environment in terms of UTAs and FTAs by K-means clustering is effective, as verified by the in-situ measurements, and (c) expansion of the green space by 1% reduces the sensible heat flux by 4.9 W/m$^2$. Thus, heat can be mitigated and spatial thermal comfort can be improved in urban areas by performing thermal spatial typification.

Among the six study areas, the highest values of both $Q_n$ and $Q_h$ were observed at Site 6, which has a high density and low-rise buildings with less green cover. Therefore, areas with spatial properties similar to those of Site 6 are considered to be the most suitable places to implement thermal mitigation in Seoul. Site 5, which is a riverside park, had the best cooling effect owing to the fact that the maximum heat decrease was noted between the peak hour and 18:00. Site 2, which is a high density area with low-rise buildings, exhibited an abrupt temporal change in $Q_h$. Unlike Site 2, Site 3, which is a neighbourhood park, showed moderate changes from the peak hour to sunset. Therefore, green areas and waterside surfaces are significant for decreasing the sensible heat flux and net radiation.

The proposed approach will eventually lead to appropriate energy consumption, and sustainable energy policies will indirectly contribute to energy usage reduction in sustainable smart cities. By enabling sustainability to be maintained via improvement of the thermal spatial environment, this method will provide a valuable tool for the implementation of smart cities. This research is expected to facilitate the establishment of a minimum land cover rate criterion to improve urban thermal environments; in addition, a standard index for implementing thermal environmental improvement can be derived. The study results offer new insights that can be utilised to develop rational methods of thermal environment improvement. In future research related to sustainable techniques for mitigating daytime radiation during heat waves, appropriate areas with adequate green coverage could be investigated in depth to control the net radiation and sensible heat flux in residential areas. More advanced studies regarding naturally emitted surplus radiation could also be performed to enable the positive reuse of energy and to achieve community energy budget goals. In the future, we expect to conduct radiation measurement research at additional sites to determine the influences of land cover factors on the net radiation over time.

In the future, the characteristics of FTAs that result in improved thermal environments should be reflected in residential areas considering the intrinsic culture. In environmental psychology, familiarity is related to place identity, which suggests that familiarity has a positive effect on the emotional stability of the citizens because it includes a unique culture and historical context [98,99]. An ideal smart city would improve the comfort of the citizens by maintaining the unique cultural identity of the place in their living environment. Thus, it is necessary to develop smart cities considering their inherent cultural identities, rather than developing such cities simply to improve quality of life [100].

To promote the development of sustainable smart cities, it is necessary to implement urban thermal environmental improvement plans and energy management policies to cope with climate change. Ultimately, smart cities, in which energy sustainability is emphasised, should be built based on a system of continuous spatial analysis and development accounting for the thermal environments and identities of the cities.

**Author Contributions:** Data curation, Y.J.K., K.L.; Conceptualization, Y.J.K.; Formal analysis, Y.J.K.; Funding acquisition, D.K.L.; Investigation, Y.J.K., K.L.; Methodology, Y.J.K., D.K.L.; Visualization, Y.J.K.; Supervision, D.K.L.; Writing—original draft, Y.J.K.; Writing—review & editing, D.K.L.

**Funding:** This research was funded by the Korea Agency for Infrastructure Technology Advancement (KAIA) grant of the Ministry of Land, Infrastructure, and Transport (Grant 19AUDP-B102406-05).

**Acknowledgments:** This work was supported by the BK 21 Plus Project in 2018 (Seoul National University Interdisciplinary Program on Landscape Architecture, Global Leadership Program toward Innovative Green Infrastructure). We truly appreciate for helping designing a colorful plot who is one of the lab colleagues at SNU, Sangjin Park. Also, we appreciate the anonymous reviewers for their constructive suggestions and elaborate comments to improve the quality of this paper and also to the editors, Tan Yigitcanlar, Md. (Liton) Kamruzzaman, Hoon Han and Elaine Chen.

**Conflicts of Interest:** The authors declare no conflicts of interest. The funding bodies had no role in the design of the study; in the collection, analyses, and interpretation of data; in the writing of the manuscript; or in the decision to publish the results.

## Appendix A

**Table A1.** CNR4 specifications of the pyranometer (details of CNR4, a net radiometer).

| Pyranometer Specification | Value (unit) | Definition |
|---|---|---|
| Sensing sensitivity | 5–20 ($\mu$V/W/m$^2$) | Calibration factor |
| Irradiance range | 0–2000 (W/m$^2$) | Measurement range |
| Net irradiance range | −250 to +250 (W/m$^2$) | |
| Shortwave radiation spectral range | 300–2800 (nm) | |
| Longwave radiation spectral range | 4500–430,000 (nm) | |
| Field of view | Upper detector: 180°, lower detector: 150° | Sensor opening angle |
| Nonlinearity | Less than 1 (%) | 0–1000 W/m$^2$ irradiance—Max. deviation from the responsivity at 500 W/m$^2$ owing to change in irradiance within the indicated range. |
| Uncertainty in daily total | Less than 5 (95% confidence level) | Achievable uncertainty |
| Temperature dependence of sensitivity | −10 to +40 (°C) | |
| Operating temperature | −40 to +80 (°C) | |
| Environmental | 0–100% RH | Relative humidity |
| Response time | less than 18 s | 95% response |
| Directional error | less than 20 (W/m$^2$) | Angles up to 80° with 1000 W/m$^2$ Beam radiation–combined zenith and azimuth errors of 0–80° with a 1000 W/m$^2$ beam |

Source: KIPP & ZONEN CNR4 Net Radiometer Instruction Manual.

## Appendix B

The following flow chart shows our overall research process.

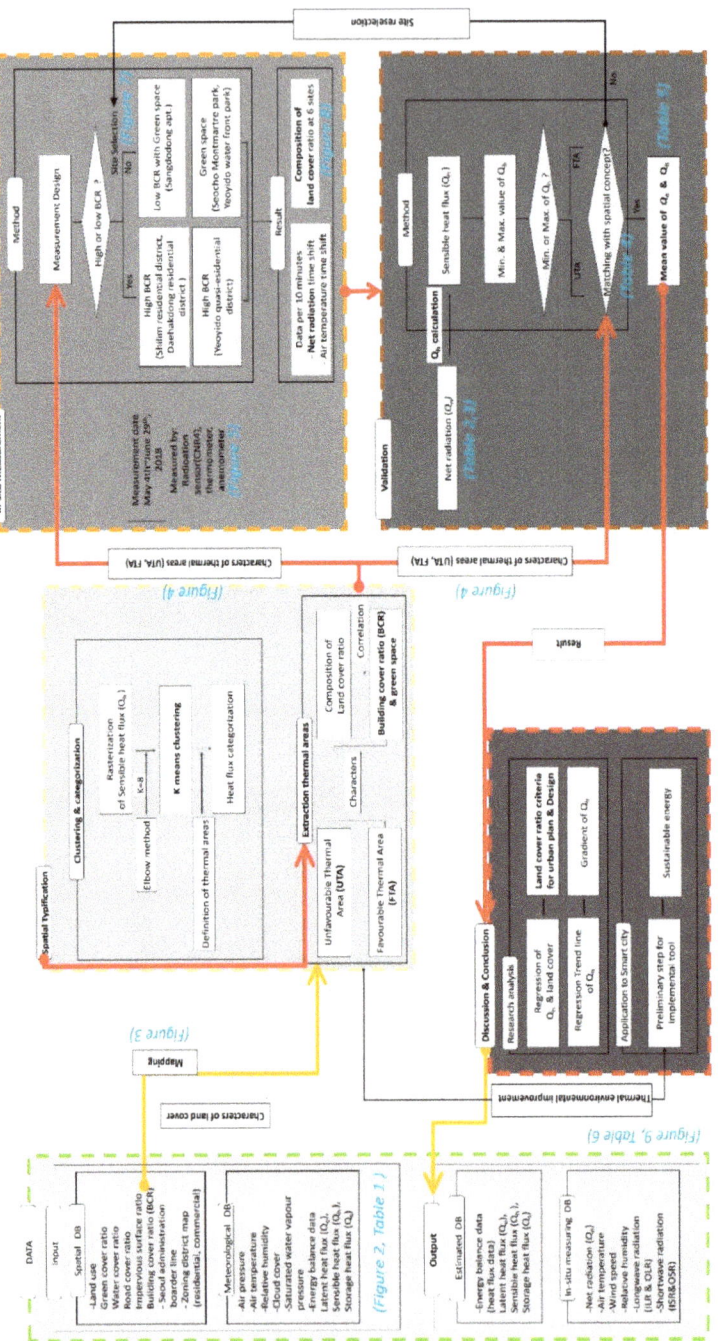

**Figure A1.** Overall research flow chart.

## Appendix C

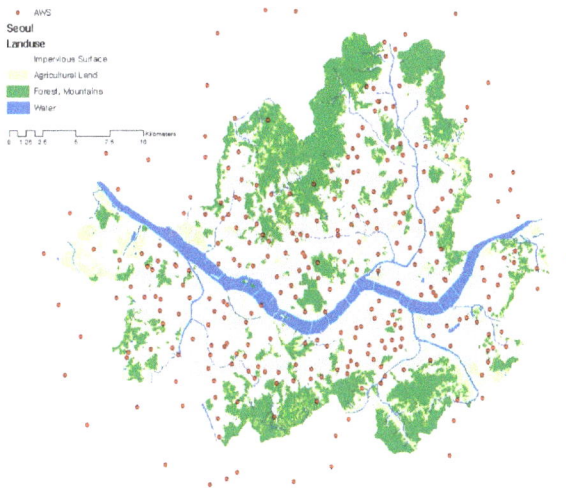

**Figure A2.** Automatic weather station locations.

We used weather data from 287 AWSs.

## Appendix D

**Table A2.** Correlations of Sensible heat flux and five land covers.

|  |  | Correlations | | | | | |
|---|---|---|---|---|---|---|---|
|  |  | $Q_h$ | gr | bd | im | water | rd |
| Pearson correlation | $Q_h$ | 1.000 | −0.727 | 0.593 | 0.209 | −0.003 | 0.637 |
|  | gr | −0.727 | 1.000 | −0.799 | −0.443 | 0.048 | −0.908 |
|  | bd | 0.593 | −0.799 | 1.000 | 0.542 | −0.610 | 0.898 |
|  | im | 0.209 | −0.443 | 0.542 | 1.000 | −0.554 | 0.493 |
|  | water | −0.003 | 0.048 | −0.610 | −0.554 | 1.000 | −0.357 |
|  | rd | 0.637 | −0.908 | 0.898 | 0.493 | −0.357 | 1.000 |
| Sig. (1-tailed) | $Q_h$ | - | 0.000 | 0.000 | 0.000 | 0.439 | 0.000 |
|  | gr | 0.000 | - | 0.000 | 0.000 | 0.003 | 0.000 |
|  | bd | 0.000 | 0.000 | - | 0.000 | 0.000 | 0.000 |
|  | im | 0.000 | 0.000 | 0.000 | - | 0.000 | 0.000 |
|  | water | 0.439 | 0.003 | 0.000 | 0.000 | - | 0.000 |
|  | rd | 0.000 | 0.000 | 0.000 | 0.000 | 0.000 | - |
| N | $Q_h$ | 3240 | 3240 | 3240 | 3240 | 3240 | 3240 |
|  | gr | 3240 | 3240 | 3240 | 3240 | 3240 | 3240 |
|  | bd | 3240 | 3240 | 3240 | 3240 | 3240 | 3240 |
|  | im | 3240 | 3240 | 3240 | 3240 | 3240 | 3240 |
|  | water | 3240 | 3240 | 3240 | 3240 | 3240 | 3240 |
|  | rd | 3240 | 3240 | 3240 | 3240 | 3240 | 3240 |

The sensible heat ($Q_h$) has the highest correlation (−0.727) with green surfaces, among the five land cover factors, but green surfaces negatively affect $Q_h$. We found that the BCR and road cover are related to $Q_h$ but that impervious surfaces and water cover are not, because the absolute ratios of impervious surfaces and water cover are equivalent to small portions of the overall land surface of Seoul.

## References

1. Stewart, I.D.; Oke, T.R. Local climate zones for urban temperature studies. *Bull. Am. Meteorol. Soc.* **2012**, *93*, 1879–1900. [CrossRef]
2. Oke, T.R. The temperature profile near the ground on calm clear nights. *Q. J. R. Meteorol. Soc.* **1970**, *96*, 14–23. [CrossRef]
3. Jochner, S.; Menzel, A. Urban phenological studies—Past, present, future. *Environ. Pollut.* **2015**, *203*, 250–261. [CrossRef] [PubMed]
4. Arnfield, A.J. Two decades of urban climate research: A review of turbulence, exchanges of energy and water, and the urban heat island. *Int. J. Climatol.* **2003**, *23*, 1–26. [CrossRef]
5. Oke, T.R. The energetic basis of the urban heat island. *Q. J. R. Meteorol. Soc.* **1982**, *108*, 1–24. [CrossRef]
6. Res, C.; Pape, R.; Wundram, D.; Löffler, J. Modelling near-surface temperature conditions in high mountain environments: An appraisal. *Clim. Res.* **2009**, *39*, 99–109. [CrossRef]
7. Johansson, E.; Thorsson, S.; Emmanuel, R.; Krüger, E. Instruments and methods in outdoor thermal comfort studies—The need for standardization. *Urban Clim.* **2014**, *10*, 346–366. [CrossRef]
8. Gluch, R.; Quattrochi, D.A.; Luvall, J.C. A multi-scale approach to urban thermal analysis. *Remote Sens. Environ.* **2006**, *104*, 123–132. [CrossRef]
9. Li, Y.Y.; Zhang, H.; Kainz, W. Monitoring patterns of urban heat islands of the fast-growing Shanghai metropolis, China: Using time-series of Landsat TM/ETM+ data. *Int. J. Appl. Earth Obs. Geoinf.* **2012**, *19*, 127–138. [CrossRef]
10. Mauree, D.; Blond, N.; Clappier, A. Multi-scale modeling of the urban meteorology: Integration of a new canopy model in the WRF model. *Urban Clim.* **2018**, *26*, 60–75. [CrossRef]
11. Liao, W.; Liu, X.; Wang, D.; Sheng, Y. The impact of energy consumption on the surface urban heat island in China's 32 major cities. *Remote Sens.* **2017**, *9*, 250. [CrossRef]
12. Ahmed Memon, R.; Leung, D.Y.; Chunho, L. A review on the generation, determination and mitigation of Urban Heat Island. *J. Environ. Sci.* **2008**, *20*, 120–128. [CrossRef]
13. Grimmond, S.U.E. Urbanization and global environmental change: Local effects of urban warming. *R. Geogr. Soc.* **2007**, *173*, 83–88. [CrossRef]
14. Cohen, J.E. Human Population: The Next Half Century. *Science* **2003**, *302*, 1172–1175. [CrossRef] [PubMed]
15. Dell, R. Energy storage—A key technology for global energy sustainability. *J. Power Sources* **2001**, *100*, 2–17. [CrossRef]
16. Kondo, H.; Kikegawa, Y. Temperature Variation in the Urban Canopy with Anthropogenic Energy Use. In *Air Quality*; Rao, G.V., Raman, S., Singh, M.P., Eds.; Springer Basel AG: Basel, Switzerland, 2003; pp. 317–324. ISBN 9783764370053.
17. Narumi, D.; Kondo, A.; Shimoda, Y. Effects of anthropogenic heat release upon the urban climate in a Japanese megacity. *Environ. Res.* **2009**, *109*, 421–431. [CrossRef] [PubMed]
18. Huang, H.; Ooka, R.; Kato, S. Urban thermal environment measurements and numerical simulation for an actual complex urban area covering a large district heating and cooling system in summer. *Atmos. Environ.* **2005**, *39*, 6362–6375. [CrossRef]
19. Angelidou, M. Smart city policies: A spatial approach. *Cities* **2014**, *41*, S3–S11. [CrossRef]
20. Grossi, G.; Pianezzi, D. Smart cities: Utopia or neoliberal ideology? *Cities* **2017**, *69*, 79–85. [CrossRef]
21. Fernandez-Anez, V.; Fernández-Güell, J.M.; Giffinger, R. Smart City implementation and discourses: An integrated conceptual model. The case of Vienna. *Cities* **2018**, *78*, 4–16. [CrossRef]
22. Baum, S.; Yigitcanlar, T.; Horton, S.; Velibeyoglu, K.; Gleeson, B. Knowledge Work(ers) in Urban Context. In *The Role of Community and Lifestyle in the Making of a Knowledge City*; Urban Research Program Griffith University: Brisbane, Australia, 2007; pp. 18–24, ISBN 9781921291012.
23. Yigitcanlar, T.; Kamruzzaman, M. Does smart city policy lead to sustainability of cities? *Land Use Policy* **2018**, *73*, 49–58. [CrossRef]
24. Rodríguez-Pose, A.; Ketterer, T.D. Do local amenities affect the appeal of regions in europe for migrants? *J. Reg. Sci.* **2012**, *52*, 535–561. [CrossRef]
25. Deller, S.C.; Tsai, T.; Marcouiller, D.W.; English, D.B.K. The Role of Amenities and Quality of Life in Rural Economic Growth. *Am. J. Agric. Econ.* **2013**, *83*, 352–365. [CrossRef]

26. Howell, B.R.; Menguc, M.P.; Siegel, R. *Thermal Radiation Heat Transfer*, 6th ed.; CRC Press: Boca Raton, FL, USA, 2015; ISBN 9780429190599.
27. Sheth, J.; Sisodia, R. Feeling the heat—Part 1 and Part 2. *Mark. Manag.* **1995**, *4*, 8–23, 19–33.
28. Stull, R.B. Chap 03: Heat. In *Meteorology for Scientists and Engineers*; Brooks/Cole: Belmont, CA, USA, 2015; pp. 53–86, ISBN 978-0-534-37214-9.
29. Zhuang, Q.; Wu, B.; Yan, N.; Zhu, W.; Xing, Q. A method for sensible heat flux model parameterization based on radiometric surface temperature and environmental factors without involving the parameter KB −1. *Int. J. Appl. Earth Obs. Geoinf.* **2016**, *47*, 50–59. [CrossRef]
30. Voogt, J.A.; Grimmond, C.S.B. Modeling Surface Sensible Heat Flux Using Surface Radiative Temperatures in a Simple Urban Area. *J. Appl. Meteorol.* **2010**, *39*, 1679–1699. [CrossRef]
31. Stern, N.; Taylor, C. Climate Change: Risk, Ethics, and the Stern Review. *Science* **2007**, *317*, 203–204. [CrossRef] [PubMed]
32. Grimm, N.B.; Faeth, S.H.; Golubiewski, N.E.; Redman, C.L.; Wu, J.; Bai, X.; Briggs, J.M. Global Change and the Ecology of Cities SUPPLEMENTARY. *Science* **2008**, *319*, 756–760. [CrossRef]
33. Ng, E.; Chen, L.; Wang, Y.; Yuan, C. A study on the cooling effects of greening in a high-density city: An experience from Hong Kong. *Build. Environ.* **2012**, *47*, 256–271. [CrossRef]
34. Oliveira, S.; Andrade, H.; Vaz, T. The cooling effect of green spaces as a contribution to the mitigation of urban heat: A case study in Lisbon. *Build. Environ.* **2011**, *46*, 2186–2194. [CrossRef]
35. Hamada, S.; Ohta, T. Seasonal variations in the cooling effect of urban green areas on surrounding urban areas. *Urban For. Urban Green.* **2010**, *9*, 15–24. [CrossRef]
36. Wong, N.H.; Yu, C. Study of green areas and urban heat island in a tropical city. *Habitat Int.* **2005**, *29*, 547–558. [CrossRef]
37. Armson, D.; Stringer, P.; Ennos, A.R. The effect of tree shade and grass on surface and globe temperatures in an urban area. *Urban For. Urban Green.* **2012**, *11*, 245–255. [CrossRef]
38. Monk, C.D.; Gabrielson, F.C. Effects of Shade, Litter and Root Competition on Old-Field Vegetation in South Carolina. *Bull. Torrey Bot. Club* **1985**, *112*, 383–392. [CrossRef]
39. Kobayashi, T.H.Y.; Nomoto, N. Effects of trampling and vegetation removal on species diversity and micro-environment under different shade conditions. *J. Veg. Sci.* **1997**, *8*, 873–880. [CrossRef]
40. Takebayashi, H.; Moriyama, M. Study on the urban heat island mitigation effect achieved by converting to grass-covered parking. *Sol. Energy* **2009**, *83*, 1211–1223. [CrossRef]
41. Onishi, A.; Cao, X.; Ito, T.; Shi, F.; Imura, H. Evaluating the potential for urban heat-island mitigation by greening parking lots. *Urban For. Urban Green.* **2010**, *9*, 323–332. [CrossRef]
42. Rowe, D.B.; Getter, K.L.; Durhman, A.K. Effect of green roof media depth on Crassulacean plant succession over seven years. *Landsc. Urban Plan.* **2012**, *104*, 310–319. [CrossRef]
43. Jim, C.Y.; Peng, L.L.H. Weather effect on thermal and energy performance of an extensive tropical green roof. *Urban For. Urban Green.* **2012**, *11*, 73–85. [CrossRef]
44. Peng, L.L.H.; Jim, C.Y. Green-roof effects on neighborhood microclimate and human thermal sensation. *Energies* **2013**, *6*, 598–618. [CrossRef]
45. Maciel, C.D.R.; Kolokotroni, M.; Paulo, S. Cool Materials in the Urban Built Environment to Mitigate Heat Islands: Potential Consequences for Building Ventilation. Available online: http://bura.brunel.ac.uk/handle/2438/15309 (accessed on 14 May 2019).
46. Santamouris, M.; Synnefa, A.; Karlessi, T. Using advanced cool materials in the urban built environment to mitigate heat islands and improve thermal comfort conditions. *Sol. Energy* **2011**, *85*, 3085–3102. [CrossRef]
47. Hong, B.; Lin, B. Numerical studies of the outdoor wind environment and thermal comfort at pedestrian level in housing blocks with different building layout patterns and trees arrangement. *Renew. Energy* **2015**, *73*, 18–27. [CrossRef]
48. Gu, Z.L.; Zhang, Y.W.; Cheng, Y.; Lee, S.C. Effect of uneven building layout on air flow and pollutant dispersion in non-uniform street canyons. *Build. Environ.* **2011**, *46*, 2657–2665. [CrossRef]
49. Blum, J. Contribution of Ecosystmes Servies to Air Quality and Climate Change Mitigation Policies: The Case of Urban Forests in Barcelona, Spain. In *Urban Forests*; Blum, J., Ed.; Apple Academic Press: New York, NY, USA, 2016; pp. 3–36, ISBN 9781315366081.

50. Tran, D.X.; Pla, F.; Latorre-Carmona, P.; Myint, S.W.; Caetano, M.; Kieu, H.V. Characterizing the relationship between land use land cover change and land surface temperature. *ISPRS J. Photogramm. Remote Sens.* **2017**, *124*, 119–132. [CrossRef]
51. Nayak, S.G.; Shrestha, S.; Kinney, P.L.; Ross, Z.; Sheridan, S.C.; Pantea, C.I.; Hsu, W.H.; Muscatiello, N.; Hwang, S.A. Development of a heat vulnerability index for New York State. *Public Health* **2018**, *161*, 127–137. [CrossRef] [PubMed]
52. Shahadat Hossain, M.; Kwei Lin, C. Land Use Zoning for Integrated Coastal Zone Management Remote Sensing, GIS and RRA Approach in Cox's Bazar Coast, Bangladesh. Available online: http://citeseerx.ist.psu.edu/viewdoc/download?doi=10.1.1.471.1264&rep=rep1&type=pdf (accessed on 14 May 2019).
53. Kim, S.; Ryu, Y. Describing the spatial patterns of heat vulnerability from urban design perspectives. *Int. J. Sustain. Dev. World Ecol.* **2015**, *22*, 189–200. [CrossRef]
54. Lunetta, R.S.; Knight, J.F.; Ediriwickrema, J.; Lyon, J.G.; Worthy, L.D. Land-cover change detection using multi-temporal MODIS NDVI data. *Remote Sens. Environ.* **2006**, *105*, 142–154. [CrossRef]
55. Pandey, P.K.; Soupir, M.L. A new method to estimate average hourly global solar radiation on the horizontal surface. *Atmos. Res.* **2012**, *114–115*, 83–90. [CrossRef]
56. Malys, L.; Musy, M.; Inard, C. Direct and indirect impacts of vegetation on building comfort: A comparative study of lawns, greenwalls and green roofs. *Energies* **2016**, *9*, 32. [CrossRef]
57. Weng, Q. Thermal infrared remote sensing for urban climate and environmental studies: Methods, applications, and trends. *ISPRS J. Photogramm. Remote Sens.* **2009**, *64*, 335–344. [CrossRef]
58. Zou, L.; Wang, L.; Li, J.; Lu, Y.; Gong, W.; Niu, Y. Global surface solar radiation and photovoltaic power from Coupled Model Intercomparison Project Phase 5 climate models. *J. Clean. Prod.* **2019**, *224*, 304–324. [CrossRef]
59. Yang, J.; Gong, P.; Fu, R.; Zhang, M.; Chen, J.; Liang, S.; Xu, B.; Shi, J.; Dickinson, R. The role of satellite remote sensing in climate change studies. *Nat. Clim. Chang.* **2013**, *3*, 875–883. [CrossRef]
60. Dousset, B.; Gourmelon, F. Satellite multi-sensor data analysis of urban surface temperatures and landcover. *ISPRS J. Photogramm. Remote Sens.* **2003**, *58*, 43–54. [CrossRef]
61. Saha, S.K. Retrieval of Agrometeorological parameters using satellite data. In *Satellite Remote Sensing and GIS Applications in Meteorology*; Sivakumar, M.V.K., Roy, P.S., Harmsen, K., Saha, S.K., Eds.; World Meteorological Organisation: Geneva, Switzerland, 2004; pp. 175–194, ISBN 978-0-415-40166-1.
62. Crank, P.J.; Sailor, D.J.; Ban-Weiss, G.; Taleghani, M. Evaluating the ENVI-met microscale model for suitability in analysis of targeted urban heat mitigation strategies. *Urban Clim.* **2018**, *26*, 188–197. [CrossRef]
63. Simon, H.; Lindén, J.; Hoffmann, D.; Braun, P.; Bruse, M.; Esper, J. Modeling transpiration and leaf temperature of urban trees—A case study evaluating the microclimate model ENVI-met against measurement data. *Landsc. Urban Plan.* **2018**, *174*, 33–40. [CrossRef]
64. Tsoka, S.; Tsikaloudaki, A.; Theodosiou, T. Analyzing the ENVI-met microclimate model's performance and assessing cool materials and urban vegetation applications—A review. *Sustain. Cities Soc.* **2018**, *43*, 55–76. [CrossRef]
65. Sentelhas, P.C.; Alvares, C.A.; Stape, J.L.; Sparovek, G.; de Moraes Gonçalves, J.L. Köppen's climate classification map for Brazil. *Meteorol. Z.* **2014**, *22*, 711–728.
66. Guo, X.; Coops, N.C.; Tompalski, P.; Nielsen, S.E.; Bater, C.W.; John Stadt, J. Regional mapping of vegetation structure for biodiversity monitoring using airborne lidar data. *Ecol. Inform.* **2017**, *38*, 50–61. [CrossRef]
67. Lechner, A.M.; Baumgartl, T.; Matthew, P.; Glenn, V. The Impact of Underground Longwall Mining on Prime Agricultural Land: A Review and Research Agenda. *Land Degrad. Dev.* **2016**, *27*, 1650–1663. [CrossRef]
68. Mcmanamay, R.A.; Bevelhimer, M.S.; Kao, S.C. Updating the US hydrologic classification: An approach to clustering and stratifying ecohydrologic data. *Ecohydrology* **2014**, *7*, 903–926. [CrossRef]
69. Jain, A.K. Data clustering: 50 years beyond K-means. *Pattern Recognit. Lett.* **2010**, *31*, 651–666. [CrossRef]
70. Janssens, F.; Glänzel, W.; De Moor, B. A hybrid mapping of information science. *Scientometrics* **2008**, *75*, 607–631. [CrossRef]
71. Ma, E.; Deng, X.; Zhang, Q.; Liu, A. Spatial variation of surface energy fluxes due to land use changes across China. *Energies* **2014**, *7*, 2194–2206. [CrossRef]
72. Lindberg, F.; Grimmond, C.S.B. The influence of vegetation and building morphology on shadow patterns and mean radiant temperatures in urban areas: Model development and evaluation. *Theor. Appl. Climatol.* **2011**, *105*, 311–323. [CrossRef]

73. Kwon, Y.J.; Ahn, S.; Lee, D.K.; Yoon, E.J.; Sung, S.; Lee, K. Spatial Typification based on Heat Balance for Improving Thermal Environment in Seoul. *J. Korea Plan. Assoc.* **2018**, *53*, 109–126. [CrossRef]
74. Parsons, K. Thermal Models. In *Human Thermal Environments*; Parsons, K., Ed.; CRC Press: Boca Raton, FL, USA, 2002; pp. 161–186. ISBN 978-1-4665-9600-9.
75. Mohajerani, A.; Bakaric, J.; Jeffrey-Bailey, T. The urban heat island effect, its causes, and mitigation, with reference to the thermal properties of asphalt concrete. *J. Environ. Manag.* **2017**, *197*, 522–538. [CrossRef]
76. Choi, H.-J.; Lee, H.-W.; Sung, K.-H. Air Quality Modeling of Ozone Concentration According to the Roughness Length on the Complex Terrain. *J. Korean Soc. Atmos. Environ.* **2011**, *23*, 430–439. [CrossRef]
77. Miller, H.J. Tobler's First Law and Spatial Analysis. *Ann. Assoc. Am. Geogr.* **2004**, *94*, 284–289. [CrossRef]
78. Janssens, W.; Wijnen, K.; De Pelsmacker, P.; Kenhove, P. Van Cluster analysis. In *Marketing Research with SPSS*; Janssens, W., Ed.; Pearson Education: London, UK, 2008; pp. 319–362, ISBN 9780273703839.
79. Offerle, B.; Grimmond, C.S.B.; Fortuniak, K. Heat storage and anthropogenic heat flux in relation to the energy balance of a central European city centre. *Int. J. Climatol.* **2005**, *25*, 1405–1419. [CrossRef]
80. Holtslag, A.A.M.; Van Ulden, A.P. A Simple Scheme for Daytime Estimates of the Surface Fluxes from Routine Weather Data. *J. Clim. Appl. Meteor.* **1983**, *22*, 517–529. [CrossRef]
81. Pigeon, G.; Legain, D.; Durand, P.; Masson, V. Anthropogenic heat release in an old European agglomeration_Toulouse, France. *R. Meteorol. Soc.* **2007**, *27*, 1969–1981. [CrossRef]
82. Ng, Y. A Study of Urban Heat Island using "Local Climate Zones"—The Case of Singapore. *Br. J. Environ. Clim. Chang.* **2015**, *5*, 116–133. [CrossRef] [PubMed]
83. Grimmond, C.S.B.; Cleugh, H.A.; Oke, T.R. an Objective Urban Heat Storage Model and Its. *Atmos. Environ.* **1991**, *25*, 311–326. [CrossRef]
84. Roberts, S.M.; Oke, T.R.; Grimmond, C.S.B.; Voogt, J.A. Comparison of four methods to estimate urban heat storage. *J. Appl. Meteorol. Climatol.* **2006**, *45*, 1766–1781. [CrossRef]
85. Grimmond, C.S.B.; Roth, M.; Oke, T.R.; Au, Y.C.; Best, M.; Betts, R.; Carmichael, G.; Cleugh, H.; Dabberdt, W.; Emmanuel, R.; et al. Climate and more sustainable cities: Climate information for improved planning and management of cities (Producers/Capabilities Perspective). *Procedia Environ. Sci.* **2010**, *1*, 247–274. [CrossRef]
86. Kwon, Y.J.; Lee, D.K. Thermal Comfort and Longwave Radiation over Time in Urban Residential Complexes. *Sustainability* **2019**, *11*, 2251. [CrossRef]
87. Park, C.Y.; Lee, D.K.; Asawa, T.; Murakami, A.; Kim, H.G.; Lee, M.K.; Lee, H.S. Influence of urban form on the cooling effect of a small urban river. *Landsc. Urban Plan.* **2019**, *183*, 26–35. [CrossRef]
88. Hart, M.A.; Sailor, D.J. Quantifying the influence of land-use and surface characteristics on spatial variability in the urban heat island. *Theor. Appl. Climatol.* **2009**, *95*, 397–406. [CrossRef]
89. Cengel, Y.; Ghajar, A. Introduction and Basic Concepts. In *Heat and Mass Transfer: Fundamentals and Applications Chapter 16: Heating and Cooling of Buildings*; Francis, C.P., Ed.; McGraw Hill: New York, NY, USA, 2015; pp. 1–60, ISBN 978-0-07-339818-1.
90. Taleghani, M.; Kleerekoper, L.; Tenpierik, M.; Van Den Dobbelsteen, A. Outdoor thermal comfort within five different urban forms in the Netherlands. *Build. Environ.* **2015**, *83*, 65–78. [CrossRef]
91. Nichol, J. Remote sensing of urban heat islands by day and night. *Photogramm. Eng. Remote Sensing* **2005**, *71*, 613–621. [CrossRef]
92. Vallati, A.; Mauri, L.; Colucci, C.; Ocłoń, P. Effects of radiative exchange in an urban canyon on building surfaces' loads and temperatures. *Energy Build.* **2017**, *149*, 260–271. [CrossRef]
93. Kramers, A.; Höjer, M.; Lövehagen, N.; Wangel, J. Smart sustainable cities—Exploring ICT solutions for reduced energy use in cities. *Environ. Model. Softw.* **2014**, *56*, 52–62. [CrossRef]
94. Granier, B.; Kudo, H. How are citizens involved in smart cities? Analysing citizen participation in Japanese "smart Communities". *Inf. Polity* **2016**, *21*, 61–76. [CrossRef]
95. Punter, J. Developing urban design as public policy: Best practice principles for design review and development management. *J. Urban Des.* **2007**, *12*, 167–202. [CrossRef]
96. Fisher, B.; Turner, R.K.; Morling, P. Defining and classifying ecosystem services for decision making. *Ecol. Econ.* **2009**, *68*, 643–653. [CrossRef]
97. Bouzarovski, S.; Simcock, N. Spatializing energy justice. *Energy Policy* **2017**, *107*, 640–648. [CrossRef]
98. Ujang, N. *Place Attachment, Familiarity and Sustainability of Urban Place*; Department of Landscape Architecture, Faculty of Design and Architecture, University Putra Malaysia: Seri Kembangan, Malaysia, 2008; p. 10.

99. Ujang, N. Place Attachment and Continuity of Urban Place Identity. *Procedia Soc. Behav. Sci.* **2012**, 156–167. [CrossRef]
100. Han, H.; Hawken, S. Introduction: Innovation and identity in next-generation smart cities. *City Cult. Soc.* **2018**, *12*, 1–4. [CrossRef]

 © 2019 by the authors. Licensee MDPI, Basel, Switzerland. This article is an open access article distributed under the terms and conditions of the Creative Commons Attribution (CC BY) license (http://creativecommons.org/licenses/by/4.0/).

*Article*

# Digital Commons and Citizen Coproduction in Smart Cities: Assessment of Brazilian Municipal E-Government Platforms

**Maurício José Ribeiro Rotta [1],\*, Denilson Sell [1,2], Roberto Carlos dos Santos Pacheco [1] and Tan Yigitcanlar [3]**

1. Department of Knowledge Engineering and Management, Federal University of Santa Catarina, Campus Universitário Reitor João David Ferreira Lima, Florianopolis 88040-900, Brazil
2. Department of Business Administration, State University of Santa Catarina, Av. Me. Benvenuta, 2007, Itacorubi, Florianopolis 88035-001, Brazil
3. School of Civil Engineering and Built Environment, Queensland University of Technology, 2 George Street, Brisbane, QLD 4000, Australia
* Correspondence: maurotta@gmail.com; Tel.: +55-(48)-99144-4747

Received: 26 June 2019; Accepted: 9 July 2019; Published: 22 July 2019

**Abstract:** Good governance practices through electronic government (eGov) platforms can be suitable instruments for strengthening the outcomes of smart city policies. While eGov is the application of information and communication technologies to public services, good governance defines how well public authorities manage public and social resources. Contemporary public management views, such as 'new public service', include citizen participation as a critical factor to sustainable government in smart cities. Public services, in the age of digital technology, need to not only be delivered through eGov platforms, but also need to be coproduced with the engagement of social players, e.g., citizens. In this sense, eGov platforms act as digital commons, and conceived as digital spaces, where citizens and public agents interact and collaborate. In this paper, we presented the Municipal eGov Platform Assessment Model (MEPA), which is a model specifically developed to evaluate eGov platforms regarding their potential to promote commons in smart cities. The study applied MEPA to 903 municipal websites across Brazil. The results revealed that the majority of investigated Brazilian eGov platforms have only a low level of digital commons maturity. This finding discloses less citizenship coproduction, and fewer opportunities for city smartness. As the MEPA model offers public authorities an instrument to depict weaknesses and strengths of municipal eGov platforms, its adoption provides an opportunity for authorities to plan and manage their platforms to act as promoters of digital commons and citizen coproduction.

**Keywords:** smart cities; commons; digital commons; governance; e-government; smart governance; new public service; Brazil

---

## 1. Introduction

When applied to cities, "smartness" refers to efficiently use human, social, natural and technological resources towards a sustainable urban living. There is an intrinsic connection between such challenges and the notion of "smart government" as public digital spaces with both authorities and citizenship participation, based on public (good) governance and efficient electronic government (eGov) [1,2]. In smart eGov platforms, public services are not only delivered by government, but include citizen participation supported by modern information technology [3]. In this sense, in developing initiatives and projects related to eGov, public administration can use a more humanistic approach, using the principles of the 'new public service' (NSP)—the focus of which is the public interest and the coproduction of the common good, and the public servants at the service of all citizens [4].

In the context of eGov and NSP, coproduction is an essential requirement for the provision of public good and quality services in a network, presupposing the engagement of citizens, government and organizations. From this perspective, citizens are the main element to define "what" should be produced and "how", and participate in the elaboration, evaluation, and accountability of the process, through the networks of state and non-state actors. Such flexibility to change according to the citizens' interests and needs, by the use of technologies as enabler to connect and engage government and citizens, is described by Marsh [5] as the 'humane smart city'. In the 'commons theory', platforms that support such coproduction can be referred to as digital commons [6]. This can be the case of smart eGov platforms. Such platforms are public digital commons when contribute to transparency, participation, accountability, effectiveness and other open governance principles.

According to Ostrom [7], commons are goods of collective use shared by individuals, and subject to social conflicts, and by: (a) Emphasis on social interaction; (b) Common objectives, rules and standards; (c) Practices of sharing and distribution of power relations; (d) Institutions for decision-making, and; (e) Governance [8,9]. In turn, the concept of commons has advanced, and has become richer and more diversified, as is the case of collective and collaborative production of content mediated by digital and open resources—such as Wikipedia or Linux [10]. This has become a recurring practice in certain organizations, thus contributing to the creation of a collective and adaptive creative intelligence [11]. These collaborative efforts and practices can be characterized as digital commons [6,12].

To complement this understanding, it is useful to refer to Pacheco's [13] view—particularly when considering knowledge as a type of commons involved in information and communication technologies (ICTs). He describes a new type of commons. That is the digital common, and defined as follows: "digital commons is a resource based on information and communication technology, shared by groups and integrated in a value chain, under principles of equity, coproduction and sustainability".

In fact, through eGov platforms or portals, public administration presents its identity, purpose and achievements, provides services and information, providing access and interaction with citizens, as well as understanding their needs, and increasing transparency and the participation of society in government actions [14]. In addition, eGov platforms can be considered tools to promote knowledge sharing, providing users with resources to promote the dissemination of knowledge and interaction between different actors and government [15]. In this sense, the importance of knowledge management is recognized in public administration, since it deals with information and knowledge about citizens, companies, market, laws and policy. Such deliverables and the level of government-to-society relationship can be analyzed according to the maturity level that eGov platforms present.

Maturity models for eGov platforms, in turn, are structured in stages (from basic to advanced). These models provide a way to classify eGov platforms (according to the services, features and functionalities offered), and can be used as a guide to help public administrations improve the quality and efficiency of their eGov portals [16]. In spite of the existence of approaches for the evaluation of different characteristics of eGov platforms [17,18], there is a need for means to evaluate elements that may characterize such platforms as digital commons and, more specifically, to assess their potential of promoting social commons. During the last years, we have developed the Municipal eGov Platform Assessment Model (MEPA) that is a model to assess eGov platforms as digital commons [18]. MEPA allows the identification and evaluation of several factors related to digital commons principals.

In this paper, we present MEPA and its application in the evaluation of 903 official websites of Brazilian cities, and discuss its potential to help public managers to use eGov as integral smart cities instruments. The results of this research underlines the limitations in the municipal eGov platforms in Brazil. Only a few municipalities effectively manage to provide services so that the population can fulfill its role of participation and coproduction. From the sample surveyed, in approximately 7% of the municipality platforms, it was possible to verify a higher maturity level for the feasibility of the commons offered by the government, since the highest levels of maturity (fifth, fourth and third levels) were identified in only 66 Brazilian municipalities. The vast majority of Brazilian eGov platforms still offer simple, easy-to-access information or services, or just online transactions, and remaining at the first and second maturity levels. Therefore, eGov's actions and projects need to be rethought in Brazil, in terms of services, infrastructure, governance and financial resources, to achieve higher levels of eGov maturity.

## 2. Research Design and Method

The MEPA model was developed by empirical research, based on 'design science research' (DSR) method [19], according to the following steps: (1) Identify problem and motivate; (2) Define the objectives of a solution; (3) Design and development; (4) Demonstration; (5) Evaluation; and (6) Communication. In the following sections the development stages of the model according to the DSR.

*2.1. Problem Identification and Motivation*

Ostrom [7] describes that eight principles are fundamental to the sustainable management of commons. These are delimitation, context appropriation, participation, monitoring, proportionate sanctions, affordable conflict resolution, autonomy, and adhocracy. These principles were originally formulated by the author from an examination and analysis of more than 5000 case studies, through which it was possible to verify why some communities or individuals organize themselves successfully to manage the commons, and others do not.

Based on these general principles, there is a need to establish the means to evaluate its presence on eGov platforms of Brazilian municipalities. The platforms offer different services in terms of scope, scope and quality, influenced interaction between stakeholders and government, knowledge sharing and the possibility of coproducing the public good in a sustainable and sustainable manner, according to the level of their maturity.

Therefore, the research problem presented in this paper seeks to answer the following questions: (1) How to relate the maturity of eGov platforms with the instrumentation of commons promotion? and; (2) What is the situation of eGov platforms in Brazilian municipalities, in relation to their potential for promoting commons principles?

The MEPA problem and motivation are, hence, contextualized in multiple domains. Citizen coproduction and public governance are public management subjects. Electronic government platforms and human smart cities are contemporaneous multidisciplinary fields, and commons a general theory suitable as a reference to several complex community-based problems. Our first step was to establish a reference concept table, with the main research construct. The result is presented in Table 1.

Our research problem was to check municipal websites maturity regarding the promotion of commons and citizen participation. In order to do so, we have combined the seven conceptual multidisciplinary constructs presented in Figure 1.

As can be seen in Figure 1, the research problem combines the notions of municipal websites as eGov solutions, that can potentially help public managers to foster citizen participation by means (or towards) digital commons. Besides the specific constructs presented in Table 1 and illustrated in Figure 1, the MEPA design was based on both commons principles [20] and New Public Service (NPS) elements [4], as indicated in Table 2.

Table 1. Conceptual references of MEPA.

| Construct | Definition | Authors |
|---|---|---|
| Humane smart city | It is a place flexible to change according to its citizens' wishes, Interests and needs, by the use of technologies as enabler to connect and engage government and citizens, aiming to rebuild, recreate, and motivate urban communities, stimulating and supporting their collaboration activities, leading to general increase of social well-being. | Marsh (2013) |
| Citizen participation | Interaction of citizens and administrators, concerned with public policy decisions and public services (Callahan, 2007). | Ostrom (1978) |
| Public governance | Formal and informal arrangements that determine how public decisions are made and how public actions are carried out. | OECD (2018) |
| New public service | A paradigm of public management that focuses on the public interest, the coproduction of the common good, transparency, accountability and the participation of society. | Denhardt (2012) |
| Knowledge management | Involving the means by which public administration mainly promotes the sharing and dissemination of knowledge through eGov platforms. | Nah et al., 2005 |
| E-Government (eGov) | It is the use of information technology to produce and distribute customer-oriented, more cost-effective, differentiated and better public services. | Holmes (2001) |
| eGov maturity assessment | It is an assessment model composed of at least four high-level eGov applications requirements: (a) current state of maturity and capability identification, (b) benchmark with other eGov applications; (c) innovation roadmap; and (d) discretion as to whether or not to follow. | Valdés et al. (2011) |
| Commons | Resource shared by a group of people attempting to solve social problems. | Fisher and Fortmann (2010) |
| Digital commons | Resources available in information and communication technology platforms (i.e., digital), shared by a group (i.e., commons), integrated in a value chain (i.e., intangible asset) and performed by agents, either as a content or as a process, valuable on a given domain (i.e., knowledge). | Pacheco (2014) |

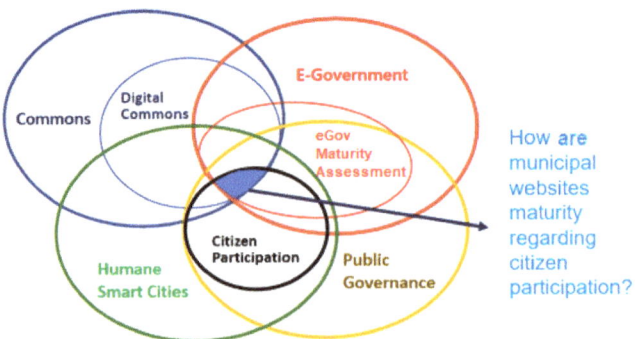

Figure 1. Research problem and conceptual constructs.

**Table 2.** Commons principles and New Public Service elements (derived from References [4,20]).

| Commons Principles | NSP Elements |
|---|---|
| Delimitation<br>Define clear boundaries | Inclusion and access<br>Access information based on education, open government, free communication, and open discussion |
| Adequacy of context<br>Match rules governing commons use to local needs and conditions | Civic engagement<br>Serve citizens, not customers |
| Participation and coproduction<br>Ensure that the ones affected by the rules can participate in regulatory changes | Coproduction<br>Promote collective efforts and collaborative processes |
| Monitoring<br>Develop a system carried out by community members for monitoring member's behavior | Transparency and publicity<br>Greater participation responds to call for greater transparency and accountability in government |
| Proportionate sanctions and rewards<br>Use graduated sanctions for rule violators | Accountability<br>Public servants must attend to law, community values, political norms, professional standards, and citizen interests |
| Resolubility<br>Provide accessible and low-cost means to dispute resolution | Shared responsibility<br>Create shared interests and shared responsibility |
| Autonomy<br>Make sure that rule-making rights are respected by outside authorities | Reaffirmation of values of democracy/citizenship |
| Adhocracy<br>Governing the common resource in based on nested tiers from the lowest level up to the entire interconnected system | Decentralization of power<br>Collaborative structures with leadership shared internally and externally |

Although the relations in Table 2 are not linear (i.e., a commons principle can be related to more than one NPS element, and vice-versa), they help to relate eGov platform services to both public commons and public decision-making. By relating commons principles to NPS elements, the MEPA model opens the possibility of eGov platform assessment in both views, first as digital commons potential promoters and, second as an instrument to support contemporaneous public management. Particularly to NPS, the MEPA analysis is concerned with:

1. Transparency and publicity: The concern of municipalities to comply with current legislation, through the publication and dissemination of information, guidelines, recommendations and open data;
2. Civic engagement: Resources to interested parties, so that they can develop activities in their communities or workplaces through social networks and media, or the services offered, seeking to assert their interests, to provide or receive common goods, or to participate in some level of the political decision-making process;
3. Inclusion and access: Services to help including citizens, public or private bodies in life in society, reducing differences;
4. Shared responsibility: Municipalities efforts to build a collective and shared notion of common, economic and socially viable good;
5. Reaffirmation of values of democracy and citizenship, power decentralization, coproduction and accountability: Services and functionalities for the concrete involvement of all stakeholders, in the definition and active participation of the decision on how the public service will be delivered, and, ultimately, how the common good will be coproduced.

Additionally, MEPA has also a conceptual relationship between the commons theory and eGov maturity assessment, as illustrated in Table 3.

**Table 3.** Relationship between eGov maturity levels and commons principles.

| eGov Maturity Level | Description | Commons Principles |
|---|---|---|
| First Level | The portal offers easy access to simple information and services | Delimitation and monitoring |
| Second Level | The portal enables online transactions | Adequacy of context |
| Third Level | The portal allows access to different sites and services, with only one authentication | Resolubility |
| Fourth Level | The portal enables interoperability between government systems and sites | Adhocracy |
| Fifth Level | The portal allows the personalization of the services offered to the users | Proportionate sanctions and rewards, autonomy, participation and coproduction |

The notion of a citizen as a client was proposed by 'new public management'. In the NPS framework, as proposed by Robert and Jane Denhardt [4], there fourth principle sets "serve citizens not customers".

We have developed MEPA considering coproduction as an essential process to do so. Citizens should participate in the creation, development and evolution of the services promoted by the public administration. In order to do so, the channels of communication and interaction must be effectively provided, so that citizens have an active and independent role. They should be able to comment, request, evaluate, and vote for public services satisfaction as well as on the effectiveness of electronic platform they are using.

In this view, citizens no longer play a secondary role, such as a solicitor at the mercy of the public administration, or as a customer who will "consume" the products and services of a menu, but rather of a co-producer, who participates and collaborates dynamically and actively in the evolution of the products and services that it is receiving from the public administration.

In summary, MEPA design was achieved by the conceptual and practical alignment of principles and procedures from commons, NPS, and eGov maturity analysis. Its ultimate goal is to allow the verification of the eGov platform maturity in relation to the common good, according to different levels of instrumentation and services. MEPA evaluations aim to help public managers to promote commons through eGov solutions.

### 2.2. Objectives of Finding a Solution

Municipal websites are, first, eGov solutions and, as such, subjected to maturity assessment. The proposal solution, however, should go further eGov common maturity analysis. More than check for technological and public services effectiveness, the maturity assessment should also reveal the potential of municipal websites promoting citizen participation. To this end, the instrument should enable the analysis of the commons' principles present in eGov platforms of Brazilian municipalities, also considering elements of NPS and knowledge management, with emphasis on how platforms promote participation and sharing of between the different actors in society.

### 2.3. Design and Development

Different approaches are proposed in the literature for the analysis of eGov platforms in order to characterize their level of maturity [11,15,17,21–28]. In common, such approaches cover the breadth of the services assembled by the platform and the level of sophistication of delivery [16,29,30] and the government's relationship with different stakeholders in society [31].

During the research of other models of maturity of e-Government platforms, we found other models that presented some level of similarity with MEPA, as is the case of the methodology presented by Fietkiewicz, Mainka and Stock [32]. Indeed, there is some similarity and points of contact between the two models. However, each model proposes to conduct a different analysis. The methodology presented by Fietkiewicz, Mainka and Stock [32], measures the maturity of e-Government, the usability of the navigation systems, and investigate the boundary documents available on the governmental websites.

The MEPA model, in turn, proposes to investigate the maturity of the municipal electronic government platforms in smart cities, considering the digital commons and citizens coproduction. MEPA was elaborated considering the following aspects: (1) Analyze the new service public [4] and knowledge sharing as factors of relationship between eGov maturity and commons principles, according to Ostrom [7]; (2) Establish a comparative framework between dimensions and maturity factors of Electronic Government (eGov), based on Holmes [17] and the principles of commons [7]; (3) Create a maturity assessment tool of Brazilian Municipal e-Government platforms, as commons promoters; (4) Apply the maturity assessment tool to Brazilian municipalities, and; (5) Analyze the results obtained, under the lens of participation and coproduction. The MEPA model does not use, primarily, criteria such as usability navigation systems, or documents that are transacted and made available between municipal sites. The two models apply two different approaches.

In this research, we have established an eGov maturity assessment including factors regarding citizen participation. This was developed based on commons general principles and by the adaptation of eGov maturity questionnaires to include checking for citizen participation related factors. The development of the work was carried out according to the following steps:

1. Systematic review of the literature for the composition of the preliminary evaluation instrument;
2. Evaluation of the preliminary version of the data collection instrument with specialists;
3. Application of the revised instrument in the pre-test stage;
4. Validation of intermediate analysis of the results achieved and of the data collection instrument with the support of specialists in the construction of items and measures;
5. Application of the evaluation instrument.

In the first step, in order to define the items that should compose the evaluation instrument, a systematic review of the literature was carried out, looking for elements in the literature that could characterize the relationship between open government data, coproduction, commons, governance, electronic government, and knowledge management. Therefore, the terms "open government", "commons", "governance", "eGov*" or "electronic government", "knowledge management" and "co*production" were combined to search the main journal bases. The search was performed on the ISI-Web of Knowledge/Web of Science databases; Scopus; Ebsco; Compendex, and ProQuest, and Google Scholar. Of the total of 54 articles found, 35 were selected after reading them and used to support the elaboration of the data collection instrument.

The literature review resulted in the definition of 56 items that formed the initial evaluation instrument. The initial instrument was evaluated by specialists and applied in the evaluation of 264 electronic platforms of Brazilian municipalities in the pre-test phase. The final version of the instrument is composed of 41 items and was configured considering the feedback from the pre-test phase. The 41 items were organized into groups by their affinity (in terms of functionality or characteristics), in order to make the filling of the questionnaire simpler and more intuitive. Table 4 presents nine groups created as well as their definition.

In Table 5 we list the items of the evaluation instrument. Each item was evaluated by means of an application instrument including the response possibilities 1 for "has the characteristic" and 0 for "does not have the characteristic".

**Table 4.** Item groups and their description.

| Group | Description |
|---|---|
| 1. Open data, information and public services are freely available to users | Citizens have the right to free access to public services and information, to exercise their participation, to improve service delivery, to monitor administration, and to expand democracy [1]. |
| 2. The platform offers open data, institutional and transparency information, and other topics | The digital environment provides greater transparency, facilitating access to information for citizens, allowing the monitoring of government actions, projects and decisions [12]. |
| 3. The platform offers features for interaction with other users or with those responsible for the platform | Citizens should be aware of the communication channels available to contact the public administration, and have access in an easy, accessible and low-cost way [33]. |
| 4. The platform provides resources for users to vote or make recommendations | Public administration should provide channels of communication, based on citizen participation, together with the assumptions of democratic decision-making processes in society [34]. |
| 5. The platform offers capabilities for downloading data (in various formats, machine-readable) | Open data must be reachable and can be physically accessed by download [35]. |
| 6. The platform provides open search/search capabilities | It should be possible to conduct research by various means to assist users in finding relevant open data [36]. |
| 7. The platform is accessible in mobile version | Citizens have the right to access public services and information, freely using ICT resources to access electronic platforms—desktop, mobile or tablet [37]. |
| 8. Quality of data and information offered by the platform | Well-informed citizens can better combat corruption, nepotism, and government inaccuracy. On the other hand, without accurate information, it is difficult to achieve effective citizen involvement in decision-making processes [38]. |
| 9. The platform provides tools for knowledge management | The GC is essential for the success of e-Government initiatives [39]. |

**Table 5.** Evaluation instrument items.

| Item | Commons Principles |
|---|---|
| 1. Does the platform require no prior registration of users? | Delimitation |
| 2. Does the platform identify the services available to each stakeholder? | Delimitation |
| 3. Does the platform provide stakeholders with guidelines for services usage? | Delimitation |
| 4. Does the platform have terms of use for the services, informing the rights and responsibilities of stakeholders? | Sanctions and rewards |
| 5. Does the platform present terms of use for the services, informing penalties in case of noncompliance? | Sanctions and rewards |
| 6. Does the platform disclose at least an index of use of the services provided? | Monitoring |
| 7. Does the platform provide updated news about the municipality? | Monitoring |
| 8. Does the platform provide additional information (economic, cultural, tourist, historical, geographic, ethnic, according to location or region)? | Monitoring |
| 9. Does the platform provide digital media - employing at least one of the following services? Podcast, interactive maps, videos, digital documents, web radio? | Context and adequacy |

**Table 5.** *Cont.*

| Item | Commons Principles |
|---|---|
| 10. Does the platform provide relevant legislation to the municipality? (It may be any type of legislation, as follows: Laws, Master Plan, Urban Zoning, Code of Works, Taxpayer's Manual, normative instructions, decree, ordinances, opinions, resolutions, etc.) | Monitoring |
| 11. Does the platform provide access to the Official Gazette? | Monitoring |
| 12. Does the platform provide access to the municipality's financial information? (availability of government documents for collection, movement of the treasury and financial application of public resources - balance sheets, financial statements) | Monitoring |
| 13. Does the platform provide content related to digital inclusion? | Delimitation |
| 14. Does the platform provide access to the municipality's transparency information? | Monitoring |
| 15. Does the platform provide access to procurement and bidding by the municipality? | Monitoring |
| 16. Does the platform provide access to at least one municipal offices website/portal? | Adhocracy |
| 17. Does the platform provide access to at least 1 website/portal of a municipal body? | Adhocracy |
| 18. Does the platform provide access to at least one website or portal of the municipality's attorney general's office? | Adhocracy |
| 19. Does the platform provide open data? | Autonomy |
| 20. Open data is available in at least one of the following formats: JSON, XML, CSV, ODS or RDF? | Context and adequacy |
| 21. Can the open data available on the portal be downloaded? | Context and adequacy |
| 22. Does the platform provide information about open data? (example: usage policies, category, identification, description, update frequency, etc.) | Delimitation |
| 23. Does the platform provide open data search? | Context and adequacy |
| 24. The platform provides a list of frequently asked questions (FAQ - Frequently Asked Questions) | Resolubility |
| 25. Does the platform provide at least one communication channel for complaints, questions, criticisms or compliments? (example: Ombudsman) | Resolubility |
| 26. Does the platform provide instant online service? (via chat or similar tool) | Resolubility |
| 27. Does the platform allow integration with social networks? (made up of groups that share common interests) | Resolubility |
| 28. Does the platform provide collaborative virtual spaces? (facilitates the meeting and interaction between people who are not physically together) | Participation and coproduction |
| 29. Does the platform provide blogs or microblogs? (example *twitter*) | Participation and coproduction |
| 30. Does the platform allow the formation of communities of practice? (e.g., to create and share common skills, knowledge, and experiences) | Autonomy |
| 31. Does the platform allow you to choose the most relevant services? (which may be due to the functionality of the platform, or through network resources or social media) | Autonomy |
| 32. Can stakeholders use the network or social media features offered by the platform? | Delimitation |
| 33. Does the platform offer features for electronic voting? | Autonomy |

Table 5. *Cont.*

| Item | Commons Principles |
|---|---|
| 34. Does the platform provide services for the composition of the decision-making agenda involving population participation? | Autonomy |
| 35. Does the platform provide features for recommending open data? | Participation and coproduction |
| 36. Does the platform offer resources for recommending services? | Participation and coproduction |
| 37. Does the platform provide resources for voting on what are the best open data? | Participation and coproduction |
| 38. Does the platform provide resources for voting on which are the best services? | Participation and coproduction |
| 39. Is the platform accessible in mobile version? | Context and adequacy |
| 40. Does the platform provide tools for knowledge management? (such as thesauri, classification schemes, taxonomies and ontologies, knowledge maps and mailing lists) | Context and adequacy |
| 41. Does the platform provide up-to-date knowledge resources? (such as lessons learned, good working practices, etc.) | Participation and coproduction |

*2.4. Demonstration*

In this phase the DSR method requires the presentation of a proof of concept in order to demonstrate the efficacy of the proposal to solve the problem. The MEPA model was built under simplicity and ease of use guidelines. The items that compose the MEPA questionnaire were developed considering the principles of the Commons [7] and the New Public Service [4].

In order to evaluate the criteria in all digital commons' maturity dimensions, the MEPA model has a questionnaire, with questions about the eGov platform and the municipal website. MEPA's questions were structured in Google Forms, used by the data researchers to register the answers they found out when checking a particular municipal website. Hence, each research data registered their answers manually, after assessing the municipal website, checking whether or not the electronic platform has the characteristic under analysis. This method of collecting data on websites is based on References [40–43].

The application of the MEPA model was performed only on the websites (electronic platforms) of the municipal public administration. Nevertheless, several items of the questionnaire aim to analyze whether the website allows the integration, interaction, voting and publication of references in relation to the services provided, through social media. It also verifies whether the website can be accessed and used on mobile devices.

After completing the data collection, the answers are organized in data sheets. MEPA researchers perform data processing (via descriptive statistics), calculate frequency of occurrence and draw graphs, classified into categories of analysis (by "size", geographical region, and municipality size) and, highlights results (findings), revealing municipality digital commons maturity rankings. For the effective application of the MEPA, the researchers underwent a training, so that everyone had knowledge of how to access the questionnaire online (using Google Forms); later, we defined a set of electronic municipal platforms for the researchers; in turn, each researcher, using the online questionnaire, accessed the municipal platforms of his responsibility, filling in the answers, and reporting findings that he considered relevant. This way, we were able to evaluate 903 electronic Brazilian municipal platforms, following the same standards and criteria of analysis, in a short period of time, with a high degree of reliability.

*2.5. Evaluation*

In this DSR phase we analyzed how well the proposed model provides solutions to the research problem by comparing the eGov maturity assessment pursued with MEPA results. The strategy was based on the comparison between expert's opinion, literature findings and MEPA results. MEPA metrics and techniques of analysis, and the way the evaluation was conducted led to results similar to cities websites situations described in the literature, as well as by the empirical observation of the researchers.

An example is the ninth publication of the eGov development benchmarking from United Nations Department of Economic and Social Affairs (UNDESA) [44]. In this survey, Brazil is in the 51st position of better eGov services. The country stands out in the basic indicators, such as the existence of a web page of the main public agencies, data supply and indicators on government websites. However, the country has unsatisfactory levels of online eGov services.

According to MEPA survey, municipal Brazilian eGov services offer little citizen participation in public policy decision making over the Internet, but is well positioned to provide information in consultations (note 93 out of 100). A large part of the municipalities visited have available electronic platform, data supply and indicators in governmental sites, in some level. However, unsatisfactory levels were identified in the offer of online services, and in the actions of digital inclusion. Comparing the results of the "online participation" carried out by the UN research with MEPA results, the eGov platforms of the Brazilian municipalities offer few services for citizen participation in the decisions, but they provide diverse information and news.

## 2.6. Communication

In this phase of DSR method, the researchers should communicate the problem, its relevance, and how their proposal presents a novelty or inedita achievements. The MEPA model was fully described in a PhD dissertation [18]. MEPA was discussed not only in the academic forum, its application has been considered beyond the municipal sphere, including the state and federal spheres in Brazil. Recently, the model was the basis of discussion with the Court of Accounts of the State of Santa Catarina, when it served as the basis for the discussion of the electronic government model for the municipalities of that state.

## 3. Application of the Model to Brazilian Municipalities

As described before, the MEPA model was developed based on different theoretical foundations and knowledge fields. In this section, we present the application of the MEPA model to Brazilian municipalities.

### 3.1. Brazilian Municipalities and Large-Scale Model Application

According to the Brazilian official national institute of geography and statistics (IBGE), the country has 5,570 cities. We have applied the MEPA model in 903 municipal websites (i.e., 16.2% of total municipalities), in cities from all five Brazilian regions, as indicated in Table 6.

Table 6. Population and sample—large-scale application.

| Brazilian Region | Quantity of Municipalities |
|---|---|
| South | 300 |
| Southeast | 300 |
| Midwest | 91 |
| Northeast | 111 |
| North | 101 |
| Total | 903 |

### 3.2. MEPA Positive Responses in the Evaluated Brazilian Municipalities

The total 903 municipal websites that were analyzed are from large (291), medium (453), and small cities (159). These websites were checked in all eight digital common maturity dimensions. In Table 6 the percentages of positive answers are presented, according to each digital commons' maturity.

As it can be seen in Table 7, around 82% of the municipal websites have positive Monitoring and 72% are adequate to its context. These positive results are followed by a good Delimitation in about 61%, and good Resolubility in almost 60% of the evaluated municipal websites.

Table 7. Dimensions and frequency of positive responses.

| Region | Frequency (%) |
|---|---|
| Monitoring | 82.77% |
| Context adequacy | 72.04% |
| Delimitation | 61.26% |
| Resolubility | 59.86% |
| Autonomy | 25.20% |
| Adhocracy | 22.22% |
| Participation and coproduction | 21.96% |
| Sanctioning and rewards | 10.96% |

The investigation revealed that the evaluated municipal eGov platforms offer the following information services: (a) Updated news, information on history, economy, tourism, and other relevant facts, using digital media; (b) Monitoring of public agents, through services aimed at the dissemination and publicity of information on Transparency and Open Data of the municipality in question, including municipal budget, laws and projects, purchases, bids and contracts, official gazette; and (c) Communication channels with the ombudsman of the municipality.

Some results in Table 7 reveal a low commitment to commons principals. The evaluated municipal websites offer low level of services related to Autonomy (25.20%), Adhocracy (22.22%), Participation and Coproduction (21.96%), and Sanctioning and Rewards (10.96%).

Therefore, fewer than one fourth of the evaluated municipal platforms provide services related to civic engagement, platform inclusion and assessment, shared accountability, regulation participation, and decentralized power. This is coherent with low levels of Autonomy (i.e., the website is limited to eGov national or regional legislation) and Adhocracy (i.e., rules and commitment is limited to eGov public owners). Additionally, a low level of sanctioning and rewarding about how parties use the platforms reveals the lack of incentives to good use or punishment when users break rules.

From the results obtained, it was possible to perceive that the electronic platforms of large municipalities presented a higher frequency of positive responses, mainly those located in the South and Southeast regions of Brazil, with higher GDP and Human Development Index (HDI) indices.

In the MEPA model, e-Gov platforms can be verified as potential instruments to promote commons and support public management. This is done by classifying MEPA answers according to Commons and NSP principles. In Table 8 we present the results of the 903 evaluated Brazilian municipalities.

According to the results in Table 8, the vast majority of Brazilian municipal eGov have Transparency and Publicity services (82.77%). This is related to Monitoring services, often related with authorities concerns about constitutional principles and laws that obligate govern to open and disseminate public data, information, guidelines, and recommendations. Around 72% of all Brazilian municipal eGov analyzed have services that help to promote Civic Engagement. In fact, most of platforms use web, social media or other services to allow users identify and/or develop activities in their communities or places of work. In time, by attending citizen interests, an eGov platform will accumulate data and information useful to support political decision-making or even to provide/or receive common goods.

Table 8. Brazilian municipal eGov regarding commons and NSP principles.

| Commons Principles | NSP Elements | Frequency (%) |
|---|---|---|
| Monitoring | Transparency and publicity | 82.77% |
| Adequacy of context | Civic engagement | 72.04% |
| Delimitation | Inclusion and access | 61.26% |
| Resolubility | Shared responsibility | 59.86% |
| Autonomy | Democracy/citizenship values | 25.20% |
| Adhocracy | Decentralization of power | 22.22% |
| Participation and coproduction | Coproduction | 21.96% |
| Proportionate sanctions and rewards | Accountability | 25.20% |

Another important finding was the fact that 61.26% of the municipal eGov websites have services related to inclusion and access to all citizens. Besides being recommended by law, digital inclusion can reduce differences between social classes, educational levels, ages, gender, disability, social prejudice or racial. Six out of 10 (59.86%) of the Brazilian municipal eGov platforms analyzed have services related to share responsibility and conflict resolution. These services help municipalities to build a collective and shared notion of common, economically and socially viable good. This includes citizens, companies, elected representatives and administrators in a broader system of governance, aimed at promoting citizenship and serving the public interest.

The lowest rates of commons and NPS principles are in services that help to promote democracy and citizenship values (25.20%), power decentralization (22.22%), coproduction (21.96%) and accountability (10.96%). Brazilian municipal governments provide an insufficient number of services and functionalities for the concrete involvement of all stakeholders. Municipal platforms can do more to engage citizens in defining and participate in decision making on how the public service will be delivered. This requires focus on deliberative democracy based on citizen participation, and shared responsibilities at levels of government and public governance.

*3.3. Brazilian Municipalities' Common Ranking*

One of MEPA model goals is to allow public managers to compare different eGov municipalities regarding their potential to promote commons. This is done by comparing commons principle eGov rates, calculated by weighted scales, where the highest levels mean more citizen participation and coproduction, as shown in Table 9.

Table 9. MEPA commons dimension weights and maturity levels.

| Maturity Level | Commons Principles | Weight | Qty Items | Range |
|---|---|---|---|---|
| First Level | Delimitation<br>Monitoring | 1 | 14 | 1 to 41 |
| Second Level | Adequacy of context | 2 | 6 | 42 to 59 |
| Third Level | Resolubility | 3 | 4 | 60 to 70 |
| Fourth Level | Adhocracy | 4 | 3 | 71 to 769 |
| Fifth Level | Proportionate sanctions and rewards<br>Autonomy<br>Participation and coproduction | 5 | 14 | 80 to 120 |

Each municipal eGov reaches a specific score calculated as follows: by summing the responses to the 41 items (positive response = 1 and negative response = 0), multiplied by the weight of the respective principle item (according to Table 9). In addition, knowing the weights assigned to each commons' principle, and the number of items per Level, it is possible to establish ranges for each Level (as shown in Table 8).

In Table 10, we present the rank results, according to each MEPA maturity level.

Table 10. Number of municipalities by level of maturity.

| Maturity Level | Quantity of Municipalities | Percentage (%) |
| --- | --- | --- |
| First Level | 4 | 0.44% |
| Second Level | 14 | 1.55% |
| Third Level | 48 | 5.32% |
| Fourth Level | 398 | 44.08% |
| Fifth Level | 439 | 48.62% |
| Total | 903 | |

Almost half of the Brazilian cities have (439 or 48.62% of the evaluated platforms) are still in the First Level of maturity, offering simple information or services, easy to access. Another significant number of cities (398 platforms or 44.08%) are in the Second Level of maturity, including users the possibility of performing online transactions. Only 48 of the evaluated municipal platforms (5.32%) are in the Third Level of maturity, adding access to different sites and services, with a single authentication. There are 14 platforms (1.55%) at the Fourth Level, enabling interoperability between systems and sites other than government. Additionally, only four platforms (0.44%) are in the Fifth Level of maturity, including the personalization of the services offered to users.

In summary, among the 903-municipal eGov platforms analyzed in Brazil, 837 (92.7%) are at the basic levels of maturity. Only a few cities provide effective services to promote citizen participation and coproduction. Brazilian municipal eGov platforms do not yet include the population, and do not provide enough means for the interested parties to participate in the elaboration and coproduction of laws, projects, budgets, as well as the services themselves and features offered by the platform, being at the mercy of the services offered by the exclusive initiative of public agencies.

In many of the platforms visited, no evidence was found to demonstrate compliance with basic requirements, such as the availability of up-to-date information and online services. For example, in terms of transparency and open data, many municipalities simply provide information that is required by legislation, often incomplete, unstructured or difficult to understand, and that does not strictly and effectively promote the transparency and publicity of actions undertaken in the public sector. The absence of services and information, or the difficulty in finding and understanding them, distances citizens from the public administration, and prevent manifestations, requests, criticisms, suggestions or compliments. The lack of inclusion of stakeholders and low understanding of the functioning, organization and execution of the actions of public services undermine citizen participation and public co-production.

*3.4. International Benchmarking*

MEPA benchmark is conducted by comparing eGov municipalities results with municipal eGov that fully comply with the commons principles proposed in its data collection instrument. We have found two eGov platforms that meet all MEPA criteria with maximum excellence grade: London/UK (see https://www.london.gov.uk/) and Singapore (see https://www.gov.sg/). By comparing the evaluated Brazilian platforms with these two international references, it is possible to recommend the following eGov good practices to improve MEPA grades:

1. Interface: Apply well-structured and elaborated interface, facilitating the access to the services of interest of the user;
2. Tutorial: Offer guidelines and easy identification of available services, based on user profiles of the interested party (citizen, company, server, tourist, etc.);
3. Content: Caveats about the services, informing rights, responsibilities and penalties to the interested parties;
4. Content: Offer municipality up-to-date news/information (economic, cultural, tourist, historical, geographic, ethnic);

5. Social eGov: Digital media (e.g., Podcast, interactive maps, videos, digital documents, web radio);
6. Openness: Provide services for government transparency;
7. Openness: Apply open government practices, providing open data, with the possibility of download and readable by machine;
8. Interoperability: Provide access to other government agencies;
9. Communication: Channels to interact with stakeholders (e.g., Ombudsman), with registration, follow-up and closing of the request;
10. Readiness: Provide instant online services;
11. Social eGov: Allow the integration and use of the social networks;
12. Interface: Allow the choice or recommendation of more relevant services;
13. Mobile eGov: Be accessible in mobile version;
14. Knowledge management: Include resources for management and knowledge sharing.

## 4. Concluding Remarks

Smart cities (particularly the humane smart cities) call for new governance models where public authorities and citizens build sustainable relationships [45]. Both commons and NPS principles relate sustainable public relations with collective governance based on citizen participation and coproduction. Additionally, smart cities also call for efficient use of information technologies, this is also a requirement for mature eGov platforms. In this study, we presented the MEPA model to accurately assess eGov platforms' performance in terms of citizen participation and coproduction to offer high quality public services. In other words, by using numerous criteria, MEPA verifies eGov platforms regarding the commons, NPS, and maturity of electronic government dimensions.

In the smart cities practice, by assessing eGov platforms as common promoters, MEPA can be a highly useful tool to evaluate the levels of citizens empowerment, collective co-creation, and public authorities' commitment to use digital technologies to develop social sense of belonging and identity. The large-scale application of MEPA in 903 Brazilian municipalities reveals the general outcome of city eGov platforms as digital commons promoters being immature/underdeveloped. There are only a few eGov municipal platforms that are in higher grades of maturity; and these are from cities with higher budgets and from more developed regions of Brazil. An international benchmark indicated several points to help public authorities to foster eGov platforms towards a higher level of commons maturity.

The MEPA model and its large-scale application indicate that in order to develop digital commons promoters, eGov platforms' authorities have to: (a) Enable citizens, public and private agencies, and government at large taking into account of their respective roles and responsibilities; (b) Develop effective mechanisms for conflict resolution (i.e., fast, affordable, and proportionate sanctioning); (c) Develop sustainable and perennial initiatives, appropriate to the context to which they refer; (d) Adopt coproduction and citizen participation as guiding principles; (e) Understand that the assets of society, more than public, are collective goods and responsibilities, and; (f) Define clear and effective rules to monitor and govern the interaction of diffuse and collective interests, considering that different communities can share the same common good.

As part of our prospective work, the MEPA model research will be expanded to its application to other levels of government (including the legislative and judicial branches). Moreover, we will adopt/develop a longitudinal data collection and analysis method to the evaluated Brazilian cities. Furthermore, we will adapt and apply the model in other countries (particularly to compare developed and developing country practices), and include eGov cross-referencing indicators such as quantity and frequency of stakeholders' eGov access. Lastly, particularly focusing on smart cities, we will relate the MEPA model with smart/knowledge-based city models (such as smart and knowledge-based urban development [27,45–48]). This will not only add another dimension to digital commons maturity analysis, but also will contribute to relate eGov platform maturity with smart cities requirements, e.g., smart governance.

**Author Contributions:** Conceptualization, M.J.R.R., D.S. and R.C.S.P.; Investigation, M.J.R.R.; Writing – Original Draft Preparation, M.J.R.R., D.S. and R.C.S.P.; Writing – Review & Editing, T.Y.

**Funding:** This study was financed in part by the Coordenação de Aperfeiçoamento de Pessoal de Nível Superior - Brasil (CAPES) - Finance Code 001.

**Acknowledgments:** The authors thank anonymous referees for their constructive comments on an earlier version of the manuscript.

**Conflicts of Interest:** The authors declare no conflict of interest.

## References

1. Bart'h, J.; Fietkiewicz, K.J.; Gremm, J.; Hartmann, S.; Ilhan, A.; Mainka, A.; Meschede, C.; Stock, W.G. Informational urbanism, A Conceptual Framework of Smart Cities. In Proceedings of the 50th Hawaii International Conference on System Sciences, Waikoloa Village, HI, USA, 4–7 January 2017; IEEE Computer Society: Washington, DC, USA, 2017; pp. 2814–2823.
2. Caragliu, A.; Del Bo, C.; Nijkamp, P. Smart Cities in Europe. *J. Urban Technol.* **2011**, *18*, 65–82. [CrossRef]
3. Anttiroiko, A.V.; Valkama, P.; Bailey, S.J. Smart cities in the new service economy: building platforms for smart services. *Ai Soc.* **2014**, *29*, 323–334. [CrossRef]
4. Denhardt, J.V.; Denhardt, R.B. *The New Public Service: Serving, Not Steering*; ME.Sharpe: London, UK; New York, NY, USA, 2003; Available online: https://epdf.pub/the-new-public-service-serving-not-steering.html (accessed on 12 July 2019).
5. Marsh, J. (Ed.) The Human Smart Cities Cookbook. 2013. Available online: www.peripheria.eu (accessed on 12 July 2019).
6. Hess, C.; Ostrom, E. A Framework for Analyzing the Knowledge Commons: A chapter from Understanding Knowledge as a Commons: From Theory to Practice. 2005. Available online: https://surface.syr.edu/sul/21 (accessed on 12 July 2019).
7. Ostrom, E. *Governing the Commons: The Evolution of Institutions for Collective Action*; Cambridge University Press: Cambridge, UK, 1990.
8. Hardt, M.; Negri, A. *Commonwealth*; Harvard University Press: Cambridge, MA, USA, 2009.
9. Chourabi, H.; Nam, T.; Walker, S.; Gil-Garcia, J.R.; Mellouli, S.; Nahon, K.; Scholl, H.J. Understanding Smart Cities: An Integrative Framework. In Proceedings of the 45th Hawaii International Conference on System Sciences, Maui, HI, USA, 4–7 January 2012; pp. 2289–2297. [CrossRef]
10. Nam, T.; Pardo, T. Conceptualizing smart city with dimensions of technology, people, and institutions. In Proceedings of the 12th Annual International Digital Government Research Conference: Digital Government Innovation in Challenging Times, College Park, MD, USA, 12–15 June 2011; ACM: New York, NY, USA, 2001; pp. 282–291. [CrossRef]
11. West, D.M. E-Government and the Transformation of Service Delivery and Citizen Attitudes. *Public Adm. Rev.* **2004**, *64*, 15–27. [CrossRef]
12. Bollier, D. The Growth of the Commons Paradigm. Understanding Knowledge as a Commons. 2007, p. 27. Available online: http://dlc.dlib.indiana.edu/dlc/bitstream/handle/10535/4975/GrowthofCommonsParadigm.pdf?sequence=1&isAllowed=y (accessed on 12 July 2019).
13. Pacheco, C.S.R. Coprodução em Ciência, Tecnologia e Inovação: Fundamentos e visões. In *Interdisciplinaridade: Universidade e Inovação Social e Tecnológica*, 1st ed.; CRV Editora: Curitiba, Brazil, 2016; pp. 21–62. Available online: https://www.researchgate.net/publication/307977522_Coproducao_em_Ciencia_Tecnologia_e_Inovacao_fundamentos_e_visoes?amp%3BenrichSource=Y292ZXJQYWdlOzMwNzk3NzUyMjtBUzo0MDUxNDExNzQ5MzE0NjdAMTQ3MzYwNDU5MTg4Ng%3D%3D&amp%3Bel=1_x_3&amp%3B_esc=publicationCoverPdf (accessed on 12 July 2019).
14. Pinho, J.A.G.D. Investigando portais de governo eletrônico de estados no Brasil: Muita tecnologia, pouca democracia. *RAP Revista de Administração Pública–Rio de Janeiro* **2008**, *42*, 471–493. Available online: http://www.scielo.br/pdf/rap/v42n3/a03v42n3 (accessed on 12 July 2019). [CrossRef]
15. Hassan, B.; Alireza, I.; Majideh, S. E- government portals: A knowledge management study. *Electron. Libr.* **2012**, *30*, 89–102. Available online: http://www.emeraldinsight.com/doi/pdfplus/10.1108/02640471211204088 (accessed on 12 July 2019).

16. Fath-Allah, A.; Cheikhi, L.; Al-Qutaish, R.E.; Idri, A. E-Government maturity models: A comparative study. *Int. J. Softw. Eng. Appl.* **2014**, *5*, 71. [CrossRef]
17. Holmes, D. *eGov: eBusiness Strategies for Government*; Nicholas Brealey Publishing: Boston, MA, USA, 2001.
18. Rotta, M.J.R. *As Plataformas de Governo Eletrônico e seu Potencial para a Promoção dos Princípios dos Commons: O caso dos Municípios Brasileiros. Doutorado em Engenharia e Gestão do Conhecimento*; Universidade Federal de Santa Catarina, UFSC: Trindade, Brazil, 2018; Available online: http://btd.egc.ufsc.br/wp-content/uploads/2018/12/Maur%C3%ADcio-Rotta.pdf (accessed on 12 July 2019).
19. Hevner, A.; Chateerjee, S. *Design Research in Information Systems Theory and Practice Forewords*; Springer: New York, NY, USA, 2012. [CrossRef]
20. Ostrom, E. Citizen Participation and Policing: What Do We Know? *J. Volunt. Action Res.* **1978**, *7*, 102–108. [CrossRef]
21. Layne, K.; Lee, J. Developing fully functional E-Government: A four stage model. *Gov. Inf. Q.* **2001**, *18*, 122–136. [CrossRef]
22. Andersen, K.V.; Henriksen, H.Z. E-Government maturity models: Extension of the Layne and Lee model. *Gov. Inf. Q.* **2006**, *23*, 236–248. [CrossRef]
23. United Nation (UN). UN E-Government Survey 2012: E-Government for the People. Available online: http://unpan1.un.org/intradoc/groups/public/documents/un/unpan048065.pdf (accessed on 12 July 2019).
24. Alonso-Muñoz, L. Transparency and Political Monitoring in the Digital Environment: Towards a Typology of Citizen-Driven Platforms. 2007. Available online: https://www.ull.es/publicaciones/latina/072paper/1223/73en.html (accessed on 12 July 2019).
25. Almazan, R.S.; Gil-Garcia, J.R. *E-Government Portals in Mexico*; University at Albany: New York, NY, USA, 2008. [CrossRef]
26. Baum, C.; Maio, A.D. *Gartner's Four Phases of e-Government Model*; Gartner Group Inc.: Stamford, CO, USA, 2000.
27. Moon, J. The Evolution of E-Government among Municipalities: Rhetoric or Reality? *Public Adm. Rev.* **2002**, *62*, 424–433. [CrossRef]
28. Toasaki, Y. *e-Government from A User's Perspective*; APEC Telecommunication and Information Working Group: Taipei, Taiwan, 2003.
29. Valdés, G.; Solar, M.; Astudillo, H.; Iribarren, M.; Concha, G.; Visconti, M. Conception, development and implementation of an e-Government maturity model in public agencies. *Gov. Inf. Q.* **2011**, *28*, 176–187. [CrossRef]
30. Cresswell, A.M.; Pardo, T.A.; Hassan, S. Assessing capability for justice information sharing. In Proceedings of the 8th Annual International Conference on Digital Government Research, Los Angeles, CA, USA, 20–23 May 2007; pp. 122–130. [CrossRef]
31. Alves, A.A.; Moreira, J.M. *Cidadania Digital e Democratização Electrónica*; SPI: Porto, Portugal, 2004; Available online: http://www.spi.pt/documents/books/inovacao_autarquia/docs/Manual_IV.pdf (accessed on 12 July 2019).
32. Fietkiewicz, K.J.; Mainka, M.; Stock, W.G. eGovernment in cities of the knowledge society. An empirical investigation of Smart Cities' governmental websites. *Gov. Inf. Quar.* **2017**, *34*, 75–83. [CrossRef]
33. Bollier, D. The commons, short and sweet. *Bollier. Org.* 2011. Available online: http://eco-literacy.net/wp-content/uploads/sites/4/2017/05/introduction-to-the-commons.pdf (accessed on 12 July 2019).
34. Gomes, W. A Democracia Digital e o Problema da Participação Civil Na Decisão Política. *Fronteiras-Estudos Midiáticos* **2005**, *7*, 214–222. Available online: http://www.revistas.unisinos.br/index.php/fronteiras/article/view/6394 (accessed on 12 July 2019).
35. Welle Donker, F.; Van Loenen, B. How to assess the success of the open data ecosystem? *Int. J. Digital Earth* **2017**, *10*, 284–306. Available online: http://www.tandfonline.com/doi/abs/10.1080/17538947.2016.1224938 (accessed on 12 July 2019). [CrossRef]
36. Máchová, R.; Lnénicka, M. Evaluating the Quality of Open Data Portals on the National Level. *J. Theor. Appl. Electron. Commer. Res.* **2017**, *12*, 21–41. Available online: http://www.scielo.cl/scielo.php?pid=S0718-18762017000100003&script=sci_arttext (accessed on 12 July 2019).
37. Brazil Ministério do Planejamento. Orçamento e Gestão. Resolução CGPAR 05, de 29 de Setembro de 2015. Available online: http://www.planejamento.gov.br/assuntos/empresas-estatais/legislacao/resolucoes/rescgpar_05.pdf (accessed on 12 July 2019).

38. Abdullah, N.N.; Rahman, M.F.A. Access to Government Information in Public Policy Making Process: A Case Study of Kurdistan. *Int. Inf. Inst.* **2015**, *18*, 3447. [CrossRef]
39. Gupta, R.; Singh, J. Knowledge Management and Innovation in (e) Government. *Int. J. Inf. Comput. Technol.* **2014**, *4*, 1637–1645. Available online: http://www.ripublication.com/irph/ijict_spl/ijictv4n16spl_04.pdf (accessed on 12 July 2019).
40. Pinterits, A.; Treiblmaier, H.; Pollach, I. Environmental websites: An empirical investigation of functionality and accessibility. *Int. J. Technol. Policy Manag.* **2006**, *6*, 2006. [CrossRef]
41. Al-Khalifa, H.S. The accessibility of Saudi Arabia government Web sites: An exploratory study. *Univ. Access. Inf. Soc.* **2010**, *10*. [CrossRef]
42. Stepchenkova, S.; Tang, L.; Jang, S.S.; Kirilenko, A.P.; Morrison, A.M. Benchmarking CVB website performance: Spatial and structural patterns. *Tour. Manag.* **2010**, *31*, 611–620. [CrossRef]
43. Tezza, R.; Bornia, A.C.; Andrade, D.F. Measuring web usability using item response theory: Principles, features and opportunities. *Interact. Comput.* **2011**, *23*, 167–175. [CrossRef]
44. United Nation (UN). *United Nations E-Government Survey 2016: E-Government in Support of Sustainable Development*; United Nations: New York, NY, USA, 2016; Available online: http://workspace.unpan.org/sites/Internet/Documents/UNPAN96407.pdf (accessed on 12 July 2019).
45. Yigitcanlar, T.; Metaxiotis, K.; Carrillo, F.J. *Building Prosperous Knowledge Cities: Policies, Plans and Metrics*; Edward Elgar Publishing: Cheltenham, UK, 2012.
46. Yigitcanlar, T.; Kamruzzaman, M.; Buys, L.; Ioppolo, G.; Sabatini-Marques, J.; da Costa, E.M.; Yun, J.J. Understanding 'smart cities': Intertwining development drivers with desired outcomes in a multidimensional framework. *Cities* **2018**, *81*, 145–160. [CrossRef]
47. Yigitcanlar, T. Position paper: Benchmarking the performance of global and emerging knowledge cities. *Expert Syst. Appl.* **2014**, *41*, 5549–5559. [CrossRef]
48. Trindade, E.P.; Hinnig, M.P.F.; Moreira da Costa, E.; Marques, J.; Bastos, R.; Yigitcanlar, T. Sustainable development of smart cities: A systematic review of the literature. *J. Open Innov. Technol. Mark. Complex.* **2017**, *3*, 11. [CrossRef]

© 2019 by the authors. Licensee MDPI, Basel, Switzerland. This article is an open access article distributed under the terms and conditions of the Creative Commons Attribution (CC BY) license (http://creativecommons.org/licenses/by/4.0/).

Article

# Modelling Interaction Decisions in Smart Cities: Why Do We Interact with Smart Media Displays?

Hoon Han [1,*], Sang Ho Lee [2] and Yountaik Leem [2]

1 City Planning Program, Faculty of the Built Environment, University of New South Wales, Sydney, NSW 2052, Australia
2 Department of Urban Engineering, Hanbat National University, Daejeon 305-719, Korea
* Correspondence: han@unsw.edu.au; Tel.: +61-2-9385-6319

Received: 3 June 2019; Accepted: 16 July 2019; Published: 23 July 2019

**Abstract:** This study examined the personal characteristics and preferences of individuals that encourage interactions with smart media displays (media façades). Specifically, it aimed to determine which key aspects of a smart display "media façade" enhance intuitive interactions. A range of smart display technologies and their effects on interaction decisions were considered. Data were drawn from a survey of 200 randomly sampled residents and/or visitors to a smart building, One Central Park, in Sydney, Australia. A binomial logistic regression analysis was undertaken to establish links between a range of design, perceptions and socio-demographic variables and individuals' decisions to interact with a smart media display. The results showed that the aesthetics of an installation, the quality of an installation's content and the safety of the operation-friendly environment significantly affected respondents' decisions to interact with the media display. Interestingly, respondents born overseas were more likely to interact with a smart display than those born in Australia. Respondents who expressed a preference for photograph-based interactions were also more likely to interact with the display. Somewhat surprisingly, age, residency and levels of familiarity with digital technology did not significantly affect respondents' decisions to interact with the display.

**Keywords:** smart cities; smart display; smart placemaking; human–computer interaction; user characteristics; media façade; intuitive interaction; living-lab

## 1. Introduction

Smart cities have increasingly encouraged human-computer intersection with a range of cutting-edge technologies. These days smart media displays/façades provide a new means of communication and creative engagement in smart cities [1,2]. In Australia, the installation of interactive media displays in public places is a recent innovation. Such installations fuse smart technology, planning, and architecture to create shared experiences between members of the public as a living-lab [3]. Recently, smart media display installations have begun to expand the potential of smart technology to encourage human–computer interactions (HCIs) in public places.

Previous research on HCIs in smart cities has adopted two distinct approaches. The demand-side approach has sought to consider the use of interaction designs for a living-lab in smart cities [4,5]. Conversely, the supply-side approach has sought to examine the application of smart technologies and technology development interactions in smart cities [6–8]. Recent HCI research has led to more interactive systems being designed that are efficient and affordable and encourage public participation in smart cities [9,10]. In particular interfaces of the smart media display are being improved and integrated into blended environments. Such interfaces enable more intuitive interactions [11] and could potentially influence patterns of social behaviour in smart cities [2].

Jacucci et al. [12] examined a range of smart interaction technologies for publicly exhibited artworks that use interactive technologies to encourage different participatory strategies. In one such artworks, lights appeared on three white canvasses in response to the external movements, sounds, pressure and electromagnetic stimuli provided by spectators. Other smart technologies have used mobile and global positioning system technologies as platforms to augment reality, transforming the role of visitors from that of passive receivers to engage in interactive and smart infrastructures [13]. Another notable smart display installation featured an evolving line-art skyline that responded to users' silhouettes while luminous creatures interacted playfully with the scaffold and users' silhouettes [14].

Despite the rapid evolution of smart media display technologies and the attendant multiplication of design possibilities in smart cities, an understanding of the interaction decision-making processes that individuals experience in response to interactive smart media displays remains limited. Specifically, very little is known about users' socioeconomic profiles or, more importantly, the perceptions, motivations and intentions of those interacting with such new smart media displays in smart cities [15]. To address these knowledge gaps, a survey was undertaken to examine the factors driving the increasing number of smart media displays using HCI. Given that a wide range of user and environmental variables could influence users' interaction decisions, this study had two key objectives: (1) to determine which user characteristics encourage interactions with a smart media display, and (2) to determine which aspects of a display's environment encourage these interactions. In this study, residents and visitors passing by a smart media display located at a commercial and retail podium of a newly developed smart precinct in central Sydney (Australia) were asked to complete a survey that had been designed to understand the unique intersection of smart technology, psychology and HCI in a public space.

## 2. Literature Review

### 2.1. Smart Media Displays: An Interactive Technology in Smart Cities

Previous research from around the world has documented the belief that media displays are merely another avenue for urban advertising. The predominant used of media displays for advertising in shopping centres has made it difficult to envisage more creative ways of these display being used [1,16]. As the incidence of an opportunity for smart media displays to be used in the smart city planning a new public interactive dimension outside the advertising sphere continue to increase [17]. For example, smart media displays could be used in historic urban areas to transform public spaces into sustainable and creative tourism destinations [18]. This agrees with Struppek [19] who claimed that media display installations can be used to enhance social sustainability of cities by boosting the livability and environmental conditions of public spaces.

The potential roles of smart displays 'interactive media façades' have only recently begun to be explored in smart cities or smart building design. Media façades (e.g., the Climate Wall in Aarhus [20] have been used to spark community discourse or to create playful spaces for interactions (e.g., the BBC Big Screen red nose game) [21]. If the media-façade concept is considered from an architectural standpoint, the value of such public displays could extend beyond the utilitarian provision of information in smart cities [16]. This tends to supported by Wiethoff and Gehring [22] that the user-centric and interactive media-façades can be developed and implemented to provide a positive experience to its users.

### 2.2. Intuitive Interaction with a Smart Media Display

Blackler and Hurtienne [11] defined intuitive interactions as the cognitive processes by which individuals use the knowledge they have gained from other experiences. Thus, the familiarity of users with smart technology is a key factor in determining the level of intuitiveness of an interface. User age is also important, as older people generally take longer to complete set tasks than their younger counterparts and experience more difficulties when engaging in new smart technology-related activities [23]. Blackler et al. [24] showed that when users are exposed to a new interface, their intuitiveness interactions are significantly related to their previous knowledge. Thus, as Blackler et al. [23] argued, it is important to understand what knowledge users have already acquired so that suitable stereotypes, features or metaphors can be used to new smart interfaces.

Individuals' cognitive processes and intuitive interactions are affected by their perceptions of new smart technology, their established user decision-making processes and their personal emotions. Consequently, effective interaction designs should seek to create emotional experiences among users. Emotion typically affects decision making and cognitive reasoning unconsciously, such that thinkers may not be aware of the irrationality of their decisions [25]. Likewise, appraisal theory posits that perceivers' needs and objectives influence their subjective appraisals of stimuli and also affects their emotions [26]. Conversely, evolutionary psychology contends that emotions represent pre-programmed responses to environmental threats and opportunities [27]; for example, anxiety is an emotional condition that prepares individuals for flight-or-fight situations in which alarms or warnings are more important than complicated words.

Fischer and Hornecker [28] explored the major trajectory (i.e., the 'activation loops') of users interacting with media façades. The activation loop describes how users change between different user roles (e.g., that of a passer-by, bystander, audience member, participant, actor or dropout) by engaging in rousing, learning, engaging, committing and dropping-out processes.

Hespanhol and Tomitsch [5] identified a recent shift in HCI research in relation to how individuals interact with smart applications in public spaces. In their research, which focused on interaction designs in public spaces, they found that in interactive public spaces: (1) the smart interactive systems may be surrounded by participants and are integrated with the surrounding built environment; and (2) the interactions can be shared by multiple people contemporaneously. In the field of HCI, the 'honeypot effect' describes an interaction whereby passers-by are passively motivated to engage in interactive activities by the people who are already interacting with the displays [29]. Thus, intuitive interactions in public spaces not only involve the individuals who are directly engaged with the interactive content, but also those in the crowd within that space [5]. Thus, installations should be accessible to a wide range of people with different socioeconomic characteristics. Such characteristics may include age, gender, income and ethnic background. In public spaces, individuals are generally self-driven towards social interactions and engage in mutual observations that are principally visual but may also involve other sensorial inputs.

## 3. Conceptual Framework: Modelling Interaction Decision with a Smart Media Display

This study adopted a conceptual framework for understanding individuals' interaction decisions in relation to behavioural intention. The theory of reasoned action states that behavioural intentions are the immediate antecedents to actual behaviour [30]. Suchman [31] developed the situated action theory of human behaviour that holds that planned behaviour and reasoned action are interleaved. Their theory undermines the notion that human action is entirely under the control of cognition. Payne et al. [32] challenged Suchman's approach, contending that the relationship between planned behaviour and reasoned action is discretionary and that a planner's intention should be viewed as a decision maker.

Payne et al. [32] further argued that the phenomena of an interactive search are the result of an individual adapting to the characteristics of an environment and human cognition (e.g., personal experiences and perceptions). As Payne et al. [32] stated (p.341): 'The phenomena are reliable not because they are themselves invariant features of human behaviour or cognition, but because they derive from decision-making strategies applied to consistently structured task environments'. The cognitions between users and computers are distributed by the users themselves. As is the cognition distribution between planning and in place responsiveness. When interacting, users consider the probability (P) of revitalising a memory to achieve a current objective, their gain (G), and the cost (C) of this cognitive process [33]. The cognition includes users' previous experiences and knowledge, and their cognitions about in situ factors (i.e., their motivations to visit a place and their awareness of their safety in a place).

Respondents' perceptions of the built environment were considered in relation to each of the following five constructs: their recognition of safety risks such as robbery, catcalling and vandalism;

their response to the socialising/communication with smart technology; their appreciation of the aesthetics of the place, their motivation to visit the place; and their behaviours as consumers. The five social constructs are developed based on the interaction design guideline for urban media facades, which emphasizes the needs of visible interactive zones of media facades [20,34]. Table 1 outlines the five social constructs of the interactive built environment considered by this study.

Table 1. Key social constructs of the interactive built environment.

| | |
|---|---|
| Recognition of safety | Slovic [35] observed that the majority of citizens have intuitive risk judgments and risk perceptions. Solvic further stated that individuals can 'sense and avoid harmful environmental conditions' and have the 'ability to codify and learn from past experience'. Jorgensen et al. [36] contended that users' perceptions of safety and preferences for public space are affected by the number and characteristics of other users, evidence of behaviours that are felt to be anti-social, the proximity of the built environment and lighting. |
| Socialising opportunity | Socialising refers to bringing a social networking opportunity under public ownership and to taking part in social activity [37]. Socialising includes social activities at which viewers of a public space meet others and exchange everyday small talk through mediated networks (e.g., local communities). It should be noted that in such activities, communication is the most important aspect in a public place [38]. |
| Appreciation of aesthetic | Eben Saleh [39] contended that excellence and aesthetic value in a public space enhances the interactions between that space and its users. Aesthetics arise as a product property and provided 'added value' to an artefact, such as a digital media display. Interactions can be induced if the aesthetics and appearance of the artefacts are attractive and pleasurable [40]. |
| Motivation to visit a place | Motivation can be measured in relation to an individual's intention to visit a public place. The value of assessing motivation predicts adherence to habitual activities, such as revisiting public spaces and may vary as a result of individual differences and socioeconomic characteristics [41]. A destination image plays a mediating role between perceived environment (i.e., risks and aesthetics) and the intention to (re)visit a public space [42]. |
| Consumer behaviour | Consumer behaviour comprises two parts: (1) how media is consumed; and (2) the effects that media consumption has on a consumer's choice. A digital media display may affect a consumer's consumption behaviour at the conscious or subconscious level. Interactions with a digital media display could potentially affect consumers' behaviour. |

The present study first examined the five aspects of a media display's interactive environment (i.e., socialising, safety recognition, aesthetic appreciation of public space, motivation to visit and consumer behaviour). Specifically, this study examined how the five aspects of a media display's interactive built environment affected individuals' perceptions and interaction decisions (see Table 1). The concept of the interactive built environment, as derived from psychometric sources, reflects individuals' particular outlooks and social perceptions of a new smart display installation [35,43]. For example, safe and secure space may increase the likelihood that users will engage with the media display [16]. Trees and walls are typical objects that evoke a protective feeling among people and create 'comfort spaces' in which people can observe interactive activities in a surrounding area [26]. A user's recognition of an interactive environment occurs within the context of a user's cognitions of the built environment whereby information, upon reception, is coded, stored and organised in the brain. A user does not merely respond reflexively to a smart media display. Thus, it was anticipated that individuals' recognition of the built environment surrounding a smart media display would first affect their individual perceptions of the digital display placement and then their interaction decisions.

In relation to the 'socialising' aspects of the interactive built environment of digital displays, respondents were asked two questions: (1) Does the interactive technology make you feel like socialising; and (2) How important is socialising in your everyday life? To assess their interaction decisions (i.e., the dependent variable), the respondents were also asked about the four other aspects of the media display's built environment. Specifically, the respondents were asked 'Are you interested in interacting with the smart media display?'

Users' perceptions of a media display are also important in shaping the manner in which users interact with the new smart display. Design illumination, interaction methods [28], the quality of the media content [34] and the frequency of displays and other visual objects in proximity to a particular media display may produce different interaction decisions [44]. Users' interaction decisions will not only be affected by their cognitions about a media display's built environment, but also by their individual socioeconomic and demographic characteristics [45]. Users' opinions of and level of familiarity with smart technology is also an important factor affecting users' interaction decisions. Different user archetypes (e.g., contributors versus passive users or non-users) will display different reactions towards the same interactive display content, and different user needs. Thus, designers should consider their target users when designing the physical appearance of a display, its content and the interaction mechanism [46]. For example, individuals' acceptance levels in relation to the content displayed on the media screen in Flagey Square, Brussels (Belgium) varied depending upon their cultural background. To increase the effectiveness of a media façade, content providers need to be sensitive to the sociocultural environment and the receivers [47]. Media surfaces with excellent usability and semantics could potentially encourage the integration of different cultures and multi-ethnic peoples [48].

Vande Moere and Wouters [47] considered four case studies and ultimately identified three aspects of the contextual integration of a media façade: (1) 'the environment' (i.e., the space where the media architecture is installed); (2) the 'content' (i.e., what the display actually delivers to the audience); and (3) 'the carrier' (i.e., who is operating the 'content' or the public image of the carrier). In the present study, a conceptual framework was developed that focused on the three factors affecting individuals' interaction decisions in relation to the media façade: (1) the 'environment'; (2) the 'carrier'; and (3) the 'content'. Tangible and intangible variables were considered in relation to each aspect of (A) the built environment where the display is installed (i.e., the environment); (B1) users' socioeconomic characteristics and (B2) familiarity with smart technology (i.e., the carriers); and (C) Satisfaction with the media display (i.e., the content) (see Figure 1).

Under this framework, a stimulus involving the interactive built environment (i.e., the Environment in Figure 1) may prompt users to interact with a media display. For instance, a friendly environment for meeting a friend (socialising) and safer place for women and children would encourage their interaction with a media display. Various inputs may inform a user's cognitive process; for example, a user may perceive the potential effects of interaction, or a lack thereof, may be registered [49]. The decision to interact with a media display is a cognitive process that involves input in the form of a user's prior knowledge, social values and experience, including a user's level of familiarity with digital technology (i.e. how many hours per day do you usually use the Internet?; how would you rate your computer technology skills?). Interactions vary according to individual's socioeconomic and demographic characteristics, such as their age, gender, ethnicity and levels of skill and experience with digital technologies [11] (see Box B2 in Figure 1). Most importantly, the quality design, media contents and technologies of the smart display affect user's interaction decision (see Box C: 'interaction' [D1] and 'no interaction' [D2]).). In this conceptual framework, we assumed that site visitors first make judgments about the current built environment around the media display based on their own experience and socioeconomic and demographic characteristics, and then to derive a certain level of satisfaction to the design and contents of media display. Subsequently, their levels of satisfaction with the media display lead interaction decisions.

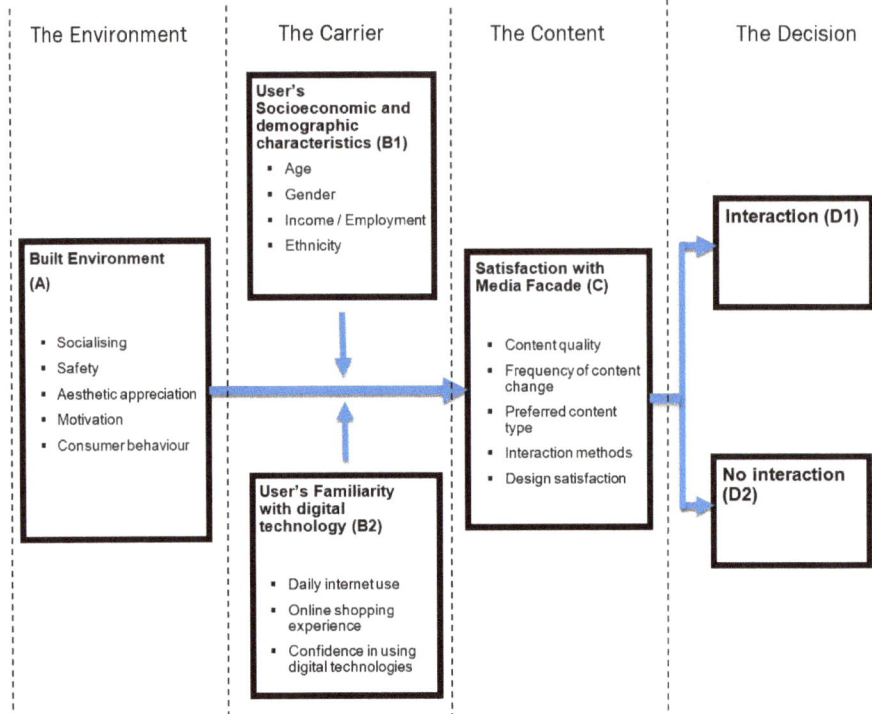

**Figure 1.** Conceptual framework of modelling the decision to interact with a media display.

## 4. Survey Design and Methods

### 4.1. Case Study: One Central Park, Sydney

A case study was conducted at the Central Park precinct, a 5.8-hectare, 2-billion-dollar urban redevelopment at Broadway, near the southern edge of the central business district in Chippendale, (Sydney, Australia). The mixed-use project is a joint venture between Frasers Property and Sekisui House to redevelop the old Carlton and United Brewery that closed in 2005 [50]. The project includes residential apartments, offices, shops and cafés. The project has won numerous awards across a range of categories, including sustainability, landscape architecture, design and innovation awards.

A 15-metre long interactive LED screen (i.e., 'a digital wall') is installed inside the retail portion of the precinct. The digital wall is located adjacent to the elevators on the ground floor. It was produced by Big Screen Projects and comprises 154 customised panels that together contain over 2 million 7 mm LED modules (Figure 2). A kiosk, microphones and cameras enable individuals to interact with the digital wall. Ramus Illumination, who acted as the creative directors of the media display, has stated that it is the largest indoor media display of its kind in a retail precinct in the world [51]. Visentin [52] noted that a media display was included at the One Central Park retail precinct to distinguish it from other modern 'sterile and soulless shopping malls'. The scale of the digital wall gives it great commercial potential; however, the project appears to be consciously trying to avoid becoming an 'advertising billboard or an incessantly-looping foyer artwork' [53] and instead offers an ever-changing range of interactive activities to shoppers. In addition to showcasing local community art and playing a library of videos, the smart digital display provides shoppers with opportunities to play games at the kiosk and engage with social media. For example, during University Orientation Week, a custom Instagram application developed by Ramus Illumination allowed students to hashtag their photos and display them on the digital wall [53]. The content of the wall can be changed by simply walking through space. Cameras and microphones around the smart digital display are mapped to the area in front of the screen. They sense people's movements and then cast those people's shadows onto the screen [53]. Ramus [51] described this as 'reflective content: it reflects the environment where the installation sits'.

**Figure 2.** Mixed Use Building and Smart Media Display at One Central Park.

*4.2. Survey Design*

The data presented in this paper was compiled after a random sample survey was administered to the respondents. Questionnaire survey (paper copy) were conducted with individuals who had seen the newly installed media display between 15 and 20 May 2015. Permission to conduct the on-site survey was granted by the Central Park One manager and research human ethics approval was obtained from the Human Research Ethics Committee (grant no. 155040). A supervisor and two trained staff administered the survey on site during the shopping centre's trading hours. They were dressed in a uniform and carried appropriate identification at all times.

It was thought that completing the survey on site might assist the respondent shoppers to understand more about the new smart technology and increase their interest in the Central Park site. The respondents were asked if they had previously seen or recognised the smart media display. Only those who responded 'yes' to this question were invited to continue to participate in the survey. The survey considered respondents' intention to interact with the media display, asking whether they like to interact or not. Regardless of their actual interaction with the media display, their interaction intention can be viewed as reasoned actions and planned behavior [32]. The conceptual model developed in this study focuses on the mediating role of the interactive built environment, design satisfaction and interaction intentions, which appear more circumscribed by behavioural theories rather than simply predicting if they will interact with the media display. In total, 200 written surveys were collected from members of the public visiting or residing in the One Central Park retail precinct. Over 330 people were asked to participate in the survey. Thus, the response rate was approximately 60%. The survey comprised a total of 26 questions. These questions were divided into three major categories: (1) interactive built environment; (2) perceptions of the media display; and (3) socioeconomics and demographics. The survey was designed to ascertain various human interactions with and perceptions of the media display. A media display was installed at One Central Park to increase visitation and support public engagement (e.g. socialising). It is important to ensure that the public is satisfied with the design of any new smart display and that they choose to interact with the screen. Consideration must also be given to how a high level of interaction with a smart media display can be sustained, especially in relation to permanent installations.

The survey questionnaire integrated the key influences of trust in a smart media interface, including the quality of the media content [54], the frequency of the displays [55], the methods of interaction [10] and users' satisfaction with the display design [56]. Users' perceptions of a media display can range from positive to negative and may include satisfaction or dissatisfaction with the content quality (five-point Likert scale), a desire for more or less frequent content change and a preference for one type of content over another. The following questions were used to measure respondents' perceptions of the media display:

- How would you rate the quality of the content shown on the media display?
- How frequently would you like to the media content/information to change?
- If you could add content using your smartphone, what would you like to display on the media display?
- How frequently would you like to the media content/information to change?
- My satisfaction with the design of media display is [ . . . .].

*4.3. Logistic Regression Model*

To examine all the potential aspects of the interaction decisions (the subject of this study), a logistic regression model was used to predict the likelihood of a respondent interacting with the smart media display. In relation to the logistic regression model, six missing cases had to be deleted; thus, a total of 194 respondents were included in the analysis. The dichotomous dependent variable selected was an individual's interaction decision (i.e., to interact [$n = 163$] or to not interact [$n = 31$]) with the smart media display. In relation to the media display's built environment, the variables chosen were socialising, safety, aesthetic appreciation, motivation to visit and consumer behaviours. To measure these variables, respondents were asked to use a five-point Likert scale (very high, high, neither high or low, low, very low satisfaction) to indicate their levels of satisfaction with the media display's environment. The 'content' variables pertaining to the media display itself included respondents' perceptions of the quality of the media display's content, their preferred frequency of content change, their content preference, their interaction method preference and their overall levels of satisfaction with the display design. At a personal level (i.e., the carrier level), a range of variables relating to respondents' levels of familiarity with smart technology (e.g., daily hours of internet usage, online shopping and confidence in technology use) and their socioeconomic and demographic characteristics (e.g., their age, gender, country of birth, residency, employment status) were selected.

The logistic regression model estimated the relative weight of each variable by the exponential of the beta coefficients (Exp β). Each coefficient shows an odds ratio as an indicator of the change in odds resulting from a unit change in an individual's interaction decision. For example, if an Exp β value is greater than 1, the likelihood of an individual interacting with a smart media display will increase, in accordance with the corresponding variable's contribution to the interaction decision. Conversely, if an Exp β value is less than 1, the likelihood of an individual interacting with a smart media display will decrease as the predictor increases. Thus, the Exp β values are interpreted as odds ratios that represent the likelihood of an individual intention of interacting with a smart media display given a particular outcome (where decisions to interact represent the independent variables in this study). The model employed a forced entry method whereby the coefficients for all selected independent variables were estimated in line with the decision-making process. The coefficients were then forced into the model simultaneously to report all coefficient estimates with a significance level for each predictor.

## 5. Findings

*5.1. Descriptive Analysis*

Table 2 outlines the socio-demographic profiles of the respondents. Of the respondents, 17.5% were residents and 82.5% were non-residents. Of those non-residents, 1.8% were first-time visitors, 23.9% were second-time visitors and 74.3% had visited the retail precinct multiple times over the past 12 months. The respondents were very culturally diverse. In relation to the country of the birth question, over 30 different countries were identified as respondents' places of birth. Of the respondents, 29.5% were born in Australia, 20.5% were born in South-East Asia, 12.5% were born in Europe and

12% were born in China. The majority of the respondents (i.e., 84.5%) were aged 18 to 30 years old and 53% were female. The following three variables were used to measure the familiarity with smart technology: (1) internet usage per day (measured in hours); (2) frequency of online shopping; and (3) level of confidence in using smart technology. Over 32% of respondents stated that they used the internet for 6 hours or more per day. Only 1.5% of respondents stated they used the internet for less than one hour per day. A large proportion of respondents indicated that they had purchased goods and services online (i.e., 94%). Further, 59% of respondents indicated that they had purchased goods or services online at least once in the last month.

*5.2. Characteristics of the Interactive Built Environment*

To examine the effects of the smart media display on respondents' interaction decisions, respondents were asked about their cognitive image in relation to five aspects of the media display's environment: (1) the smart screen's effect upon respondents' enjoyment of nearby social activities (i.e., socialising); (2) the smart screen's effect upon local safety and security; (3) the smart screen's contribution to respondents' aesthetic appreciation of the precinct; (4) the extent to which the smart screen motivated shoppers to visit the precinct; and 5) the extent to which the smart screen encouraged their consumption.

Of the respondents, 52% felt that the smart media display neither enhanced nor detracted from their social activities at One Central Park, 26% indicated that they would enjoy their social activities a little less if the media display did not exist and 6% said they would enjoy their social activities a lot less if the media display did not exist. Only 16% of respondents indicated that they would enjoy their social activities more if the media display did not exist. Thus, for almost a third of the respondents, the smart media display positively contributed to their enjoyment of social activities at One Central Park.

Of the respondents, 65% indicated that the level of safety and security they experienced would be the same if the media display did not exist. However, almost a quarter of the respondents viewed the smart media display could promote a feeling of safety. 18.5% of respondents were of the view that safety and security would be a little worse if the media display did not exist and 4.5% were of the view that it would be much worse.

A response of 'the same' attracted the highest number of responses from respondents in relation to four of the five questions about the media display's built environment (i.e., socialising, local safety and security, motivation to visit and ease of spending money). Thus, it appears that a majority of the respondents felt that the presence of the media display did not affect these aspects of their visit. However, when asked about aesthetics and appearance, 68.5% of respondents stated that the aesthetic of the place would be a little worse or much worse if the media display did not exist. This highlights the importance of the media displaying improving respondents' aesthetic and visual experience of the Central Park retail precinct.

Of the respondents, 69.5% were of the view that their motivation to visit One Central Park would be the same if the media display did not exist and 23% indicated that their motivation to visit would be a little less or much less if the media display did not exist. However, the results showed that for almost a quarter of the respondents to One Central Park, the media display informed their motivation to visit the retail precinct. The existence of the media display had the least effect on respondents' spending habits. Of the respondents, 75% were of the view that their spending at the Central park retail precinct would be the same whether or not the media display existed.

Table 2. Respondents' personal characteristics ($n = 200$).

| Attribute | Categories | Count | Percentage |
|---|---|---|---|
| Age (years) | 18–20 | 28 | 14% |
| | 21–30 | 141 | 70.5% |
| | 31–40 | 19 | 9.5% |
| | 41–50 | 7 | 3.5% |
| | 51–60 | 2 | 1% |
| | Above 60 | 1 | 0.50% |
| | Missing | 2 | 1% |
| Gender | Male | 94 | 47% |
| | Female | 106 | 53% |
| Country of birth | Australia | 59 | 29.5% |
| | South-East Asia | 41 | 20.5% |
| | Europe | 25 | 12.5% |
| | China | 24 | 12% |
| | UK | 13 | 6.5% |
| | South Korea | 11 | 5.5% |
| | New Zealand | 6 | 3% |
| | USA | 4 | 2% |
| | Middle East | 4 | 2% |
| | Other | 12 | 6% |
| Residency | Residents | 35 | 17.5% |
| | Non-residents | 165 | 82.5% |
| Employment | Full-time | 73 | 36.5% |
| | Part-time or casual | 59 | 29.5% |
| | Self-employed | 8 | 4.0% |
| | other | 60 | 30.0% |
| Visit frequency (visitors only $n = 165$) | First Time | 3 | 1.5% |
| | Second Time | 39 | 19.5% |
| | Multiple Times | 121 | 60.5% |
| | Missing | 2 | 1% |
| Time since last visit | In the last 6 months | 160 | 80% |
| | 6 to 12 months ago | 12 | 6% |
| | 1 to 2 years ago | 7 | 3.5% |
| | More than 2 years ago | 3 | 1.5% |
| | Do not know | 16 | 8% |
| | Missing | 2 | 1% |
| Daily internet usage (hours) | Less than one hour | 3 | 1.5% |
| | 1–2 hours | 18 | 9.0% |
| | 2–3 hours | 44 | 22.0% |
| | 4–5 hours | 50 | 25.0% |
| | 5–6 hours | 20 | 10.0% |
| | 6 hours or more | 65 | 32.5% |
| Frequency of online shopping | Never | 12 | 6.0% |
| | Less than once a month | 70 | 35.0% |
| | Once or twice a month | 71 | 35.5% |
| | Three to five times a month | 33 | 16.5% |
| | Six or more times a month | 14 | 7.0% |
| Confidence in computer technology | Very confident | 77 | 38.5% |
| | Somewhat confident | 78 | 39.0% |
| | Neutral | 39 | 19.5% |
| | Somewhat less confident | 3 | 1.5% |
| | Not at all confident | 2 | 1.0% |

Respondents' satisfaction levels with the design of the media display were very high. Indeed, 62.5% of respondents classed their levels of satisfaction as a very high or high. A third of respondents felt neutral about the media display and 4.5% of respondents surveyed felt low or very low levels of satisfaction. Respondents had high regard for the content shown on the media display. Specifically, 23% of respondents were of the view that mostly good content was available and 63% thought that some good content was available. Most of the respondents were of the view that the content should be changed frequently. Of the respondents, 60.5% stated they would like to see the content on the media display change every week while 19.5% stated they would like to see the content change every day. As the retail precinct has been designed to cater mostly to local residents, including those who live on site, this high demand for content change may reflect the frequency with which residents use this area of the building.

### 5.3. Modelling Interaction Decisions: A Logistic Regression Analysis

Guided by the conceptual framework, a binomial logistic regression analysis was undertaken. Table 3 sets out the results of the analysis that used the pre-defined variables in the research framework (see Figure 1).

Table 3. Odd ratios of logistic regression models (Exp β).

| Independent Variables | Model 1: Environment & Content (A + C) | Model 2: Carrier (B1 + B2) | Model 3: Composite (A + B1 + B2 + C) |
|---|---|---|---|
| Social activities | 1.079 | | 1.225 |
| Safety and security | 0.975 | | 1.210 |
| Aesthetic appearance | 0.378 ** | | 0.259 *** |
| Motivation to visit | 1.318 | | 1.274 |
| Consumer behaviour | 1.279 | | 1.553 |
| Quality of content | 4.410 ** | | 9.725 ** |
| Frequency of change in content | 0.789 | | 0.698 |
| Preferred content: advertising | 0.475 | | 0.424 |
| Preferred content: news | 1.375 | | 1.173 |
| Preferred content: video clips | 1.037 | | 0.954 |
| Preferred content: information | 0.789 | | 0.931 |
| Interaction methods: by messages/texts | 0.326 * | | 0.314 |
| Interaction methods: by images/graphics | 0.976 | | 0.794 |
| Interaction methods: by photographs | 1.346 | | 5.171 * |
| Interaction methods: by advertising | 0.456 | | 0.242 * |
| Satisfaction with design | 6.395 *** | | 10.816 *** |
| Hours of internet usage per day | | 0.986 | 0.955 |
| Experience with online shopping | | 0.876 | 0.417 * |
| Confidence with digital technology | | 0.819 | 0.637 |
| Age | | 0.997 | 1.066 |
| Country of Birth (non-Australian) | | 2.777 ** | 10.192 ** |
| Sex (male) | | 1.212 | 0.322 |
| Residency (resident) | | 1.391 | 7.061 |
| Employment status | | | |
| full-time | | 0.438 | 0.648 |
| part-time/casual | | 0.164 *** | 0.788 |
| self-employed | | 0.533 | 5.145 * |
| Constant | 0.003 | 0.464 | 0.001 |
| $n = 195$ | −2 Log likelihood = 103.208 | −2 Log likelihood = 159.312 | −2 Log likelihood = 82.475 |

Note: *** $p < 0.001$, ** $p < 0.01$, * $p < 0.05$.

The results of the three logistic regression models indicated odd ratios for all the independent variables affecting individuals' interaction decisions in relation to the built environment (i.e., the environment), individuals' perceptions of the media display (i.e., the content) and individuals' personal characteristics (i.e., the carrier). Model 1 (A + C) included the independent variables of the built environment (A) and the contents/design of media display (C) only (refer to Figure 1). Several intangible factors were closely associated with individuals' interaction decisions. Specifically, aesthetic appreciation, the quality of the content, interaction methods and levels of satisfaction with the media display's design significantly affected respondents' interaction decisions. Within the built environment group of variables (Model 1), the perception of aesthetic appearance was the most significant factor in determining interaction decisions. Respondents who were not of the view that the media display enhanced the aesthetic appearance of the public place were less likely to interact with the media display (Exp β = 0.378).

Design satisfaction with the media display was also a main driver of respondents" interaction decisions. Respondents who were satisfied with the media display's design were 10 times (Exp B = 10.816) more likely to interact with the media display than those were unsatisfied. Respondents who chose personal or text messaging as their preferred method of interaction were less likely to interact with the media display (Exp β = 0.269) than others. This may be attributable to these respondents' perception that the media display had only simplistic uses and they are being mainly interested in verbal communication.

In Model 2 (see Figure 1(B1 + B2)), several socio-demographic and technology familiarity variables were associated with the respondents who interacted with the media display. The variable of the ethnic background was significantly associated with interactions. Compared to Australian born respondents (the reference category), respondents who were born overseas were 2.7 times (Exp β = 0.269) more likely to interact with the media display. However, other socio-demographic characteristics (e.g., age, gender, residency and employment status) did not significantly affect the likelihood that respondents would interact with the media display. Most significantly, the likelihood of interaction was lower for part-time and casual workers (Exp β = 0.164) than for those with other employment statuses.

The overall model predictability of the composite model, which considered the variables of environment, personal characteristics and perceptions (Model 3, see Figure 1(A+B1+B2+C)), showed a significant improvement. The log likelihood revealed the extent to which observations remained unexplained after each model had been applied; a larger value of the log likelihood ratio indicated higher unexplained interaction decisions. The result of the –2 LL ratio in the composite model (Model 3) was 82.475, much lower than that of the other models (Model 1 and 2), indicating an improvement in the model's predictivity when the other two models were combined in the interaction decision process. This showed that individuals' socioeconomic and demographic characteristics and their familiarity with smart technology were also important predictors in their interaction decisions.

Respondents who believed that the precinct's aesthetic appearance would be significantly worse without the media display were also significantly more likely to interact with it (Exp β = 0.259). The quality of content available on the media display was another important contributor to the likelihood of interaction (Exp β = 9.725). Respondents who were born overseas, especially those born in South-East Asia, were 7.8 times more likely to interact with the smart media display than those born in Australia. Surprisingly, age and residency did not significantly affect respondents' decisions to interact with the display.

Decisions to display personal photographs/images or work-related advertisements as a preferred interaction method proved to be a significant predictor of respondents' decisions to interact with the media display. The interaction method that included digital photographs increased the likelihood of interactions; however, the respondents who chose to display an advertisement were less likely to interact with the media display. Notably, all the independent variables relating to familiarity with technology (e.g., daily hours of internet usage, levels of confidence with digital technology and online shopping experience) did not increase the likelihood that the respondents would interact with the media display. The results suggest that content quality, levels of satisfaction with the display design of the media display, its aesthetic appearance, respondents' preferred interaction methods and their ethnic backgrounds, were the key independent variables that differentiated between the respondents who interacted with the media display and those who did not (at $p < 0.05$).

## 6. Discussion and Conclusions

The complexities of public display designs and HCI interfaces appear to be hindering the development of smart media display or façades in smart cities; however, opportunities still exist for better understanding individuals' perceptions of media display and the variables that influence individuals' interaction decisions. Documentation of such encounters have raised an important question: Would new smart media display be more successful and socially accepted if they presented more interactive opportunities in smart cities?

Decisions to interact with media display necessitate a new standard for smart media display design and media content development in smart cities. Specifically, consideration must be given to the built environment (i.e., the environment) and the socio-demographic characteristics of individuals (i.e., the carriers). It should not be assumed that new smart technologies will resolve or address discontinuities of HCIs in smart cities. Indoor media displays adjacent to main building entries and gathering points can effectively leverage HCI based on their proximity to retail shops; however, the smart media display may not be able to remove the influence of the perceptions created by the surrounding built environment (i.e., perceptions about safety and the aesthetics of a place) in the precinct.

The perceived built environment (i.e., individuals' perceptions of safety and aesthetic quality) can influence individuals' interaction decisions in smart cities. Thus, the socioeconomic and demographic characteristics of engagers should influence both the cognitive and affective images displayed in any smart media display (i.e., the content). Individuals' personal familiarity with smart technology was not found to significantly affect their interaction decisions. Further, individuals were more likely to interact with the smart media display if a variety of interaction methods were offered (e.g., the possibility of using digital images or digital twins rather than simple text messages).

Socio-demographic factors still play a mediating role in encouraging or discouraging human interactions with smart media displays. Gender, age and residency are commonly believed to play an important role in determining interaction decisions, but the related independent variables considered by this study's models proved non-significant. It may be attributed to such structural variables (gender, age and residency) largely play in an independent manner to predict interaction behaviours while relationships between satisfaction and interaction are closely related. The relative significance of specific ethnicities was not explored in this study; however, in the descriptive analysis respondents with South-East Asian backgrounds would interact with the new smart media display considered in the present study. As the majority of South-East Asian (90.2%) is a visitor to the site the underlying cause might be that they are a tourist who are more likely to interact with the smart technologies.

An individual's digital familiarity was considered in relation to particular individual attributes; however, it could arguably be better established by considering an individual's social network interactions on Facebook, Twitter, LinkedIn, etc. Given that digitally familiar people now spend a large amount of time interacting with digital screens, the issue of saturation arises, as such individuals may lack the desire to interact with other devices. Thus, the extent of individuals' screen exhaustion may negate other effects relating to different levels of technological experience. This issue should be the subject of an extended behavioral study in smart cities.

This study is one of the first studies empirically identified the key aspects of a smart display "media façade" in relations to the enhancement of intuitive interactions with a new generation of public display. This study, therefore, has important implications for planners, architects, engineers and policymakers seeking to deploy creative media façade and innovative HCI methods in smart cities. We argued that individuals' perceptions of the built environment (i.e., safety) and their personal characteristics significantly mediate the relationship between the perceived image of a media display and individuals' interaction decisions. However, the implication of this intention for actual interaction may differ if the respondents have interacted in different time, date and weather. Further studies therefore need to be gained of actual interaction behaviors with newly implemented smart technologies and media contents and investigate their social interaction patterns if smart media displays are to become more effective at achieving their intended objectives. A semi-structured interview with end-users helps understand which the attributes of 'environment', 'carrier' and 'content' affect the extent to which intentions are translated into actual interaction.

This paper found addressing the needs of users with particular socio-demographic characteristics and attracting visitors to a smart building represent particular challenges. Interestingly, cultural differences (defined in this study as country of birth) had a varying effect on individuals' interaction decision, indicating that more cultural content should be featured in the displays. Many existing media display programs do not consider cultural differences in interaction decisions. However, the inclusion of multicultural content and international language services provided by smart technology (real-time translation), not least for Chinese, Indonesian and French visitors to One Central Park, could significantly enhance interaction decisions, especially for visitors or tourists. The findings suggest that a one-size-fits-all approach to content should not be adopted when decisions are being made to install or vary the multicultural content of a media display. Clearly, smart media display installations can be successful in culturally mixed communities and may also be successful at international tourist destinations in Sydney.

Media displays are typically developed for specific temporal and spatial contexts. Thus, many opportunities and challenges arise in exploring when they should be implemented. Issues of spatial organisation and built environment will be unique to each individual smart display project; however, guidelines or principles could be developed to ensure that a better approach to the design and content development of specific media façades is adopted (Dalsgaard and Halskov [20]). Higher levels of satisfaction with the design of a smart media display may also motivate individuals to (re)visit a place while the quality of the display content could also encourage viewers to remain in a space, and thus affect their interaction decisions.

The findings of this paper further suggest that a need for smart technologies for safe and secure place and an appreciation of the aesthetics of a smart media display affects people's decisions to interact with a smart installation. Thus, the creation of a safe built environment in the display zone and an aesthetically appealing display should be a design priority rather than providing socialising place. Sustaining user interest through the provision of high-quality content and interaction opportunity, especially if a media screen is intended to be displayed more permanently, is likely to present an ongoing challenge for a smart city planning.

**Author Contributions:** The research is initially designed and proposed by H.H.; The survey design and data collection are conducted by H.H. and S.H.L.; The manuscript is further improved by S.H.L. and Y.L.

**Funding:** This research is funded by the Architecture & Urban Development Research Program Gant (13AUDP-B070066-01), the Ministry of Land, Infrastructure and Transport of the Korean Government.

**Conflicts of Interest:** The authors declare no conflict of interest.

## References

1. Barker, T.; Haeusler, M.; Beilharz, K.A. *Interactive Polymedia Pixel and Protocol for Collaborative Creative Content Generation on Urban Digital Media Displays*; Marmara University: Istanbul, Turkey, 2010.
2. Hespanhol, L.; Dalsgaard, P. Social Interaction Design Patterns for Urban Media Architecture. In Proceedings of the INTERACT 2015 15th IFIP TC 13 International Conference on Human-Computer Interaction, Bamberg, Germany, 14–18 September 2015; Abascal, J., Barbosa, S., Fetter, M., Gross, T., Palanque, P., Winckler, M., Eds.; Springer: Berlin/Heidelberg, Germany, 2015; pp. 596–613, ISBN 978-3-319-22697-2.
3. Haeusler, M.H.; Tomitsch, M.; Tscherteu, G.; van Berkel, B. *New Media Facades. A Global Survey*; Avedition: Ludwigsburg, Germany, 2012; ISBN 3899861701.
4. Brignull, H.; Rogers, Y. Enticing People to Interact with Large Public Displays in Public Spaces. In Proceedings of the INTERACT '03, IFIP TC13 International Conference on Human-Computer Interaction, Zurich, Switzerland, 1–5 September 2003; Rauterberg, M., Menozzi, M., Wesson, J., Eds.; IOS Press: Amsterdam, The Netherlands; Oxford, UK, 2003; pp. 17–24, ISBN 4274906140.
5. Hespanhol, L.; Tomitsch, M. Strategies for Intuitive Interaction in Public Urban Spaces. *Interact. Comput.* **2015**, *27*, 311–326. [CrossRef]

6. Behrens, M.; Valkanova, N.; Gen Schieck, A.F.; Brumby, D.P. Smart Citizen Sentiment Dashboard: A Case Study into Media Architectural Interfaces. In Proceedings of the Third International Symposium on Pervasive Displays, Copenhagen, Denmark, 3–4 June 2014; Boring, S., Ed.; ACM: Copenhagen, Denmark, 2014; pp. 19–24, ISBN 978-1-4503-2952-1.
7. Michelis, D.; Müller, J. The Audience Funnel: Observations of Gesture Based Interaction with Multiple Large Displays in a City Center. *Int. J. Hum. Comput. Interact.* **2011**, *27*, 562–579. [CrossRef]
8. Valkanova, N.; Walter, R.; Vande Moere, A.; Müller, J. My Position: Sparking Civic Discourse by a Public Interactive Poll Visualization. In Proceedings of the 17th ACM Conference on Computer Supported Cooperative Work & Social Computing, Baltimore, MD, USA, 15–19 February 2014; Fussell, S., Ed.; ACM Press: New York, NY, USA, 2014; pp. 1323–1332, ISBN 9781450325400.
9. Antle, A.N.; Corness, G.; Droumeva, M. Human-computer-intuition? Exploring the cognitive basis for intuition in embodied interaction. *Int. J. Arts Technol.* **2009**, *2*, 235. [CrossRef]
10. Hespanhol, L.; Tomitsch, M. Designing for Collective Participation with Media Installations in Public Spaces. In Proceedings of the 4th Media Architecture Biennale Conference Participation, Aarhus, Denmark, 15–17 November 2012; Dalsgaard, P., Brynskov, M., Schieck, A.F., Eds.; ACM Press: New York, NY, USA, 2012; pp. 33–42, ISBN 9781450317924.
11. Blackler, A.L.; Hurtienne, J. Towards a unified view of intuitive interaction: definitions, models and tools across the world. *MMI-Interakt.* **2007**, *13*, 36–54.
12. Jacucci, G.; Wagner, M.; Wagner, I.; Giaccardi, E.; Annunziato, M.; Breyer, N.; Hansen, J. ParticipArt: Exploring participation in interactive art installations. In Proceedings of the 2010 IEEE International Symposium on Mixed and Augmented Reality—Arts, Media, and Humanities, Seoul, Korea, 13–16 October 2010.
13. Dindler, C. Designing infrastructures for creative engagement. *Digit. Creat.* **2014**, *25*, 212–223. [CrossRef]
14. Brynskov, M.; Dalsgaard, P.; Ebsen, T.; Fritsch, J.; Halskov, K.; Nielsen, R. Staging Urban Interactions with Media Façades. In Proceedings of the 12th IFIP TC 13 International Conference on Human-Computer Interaction, Uppsala, Sweden, 24–28 August 2009; Gross, T., Ed.; Springer: Berlin, Germany, 2009; pp. 154–167, ISBN 978-3-642-03654-5.
15. Yigitcanlar, T.; Sabatini-Marques, J.; Da-Costa, E.; Kamruzzamana, M.; Ioppoloc, G. Stimulating technological innovation through incentives: Perceptions of Australian and Brazilian firms. *Technol. Forecast. Soc. Chang.* **2017**. [CrossRef]
16. Fischer, P.T.; Zöllner, C.; Hoffmann, T.; Piatza, S.; Hornecker, E. Beyond information and utility: Transforming public spaces with media facades. *IEEE Eng. Med. Biol. Mag.* **2013**, *33*, 38–46. [CrossRef]
17. Park, J.W. Interactive Kinetic Media Facades: A Pedagogical Design System to Support an Integrated Virtual-Physical Prototyping Environment in the Design Process of Media Facades. *J. Asian Arch. Build. Eng.* **2013**, *12*, 237–244. [CrossRef]
18. Javadi, N.; Dağlı, U. Media Facades Utilization for Sustainable Tourism Promotion in Historic Places: Case Study of the Walled City of Famagusta, North Cyprus. *Int. J. Humanit. Soc. Sci.* **2016**, *10*, 431–438. [CrossRef]
19. Struppek, M. The social potential of Urban Screens. *Vis. Commun.* **2006**, *5*, 173–188. [CrossRef]
20. Dalsgaard, P.; Halskov, K. Designing Urban Media Façades: Cases and Challenges. In Proceedings of the CHI 2010 the 28th Annual CHI Conference on Human Factors in Computing Systems, Atlanta, GA, USA, 10–15 April 2010; Mynatt, E., Schoner, D., Fitzpatrick, G., Hudson, S., Edwards, K., Rodden, T., Eds.; Association for Computing Machinery: New York, NY, USA, 2010; p. 2277, ISBN 9781605589299.
21. O'Hara, K.; Glancy, M.; Robertshaw, S. Understanding Collective Play in an Urban Screen Game. In Proceedings of the 2008 ACM Conference on Computer Supported Cooperative Work, San Diego, CA, USA, 8–12 November 2008; Begole, B., McDonald, D.W., Eds.; Association for Computing Machinery: New York, NY, USA, 2008; p. 67, ISBN 9781605580074.
22. Wiethoff, A.; Gehring, S. Designing Interaction with Media Façades: A Case study. In Proceedings of the Designing Interactive Systems Conference, Newcastle Upon Tyne, UK, 11–15 June 2012; ACM: New York, NY, USA, 2012.
23. Blackler, A.L.; Popovic, V.; Mahar, D.P.; Reddy, R.; Lawry, S. Intuitive Interaction and Older People. In Proceedings of the Design Research Society (DRS) 2012 Conference, Bangkok, Thailand, 1–4 July 2012; Israsena, P., Tangsantikul, J., Durling, D., Eds.; Chulalongkorn University: Bangkok, Thailand, 2012; pp. 560–578, ISBN 6165515746.

24. Blackler, A.L.; Popovic, V.; Mahar, D.P. Applying and testing design for intuitive interaction. *Int. J. Des. Sci. Technol.* **2014**, *20*, 7–26.
25. Lottridge, D.; Chignell, M.; Jovicic, A. Affective Interaction: Understanding, Evaluating, and Designing for Human Emotion. *Rev. Hum. Factors Ergon.* **2011**, *7*, 197–217. [CrossRef]
26. Arnold, M.B. Emotion and Personality. In *Volume I. Psychological Aspects*; Columbia University Press: Oxford, UK, 1960.
27. Marks, I.F.; Nesse, R.M. Fear and fitness: An evolutionary analysis of anxiety disorders. *Evol. Hum. Behav.* **1994**, *15*, 247–261. [CrossRef]
28. Fischer, P.T.; Hornecker, E. Urban HCI: Spatial Aspects in the Design of Shared Encounters for Media Façades. In Proceedings of the 30th ACM Conference on Human Factors in Computing Systems (CHI), Austin, TX, USA, 5–10 May 2012; Konstan, J.A., Chi, E.H., Höök, K., Eds.; Association for Computing Machinery: New York, NY, USA, 2012; p. 307, ISBN 9781450310154.
29. Wouters, N.; Downs, J.; Harrop, M.; Cox, T.; Oliveira, E.; Webber, S.; Vetere, F.; Vande Moere, A. Uncovering the Honeypot Effect. In Proceedings of the 2016 ACM Conference on Designing Interactive Systems (DIS 2016), Brisbane, Australia, 4–8 June 2016; Foth, M., Ju, W., Schroeter, R., Viller, S., Eds.; ACM: New York, NY, USA, 2016; pp. 5–16, ISBN 9781450340311.
30. Ajzen, I.; Fishbein, M. *Understanding Attitudes and Predicting Social Behavior*; Prentice-Hall: Englewood Cliffs, NJ, USA; London, UK, 1980; ISBN 0139364439.
31. Suchman, L.A. *Plans and Situated Actions. The Problem of Human-Machine Communication*; Cambridge University Press: Cambridge, UK, 1987; ISBN 0521331374.
32. Payne, S.J.; Howes, A.; Reader, W.R. Adaptively distributing cognition: A decision-making perspective on human— Computer interaction. *Behav. Inf. Technol.* **2001**, *20*, 339–346. [CrossRef]
33. Anderson, J.R. *The Adaptive Character of Thought*; Psychology Press: London, UK, 1990; ISBN 9780203771730.
34. Bless, H.; Fiedler, K.; Strack, F. Social Cognition. In *How Individuals Construct Social Reality*; Psychology Press: Hove, UK, 2002; ISBN 0863778291.
35. Slovic, P. *The Perception of Risk*; Earthscan Publications: London, UK; Sterling, VA, USA, 2000; ISBN 9781853835278.
36. Jorgensen, A.; Hitchmough, J.; Calvert, T. Woodland spaces and edges: their impact on perception of safety and preference. *Landsc. Urban Plann.* **2002**, *60*, 135–150.
37. Tacon, P. Socialising Landscapes: The Long-Term Implications of Signs, Symbols and Marks on the Land. *Archaeol. Ocean.* **1994**, *29*, 117–129. [CrossRef]
38. Enli, G.; Syvertsen, T. Participation, play and socializing in new media environments. *New Media Worlds Chall. Converg.* **2007**, 147–162.
39. Eben Saleh, M.A. The architectural form and landscape as a harmonic entity in the vernacular settlements of Southwestern Saudi Arabia. *Habitat Int.* **2000**, *24*, 455–473. [CrossRef]
40. Locher, P.; Overbeeke, K.; Wensveen, S. Aesthetic Interaction: A framework. *Des. Issues* **2010**, *26*, 70–79. [CrossRef]
41. Dishman, R.K.; Ickes, W.; Morgan, W.P. Self-motivation and adherence to habitual physical activity. *J. Appl. Soc. Psychol.* **1980**, *10*, 115–132.
42. Chew, E.; Jahari, S. Destination image as a mediator between perceived risks and revisit intention: A case of post-disaster Japan. *Tour. Manag.* **2014**, *40*, 382–392. [CrossRef]
43. Wadley, D.; Elliott, P.; Han, H. Installing large-scale community infrastructure: Homeowners' preferences toward notification and recourse. *Community Dev.* **2017**, *21*, 1–17. [CrossRef]
44. Cikic-Tovarovic, J.; Sekularac, N.; Ivanovic-Sekularac, J. Specific problems of media facade design. *Facta Univ. Ser. Archit. Civ. Eng.* **2011**, *9*, 193–203. [CrossRef]
45. Bless, H.; Fiedler, K.; Strack, F. *Social Cognition: How Individuals Construct Social Reality*; Psychology Press: Hove, UK, 2003.
46. Schroeter, R. Engaging new digital locals with interactive urban screens to collaboratively improve the city. In Proceedings of the ACM 2012 Conference on Computer Supported Cooperative Work (CSCW'12). ACM 2012 Conference, Seattle, WA, USA, 11–15 February 2012; Poltrock, S., Simone, C., Grudin, J., Mark, G., Riedl, J., Eds.; Association for Computing Machinery: New York, NY, USA, 2012; p. 227, ISBN 9781450310864.

47. Vande Moere, A.; Wouters, N. The Role of Context in Media Architecture. In Proceedings of the 2012 International Symposium on Pervasive Displays 2012, Porto, Portugal PerDis, 4–5 June 2012; Huang, E.M., José, R., Eds.; ACM Press: New York, NY, USA, 2012; pp. 1–6, ISBN 9781450314145.
48. Gasparini, K. Media-surface design for urban regeneration: the role of colour and light for public space usability. *J. Int. Colour Assoc.* **2017**, *17*, 38–49.
49. Kim, Y. Impacts of the perception of physical environments and the actual physical environments on self-rated health. *Int. J. Urban Sci.* **2016**, *20*, 1–15. [CrossRef]
50. Frasers Property. Central Park Sydney: Once Upon a Time. Available online: https://www.centralparksydney.com/explore/chippendale-a-rich-heritage (accessed on 7 July 2019).
51. Ramus Illumination. The Digital Wall. Available online: http://ramus.com.au/project/portfolio-post-with-video/ (accessed on 3 June 2019).
52. Visentin, L. Digital Canvas Brings Art to Commerce at Shopping Centre. Sydney Morning Herald. Available online: https://www.smh.com.au/entertainment/art-and-design/digital-canvas-brings-art-to-commerce-at-shopping-centre-20140604-zrww2.html (accessed on 4 June 2019).
53. Holder, C. Digital Dexterity: The Digital Wall at Central Park is not your average shopping mall digital signage. Available online: http://www.digitalsignagemagazine.com.au/wp/index.php/digital-dexterity/ (accessed on 6 July 2019).
54. Al-Azhari, W.; Haddad, L.; Al Absi, M. Large Interactive Media Display and Its Influence on Transformation Urban Spaces from Neglecting to Action: The Case of Al-Thaqafa Street in Amman City. *J. Softw. Eng. Appl.* **2014**, *7*, 817–827. [CrossRef]
55. Sato, M.; Suzuki, Y.; Hiyama, A.; Tanikawa, T.; Hirose, M. Particle Display System-A Large Scale Display for Public Space. In Proceedings of the Joint Virtual Reality Conference of EGVE—the 15th Eurographics Symposium on Virtual Environments—ICAT-EuroVR, Lyon, France, 7–9 December 2009; pp. 29–36.
56. Lindgaard, G.; Dudek, C. What is this evasive beast we call user satisfaction? *Interact. Comput.* **2003**, *15*, 429–452. [CrossRef]

© 2019 by the authors. Licensee MDPI, Basel, Switzerland. This article is an open access article distributed under the terms and conditions of the Creative Commons Attribution (CC BY) license (http://creativecommons.org/licenses/by/4.0/).

Article

# Systematic Integration of Energy-Optimal Buildings With District Networks

Raluca Suciu *, Paul Stadler, Ivan Kantor, Luc Girardin and François Maréchal

Industrial Process and Energy Systems Engineering (IPESE), École Polytechnique Fédérale de Lausanne, CH-1951 Sion, Switzerland
* Correspondence: raluca-ancuta.suciu@epfl.ch

Received: 21 June 2019; Accepted: 25 July 2019; Published: 31 July 2019

**Abstract:** The residential sector accounts for a large share of worldwide energy consumption, yet is difficult to characterise, since consumption profiles depend on several factors from geographical location to individual building occupant behaviour. Given this difficulty, the fact that energy used in this sector is primarily derived from fossil fuels and the latest energy policies around the world (e.g., Europe 20-20-20), a method able to systematically integrate multi-energy networks and low carbon resources in urban systems is clearly required. This work proposes such a method, which uses process integration techniques and mixed integer linear programming to optimise energy systems at both the individual building and district levels. Parametric optimisation is applied as a systematic way to generate interesting solutions for all budgets (i.e., investment cost limits) and two approaches to temporal data treatment are evaluated: monthly average and hourly typical day resolution. The city center of Geneva is used as a first case study to compare the time resolutions and results highlight that implicit peak shaving occurs when data are reduced to monthly averages. Consequently, solutions reveal lower operating costs and higher self-sufficiency scenarios compared to using a finer resolution but with similar relative cost contributions. Therefore, monthly resolution is used for the second case study, the whole canton of Geneva, in the interest of reducing the data processing and computation time as a primary objective of the study is to discover the main cost contributors. The canton is used as a case study to analyse the penetration of low temperature, $CO_2$-based, advanced fourth generation district energy networks with population density. The results reveal that only areas with a piping cost lower than 21.5 k€/100 $m^2_{ERA}$ connect to the low-temperature network in the intermediate scenarios, while all areas must connect to achieve the minimum operating cost result. Parallel coordinates are employed to better visualise the key performance indicators at canton and commune level together with the breakdown of energy (electricity and natural gas) imports/exports and investment cost to highlight the main contributors.

**Keywords:** optimal cities; energy autonomy; low-carbon resources; multi-energy networks; parametric optimisation; $CO_2$ networks

## 1. Introduction

Increasing population, urbanization and rapid industrialization corresponds to parallel and continuous increases in world energy demand, where up to 65% of the energy consumption comes from urban areas [1]. While the consumption of major sectors, such as commercial, industrial, transportation and agriculture are relatively well-understood due to their centralized ownership, self-interest in reducing the energy consumption and high level of regulation, the residential sector is an energy sink which is difficult to characterize, since it encloses a large variety of geometries, structure sizes and envelope materials. At the same time, privacy concerns restrict energy consumption data collection and distribution and detailed metering of households bears high costs. Nevertheless, Pachauri et al. reports that there is a great potential to achieve significant reductions in energy consumption, mainly in the building sector, at a relatively modest cost [2], which highlights the requirement to better understand the defining characteristics of energy consumption in this sector.

Major end-use energy consumption groups in the residential sector are: space heating and cooling, energy required to overcome thermal flows through the building envelope, by conduction, radiation and through air infiltration/ventilation; domestic hot water-energy consumed to heat water to the comfort temperature; appliances and lighting-energy needed to operate appliances (e.g., refrigerator, electronics) and for supplying appropriate lighting. Fossil fuels are currently the main energy sources to supply these demands [3]; however, they have a high environmental impact and limited reserves which also correspond to fluctuating prices, which affects national economies and results in a prominent interest in using renewable energy sources. Renewable energy comes from a variety of sources, such as biomass, geothermal heat, ocean waves, sun, tides, water and wind. Hybrid (i.e., multi-source) renewable energy systems are favored over single sources since they are more reliable, more efficient, require less energy storage capacity and have lower levelized life cycle electricity generation cost under optimum design [4]. Multi-source generation makes hybrid system solutions complex, thus a techno-economic analysis of these systems is essential to ensure the optimal use of renewable sources. This, in turn, requires models and software which can be employed for design, optimization and techno-economic planning.

Another dilemma that arises with integration of renewable energies is the mismatch between renewable energy supply and demand profiles in the residential sector, which is often pronounced and requires extensive storage solutions [5]. Heat storage solutions already exist at small scale in individual buildings and via district heating networks (DHNs) in large bore-hole storage systems. Alternative solutions exist for multi-energy systems, such as power-to-gas, fuel cells, electric/hydrogen mobility and large scale batteries [6,7].

Balancing energy demand and supply both spatially and temporally can be modeled using computational methods, such as mathematical programming, among which linear programming techniques have been used to optimize multi-energy systems for more than thirty years [8]. Generally, there is a separation of topics in residential energy system analysis based on the scale, namely: individual building scale and urban scale. The former focuses principally or solely on the building itself and omits any relationship with the urban environment. It treats a building as an independent object, isolated from the built environment; however, real buildings are connected to their surroundings through physical means (infrastructure) and users (residents, workers). The latter scale focuses on the entire system, often without details at the building scale. Therefore, there are improvements to be made by coupling building-level models with those at the urban level while also using detailed equipment models (e.g., energy conversion technologies, heat pumps (HP)). Linking buildings with district systems requires tools for design, sizing, operation and control of energy system components, buildings and district networks. An even larger challenge, though, is to provide simple tools, which can aid decision-makers at an early stage in the design process at both the building and urban levels.

This paper proposes a double-optimisation approach with meta-models [9,10] for the design and optimisation of urban systems at building and urban levels, with interaction between the two scales, including renewable energy integration and long-term energy storage solutions. The connection between the building and the urban level is realised through a low-temperature $CO_2$ district energy network and meta-models models are used to integrate building solutions into the district optimisation. Therefore, this paper contributes a novel approach for optimal design of urban energy systems, coupling optimal solutions for individual buildings with the larger energy system to provide guidance for holistic urban energy system design. Additionally, this work provides unique insights into various objectives of such systems and the inherent balance between them, providing a set of optimal solutions to be ultimately selected by decision-makers. Section 2 reviews the main tools and approaches currently employed for this purpose and their limitations, Section 3 presents the mathematical formulation and the case studies considered, Section 4 shows the results and conclusions are drawn in Section 5.

## 2. State of the Art

Energy use in the residential sector has been studied extensively, across a variety of fields, such as civil engineering, architecture, economy, environmental assessment, sociology, transport, city and regional planning. Energy consumed in this sector is generally classified as either embodied or operational. Embodied energy is the energy required to produce and transport materials to the construction site and for the construction process itself, while operational energy is consumed for the daily use of the building to provide electricity, water, hot water, ventilation, heating and cooling.

A clear distinction in the scale of the analysis arises when trying to summarise the research in the area, namely at the individual building and urban scales [1]. Research at the individual building scale usually covers topics such as building materials used, architectural design, structural and operational system and construction. Developments in the area include improving the accuracy of the models and reducing the computation time of the assessment [11], analysing the results with different objectives [12] techniques to reduce energy and $CO_2$ emissions. Kofoworola et al. showed a combination of energy savings measures to reduce the electricity consumption in a typical office building in Thailand by 40–50% [13]. Ochoa et al. stated that the usage phase of buildings accounts for the largest share of the energy use and environmental impact, followed by the construction phase, while the disposal phase is negligible from both perspectives [14]. Junilla et al. presented the elements in the life-cycle assessment of office buildings which cause the highest emissions and should therefore be targeted for improvement [15] and in a similar study concluded that lighting, HVAC systems and outlets, manufacturing and maintenance of steel, manufacturing of concrete and paint and water use have the largest environmental impacts in office buildings [16].

The second scale of analysis for energy use in the residential sector is the urban scale. Research at this scale typically covers topics such as urban form, density, transportation, infrastructure and consumption. Studies in the field focus mainly on quantification of energy use, transportation infrastructure, water infrastructure, construction, and modeling of energy use in urban systems. Glaeser and Kahn studied the energy use and environmental impact due to driving, public transit, providing heating and electricity in households and found a strong negative correlation between emissions and land use regulations, leading them to conclude that cities have significantly lower emissions compared to suburban areas[17]. Kennedy et al. performed a study on ten global cities, showing correlations between public transit quality and personal income, and between heating and industrial fuel use [18]. Troy et al. quantified the embodied energy in urban areas and found it to be more significant than previously supposed and suggested that knowing the embodied energy consumed can be used for control tool development [19].

Jones et al. assessed energy consumption and environmental impact in urban areas due to transportation, energy, water, waste, food, goods and services, and suggested that results were highly dependent on the basic demographic characteristics of the area studied [20]. Regarding energy use modeling, Howard et al. developed a model to estimate end-use energy intensity in New York, as a tool for cost-efficient policies regarding renewable energy efficient solutions [21]. Gurney et al. used simulation tools, traffic data, power production reporting and local air pollution reporting to build a model which quantified $CO_2$ emissions across the city of Indianapolis [22]. Keirstead et al. reviewed approximately 220 papers on urban energy system modeling and concluded that the four most common challenges are data quality and uncertainty, model integration, model complexity and policy relevance [23]. They also concluded that urban energy system models have a significant potential of moving toward a more integrated perspective, which could capture their intricacies.

While these references offer a first insight into multi-scale integration analysis, additional methodological developments are required to directly address the interaction between scales. In view of that, this paper proposes a method which combines the work of Stadler et al. [24] on building optimisation at building level with the work of Suciu et al. [25] on optimisation at district level, to perform a detailed multi-level energy integration optimisation. The link between the building and the urban level is realised through a low-temperature district heating and cooling network and meta-models are employed to embed the building solutions into the district level optimisation.

*Low Temperature DHC Networks*

Low temperature district energy networks (DENs) provide a low temperature source, which can be used for heating via decentralized heat pumps, directly for cooling, indirectly as a low temperature source for chillers and can recover waste heat from processes and other buildings in the proximity; they are also often linked to large seasonal storage in the form of borehole fields [26].

Low-temperature networks have been discussed in the literature, for example, De Carli et al. performed an energo-economic analysis of a small-scale, low-temperature district heating and cooling network in Italy [27], Bestenlehner compared a low-temperature and a conventional district heating network in a quarter of Stuttgart [28], Ruesch modeled the time evolution of large borehole fields connected to low temperature district heating networks [26], Kräuchi et al. modelled a low-temperature district heating and cooling network using the IDA indoor climate and energy (IDA ICE) simulation software [29] and Molyneaux et al. performed an enviro-economic optimisation for low-temperature heat networks with heat pumps [30].

This work analyzes both conventional networks and low temperature refrigerant ($CO_2$)-based networks. Weber and Favrat introduced the idea of distributing $CO_2$ in the district energy networks at a temperature below the critical pressure of 74 bar. $CO_2$ networks (Figure 1) use a double-pipe system to deliver both heating and cooling services. A pressure of 50 bar is suggested for use in the network to remain within the saturation temperature range of 12–18 °C, which allows network operations to leverage the latent heat and small pressure difference between liquid and gas phases to provide cooling services by gas expansion. Unlike water-based networks currently in place in several cities, $CO_2$ networks use phase change to realize the heat transfer and allow cooling services to provide heating, which is not possible with conventional systems. The approach is based on a $CO_2$ "closed-loop" concept, i.e., except for leaking (considered negligible) no $CO_2$ enters/leaves the network.

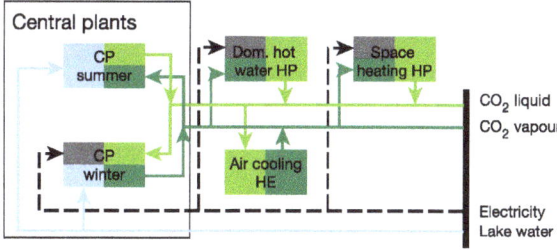

**Figure 1.** $CO_2$ network schematic representation.

$CO_2$ networks have also been integrated with advanced technologies for energy storage and heat integration, such as power-to-gas [25]. Power-to-gas systems use electricity in periods of high production (summer) to produce hydrogen and oxygen by water electrolysis and then methane in a Sabatier reaction, which is stored to provide electricity and heat during cold periods or periods of low electricity production (Figure 2). The waste heat of the co-generation system is first used in a steam network to produce electricity with the remaining low temperature heat used to vaporize $CO_2$, which is used to provide heating services.

This work proposes a method which links analysis and optimisation in individual buildings with urban-level systems through low temperature $CO_2$ networks and long term power-to-gas storage systems. The method proposed uses a double optimisation approach with surrogate models, using two different time scales: monthly averages and typical days.

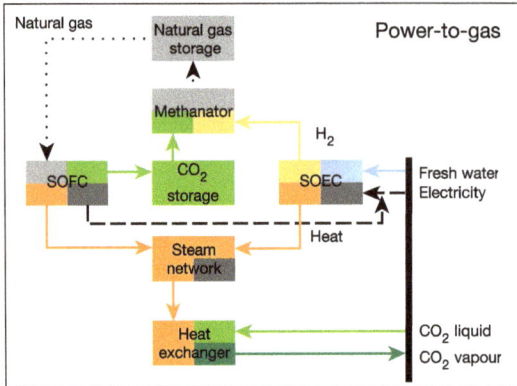

**Figure 2.** P2G schematic representation.

## 3. Materials and Methods

*3.1. Methodology Overview*

The proposed method models energy systems using a double optimisation approach with meta-models (Figure 3). The first optimization is performed at the building level, where different utilities can be chosen, such as photo-voltaic (PV) panels with short-term electricity storage (batteries), $CO_2$ and air-water heat pumps, co-generation units, heat storage tanks, domestic hot water tanks, heat exchangers (HEs) for cooling and electrical heaters as back up systems. The PV panels and co-generation units are described in detail in Appendix A.2, while the other units are described in Appendix A.1. Further details on the formulation can be found in [24]. This optimization is performed to ensure that each building is operated optimally, e.g., all the controllable loads are shifted to decrease the operating cost.

The buildings considered are residential (single- and multi-family houses), mixed (residential and administrative), administrative, commercial, education and hospitals. The buildings are also grouped according to the renovation stage, as existing (built before 2005), new (built after 2005) and renovated (built before 2005 but improved to meet modern standards) [31]. The pool of building meta-models is enriched by including two energy conversion technology configurations, one with and one without $CO_2$ network utilities. Within each scenario, parametric optimisation is implemented on the investment cost (minimum operating cost, minimum investment cost and five intermediary scenarios, see Figure 4) to obtain a systematic approach for generating interesting solutions in cities and explore options for optimal utilities and connections to optimal buildings.

Figure 5 illustrates sample results of the building-level optimisation. More specifically, it depicts the operating-investment cost Pareto frontier and self-sufficiency of residential single-family houses of different renovation stages, with and without $CO_2$ network utilities. The concept of self-sufficiency is further defined in Section 3.5 and is used to evaluate the autonomy of the energy systems studied, but is defined in simple terms as the percentage of electricity consumption supplied by self-production. For all solutions with an increase in investment cost, the operating cost decreases and renewable energy sources penetrate, leading to higher values of self-sufficiency. New and renovated buildings have reduced overall demands, and therefore lower operating cost. While the solutions connected to the $CO_2$ network yield no difference for low investment cost limits, they result in lower operating costs whenever the capital expenditure (capex) limit is high enough for these technologies. However, the piping cost of the $CO_2$ network is not considered at this stage, being included only at the canton/commune level (Switzerland has 26 cantons, each of them being divided in several communes. Cities can be comprised of several communes, e.g., Geneva has four communes).

The building-level solutions are then integrated in the main optimization, where each building is represented by its resource ($CO_2$ liquid and vapor, natural gas, electricity) import and export. Decision variables and constraints are used to permit selection of any number of buildings from any type, age and utility configuration as long as the overall mix is consistent with that of the case study considered. At the city level the optimiser chooses not only the best configuration of buildings, but also additional utilities to create an optimal city. The additional utilities at the upper level include PV panels, central plants which provide $CO_2$ liquid and vapor, and a power-to-gas storage system (Figure 3). The PV panel and $CO_2$ and $CH_4$ storage unit models are described in detail in [25], the co-generation solid oxide fuel cell-gas turbine (SOFC-GT) unit is modeled according to [32] and the co-generation solid oxide electrolysis cell (SOEC) unit according to [33]. A detailed description of the unit models can be found in Appendix A.2 and in [25].

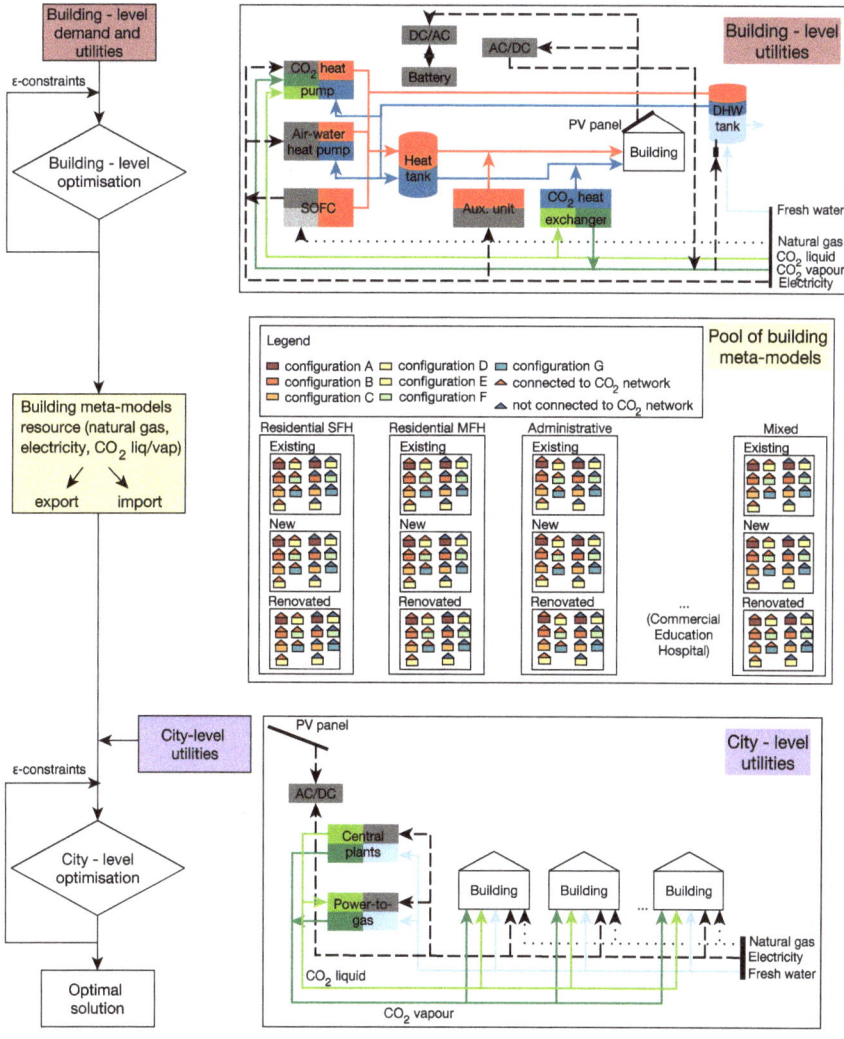

**Figure 3.** Methodology overview.

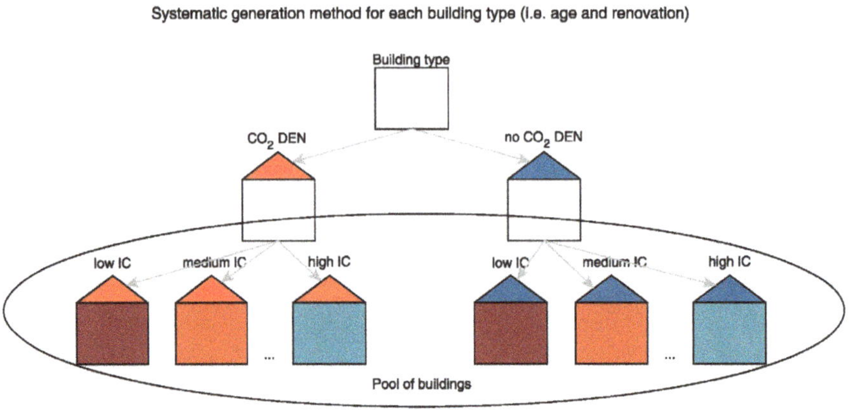

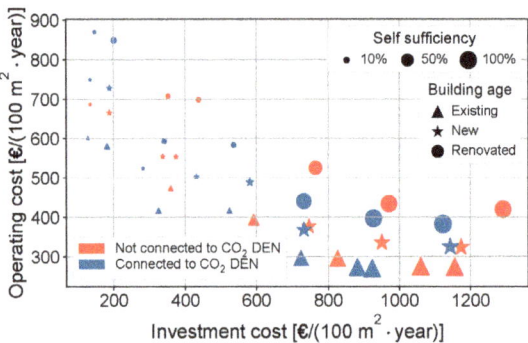

Figure 4. Systematic generation method for each building type (i.e., age and renovation).

Figure 5. Pareto frontier of residential SFH with different utility configurations and renovation stages.

A mixed integer linear programming (MILP) framework is used to find the optimal utility configurations and to integrate different technologies which satisfy the urban demand.

*3.2. Mathematical Formulation*

The building and city-level optimisations are formulated using mixed integer linear programming [34–37]. This framework was chosen to represent building energy systems, since it can model both the discrete and the continuous behavior of the units. An additional benefit is that this formulation always results in a global optimum and does not require extensive effort for problem initialization.

3.2.1. Definition of Sets

Given that energy demand is time-dependent, the problem is defined using discrete time intervals (e.g., $p \in \mathbf{P} = \{1\}$ (1 year), $t \in \mathbf{TOP}_p = \{1, 2, ..,14\}$ (12 months and two extreme days)). The system to be optimized is represented through several units, belonging to the set $\mathbf{U}$. The units are grouped in two subsets: the set of utility units ($\mathbf{UU}$ = {PV panels, batteries, heat pumps, CHPs, storage tanks, heat exchangers}) and the set of process units ($\mathbf{PU}$ = {building demands: space heating, domestic hot water, air cooling, utilities}). The process units represent the demand and hence have a fixed size, while the utility units represent the energy technologies used to satisfy the demand, with variable sizes, which are to be optimized. Units supply, demand, or convert resources ($r \in \mathbf{R}$) (electricity and material) and heat (at different temperature intervals $k \in \mathbf{K}$).

3.2.2. Objective Function and Constraints

The objective function of the problem is the minimization of the operating cost (Equation (1)), with $\epsilon$-constraints on the investment cost (Equation (2)) [38]. The objective function accounts for both the fixed ($C_u^{op,1}$) and variable ($C_u^{op,2}$) operating costs. The additional terms in the objective function are the binary variables ($y_{u,p,t}, y_u$) which decide whether a unit is used or not, the continuous variables ($f_{u,p,t}, f_u$) which determine the size of a unit, the operating time parameter ($t_t^{op}$) and the period occurrence ($p_p^{occ}$). $\epsilon$-constraints consider the fixed ($C_u^{inv,1}$) and variable ($C_u^{inv,2}$) investment costs.

$$\min_{y_u, f_u} \sum_{u \in \mathbf{U}} \left( \sum_{p=1}^{P} \left( \sum_{t=1}^{TOP} \left( C_u^{op,1} \cdot y_{u,p,t} + C_u^{op,2} \cdot f_{u,p,t} \right) \cdot t_t^{op} \right) \cdot p_p^{occ} \right) \tag{1}$$

$$\sum_{u \in \mathbf{U}} \left( C_u^{inv,1} \cdot y_u + C_u^{inv,2} \cdot f_u \right) \leq \epsilon \qquad \epsilon \in \{IC_{min} : IC_{max}\} \tag{2}$$

The main constraints of the problem include the energy conversion technology sizing and selection. Equations (3)–(6) bound the size of the unit in each time step $t$ and period $p$ to be smaller than the purchase size of the equipment, Equation (7) ensures that the purchase size of the equipment is between the minimum and maximum boundaries set ($f_u^{min}$, $f_u^{max}$), and Equations (8) and (9) fix the size of the process units.

$$y_{u,p} \leq y_u \qquad \forall u \in \mathbf{U}, \forall p \in \mathbf{P} \tag{3}$$

$$y_{u,p,t} \leq y_{u,p} \qquad \forall u \in \mathbf{U}, \forall p \in \mathbf{P}, \forall t \in \mathbf{TOP}_p \tag{4}$$

$$f_{u,p} \leq f_u \qquad \forall u \in \mathbf{U}, \forall p \in \mathbf{P} \tag{5}$$

$$f_{u,p,t} \leq f_{u,p} \qquad \forall u \in \mathbf{U}, \forall p \in \mathbf{P}, \forall t \in \mathbf{TOP}_p \tag{6}$$

$$f_u^{min} \cdot y_{u,p,t} \leq f_{u,p,t} \leq f_u^{max} \cdot y_{u,p,t} \qquad \forall u \in \mathbf{U}, \forall p \in \mathbf{P}, \forall t \in \mathbf{TOP}_p \tag{7}$$

$$y_{u,p,t} = 1 \qquad \forall u \in \mathbf{PU}, \forall p \in \mathbf{P}, \forall t \in \mathbf{TOP}_p \tag{8}$$

$$f_u^{min} = f_u^{max} = 1 \qquad \forall u \in \mathbf{PU} \tag{9}$$

The heat cascade equations ensure that heat is transferred from higher temperature intervals to lower temperature intervals and close the energy balance in each temperature interval $k$ (Equation (10a)). This is achieved using the residual heat $\dot{R}_{p,t,k}$, which cascades excess heat from higher temperature intervals ($k$) to lower temperature intervals ($k-1$). The minimum residual heat is zero, when heat cannot be transferred from the corresponding temperature interval to lower ones (Equation (10b)). Similarly, residual heat in the first interval ($\dot{R}_{t,1}$) is zero, as lower temperature intervals do not exist to accept a transfer of heat. Logically, heat cannot be cascaded to the $k^{th}$ interval as it is the highest, so $\dot{R}_{t,k+1}$ is also zero (Equation (10c)). $\dot{Q}_{u,p,t,k}$ represents the reference heat load of a unit $u$ in period $p$, time step $t$ and temperature interval $k$.

$$\sum_{u \in \mathbf{U}} f_{u,p,t} \cdot \dot{Q}_{u,p,t,k} + \dot{R}_{p,t,k+1} - \dot{R}_{p,t,k} = 0 \qquad \forall p \in \mathbf{P}, \forall t \in \mathbf{TOP}_p, \forall k \in \mathbf{K} \tag{10a}$$

$$\dot{R}_{p,t,k} \geq 0 \qquad \forall p \in \mathbf{P}, \forall t \in \mathbf{TOP}_p, k \in \mathbf{K} \qquad (10b)$$

$$\dot{R}_{p,t,1} = 0 \quad \dot{R}_{p,t,k+1} = 0 \qquad \forall p \in \mathbf{P}, \forall t \in \mathbf{TOP}_p \qquad (10c)$$

For each unit $u$, the supply $\dot{M}^{out}_{r,u,p,t}$ and demand $\dot{M}^{in}_{r,u,p,t}$ of a specific resource $r \in \mathbf{R}$ are computed (Equations (11a) and (11b)) and the balance of each resource is closed for each period $p$ and time step $t$ (Equation (11c)). $\dot{m}^-_{r,u,p,t}$ and $\dot{m}^+_{r,u,p,t}$ are the reference supply and demand flows of a unit.

$$\dot{M}^-_{r,u,p,t} = \dot{m}^-_{r,u,p,t} \cdot f_{u,p,t} \qquad \forall r \in \mathbf{R}, \forall u \in \mathbf{U}, \forall p \in \mathbf{P}, \forall t \in \mathbf{TOP}_p \qquad (11a)$$

$$\dot{M}^+_{r,u,p,t} = \dot{m}^+_{r,u,p,t} \cdot f_{u,p,t} \qquad \forall r \in \mathbf{R}, \forall u \in \mathbf{U}, \forall p \in \mathbf{P}, \forall t \in \mathbf{TOP}_p \qquad (11b)$$

$$\sum_{u \in \mathbf{U}} \dot{M}^-_{r,u,p,t} = \sum_{u \in \mathbf{U}} \dot{M}^+_{r,u,p,t} \qquad \forall r \in \mathbf{R}, \forall p \in \mathbf{P}, \forall t \in \mathbf{TOP}_p \qquad (11c)$$

### 3.2.3. Constraint Linking Individual Building and Urban Scale

Specific constraints at the building scale are presented in detail in [24]. Additional variables and sets are introduced at the urban scale, which aid the formulation of the constraints, such as building types ($bt \in \mathbf{BT}$ = {residentialSFH, residentialMFH, administrative, education, commercial, hospital, mixed}), building units of type $bt$ ($bu \in \mathbf{BUT}_{bt}$), renovation stages ($rs \in \mathbf{RS}$ = {existing, new, renovated}), building units of renovation stage $rs$ ($bu \in \mathbf{BUR}_{rs}$), building units connected to the $CO_2$ network ($bu \in \mathbf{BUC}$) and the set of cities/communes ($c \in \mathbf{C}$). The extra constraints include fixing the number of buildings of a given type to the one of the case studies considered ($N_{bt}$):

$$\sum_{p=1}^{P} \sum_{t=1}^{TOP} \sum_{bu \in \mathbf{BUT}_{bt}} f_{bu,p,t} = N_{bt} \qquad \forall bt \in \mathbf{BT} \qquad (12)$$

And making the number of buildings at each renovation stage equal to that of the urban system studied ($N_{rs}$):

$$\sum_{p=1}^{P} \sum_{t=1}^{TOP} \sum_{bu \in \mathbf{BUR}_{rs}} f_{bu,p,t} = N_{rs} \qquad \forall rs \in \mathbf{RS} \qquad (13)$$

The investment cost for the $CO_2$ network in each city/commune is computed according to [31] (for details see Appendix A.3). The commune has the choice of investing in the $CO_2$ network or not ($y_{u_{CO_2,c}}$), which translates in optimisation terms as a big $M$ constraint:

$$y_{u_{CO_2,c}} \geq f_{bu,p,t}/M \qquad \forall bu \in \mathbf{BUC}_c, \forall c \in \mathbf{C}, \forall p \in \mathbf{P}, \forall t \in \mathbf{TOP}_p \qquad (14)$$

I.e., if the commune activates a building with $CO_2$ network utilities, it must invest in piping. The size/length of the network piping is fixed for all periods and times.

### 3.2.4. Long-Term Energy Storage Model with Typical Day Resolution

To model the long-term storage units with typical day resolution, a series of new sets must be introduced (or re-defined). The equations here are based on using eight periods or typical days:

- **P**: periods, or typical days of the year, e.g., {1, 2, 3, 4, 5, 6, 7, 8};
- $\mathbf{TOP}_p$, $\forall p \in \mathbf{P}$: time steps in each period $p$, e.g., $\{\underbrace{\{1, 2, ..., 24\}, \{25, 26, ..., 48\}, ..., \{169, 170, ..., 192\}}_{\text{8 typ. days}}\}$;
- **RD**: real days of the year, e.g., {1, 2, ..., 365};

- $PORD_{rd}$, $\forall rd \in RD$: typical day corresponding to each real day of the year, e.g., $\underbrace{\{2, 2, 4, ...6\}}_{365 \text{ days}}$;

- $TORD_{rd} = t$, $\forall t \in TOP_{pr}$, $\forall pr \in PORD_{rd}$, $\forall rd \in RD$: time steps in each real day of the year, e.g., $\{\underbrace{\{25, 26, ..., 48\}}_{\text{time steps in typ. day 2}}, \underbrace{\{25, 26, ..., 48\}}_{\text{time steps in typ. day 2}}, \underbrace{\{73, 74, ..., 96\}}_{\text{time steps in typ. day 4}}, ..., \underbrace{\{121, 122, ..., 144\}}_{\text{time steps in typ. day 6}}\}$;
  
  $\underbrace{\phantom{xxxxxxxxxxxxxxxxxxxxxxxxxxxxxxxxxxxxxxxxxxxxxxxxxxxxxxxxxxxxxxxxxxxxxxxxxxxxxx}}_{365 \text{ days}}$

- $TOPNC_p = \{1, 2, ..., card(TOP_p)\}$, $\forall p \in P$: non cumulative time steps in each typical day $p$, e.g.,
  
  $\underbrace{\{\{1, 2, ..., 24\}, \{1, 2, ..., 24\}, ..., \{1, 2, ..., 24\}\}}_{8 \text{ typ. days}}$;

- $RTORD_{rd,pr,t} = \sum_{i=1}^{rd-1} card(TORD_i) + t - \sum_{j=1}^{pr-1} card(TOPNC_j)$ $\forall t \in TOP_{pr}$, $\forall pr \in PORD_{rd}$, $\forall rd \in RD$: real time of each real day of the year, e.g.,

  $\{\underbrace{\{1, 2, ..., 24\}}_{\text{time steps in real day 1}}, \underbrace{\{25, 26, ..., 48\}}_{\text{time steps in real day 2}}, \underbrace{\{49, 50, ..., 72\}}_{\text{time steps in real day 3}}, ..., \underbrace{\{8737, 8738, ..., 8760\}}_{\text{time steps in real day 365}}\}$

  $\underbrace{\phantom{xxxxxxxxxxxxxxxxxxxxxxxxxxxxxxxxxxxxxxxxxxxxxxxxxxxxxxxxxxxxxxxxxxxxxxxx}}_{365 \text{ days}}$

- $RT = 1, ..., \sum_{rd \in RD} card(TORD_{rd})$, (ordered set): real times of the year, e.g., $\{1, 2, ..., 8760\}$.

Given these sets above, the long-term storage units ($u \in SU$) are represented by the constraint:

$$SL_{rt} = \begin{cases} \text{if } rt = \text{first}(RT): & \sigma \cdot SL_{\text{last}(RT)} + \eta_{ch} \cdot M^+_{r,u,pr,t} - \frac{1}{\eta_{dch}} \cdot M^-_{r,u,pr,t} \\ \text{else}: & \sigma \cdot SL_{rt-1} + \eta_{ch} \cdot M^+_{r,u,pr,t} - \frac{1}{\eta_{dch}} \cdot M^-_{r,u,pr,t} \end{cases}$$

$\forall r \in R$, $\forall u \in SU$, $\forall rd \in RD$, $\forall pr \in PORD_{pr}$, $\forall t \in TOP_{pr}$, $\forall rt \in RTORD_{rd,pr,top}$  (15)

with $SL_{rt}$ as the storage level of the unit at each real time step of the year $rt \in RT$, $\sigma = 0.9992$ [39] the self-discharge rate of the unit, and $\eta_{ch} = \eta_{dch} = 0.9$ [39] as the charging and discharging efficiencies of the unit. A summary of all the sets used in the problem formulation is given in Table 1.

Table 1. Sets used in the mathematical formulation.

| Set Symbol | Name | Index | Increment | Cyclicity |
|---|---|---|---|---|
| P | periods | - | day | no |
| TOP | times of period | p | hour | no |
| U | units | - | - | no |
| UU | utility units | - | - | no |
| PU | process units | - | - | no |
| SU | storage units | - | - | no |
| R | resources | - | - | no |
| K | temperature intervals | - | - | no |
| BT | building types | - | - | no |
| BUT | building units of type | bt | - | no |
| RS | renovation stages | - | - | no |
| BUR | building units of renovation | rs | - | no |
| C | communes | - | - | no |
| BUC | building units connected to $CO_2$ DEN | c | - | no |
| RD | real days | - | day | no |
| PORD | periods of real day | rd | day | no |
| TORD | times of real day | rd | hour | no |
| TOPNC | times of period non cummulative | pr | hour | no |
| RTORD | real times of real day | rd, pr, t | hour | no |
| RT | real times | - | hour | yes |

171

*3.3. Case Study*

The case studies considered are Geneva city center (four communes: Genève-Cité, Genève-Plainpalais, Genève-Eaux-Vives and Genève-Petit-Saconnex) and the canton of Geneva (all 48 communes, Figure 6).

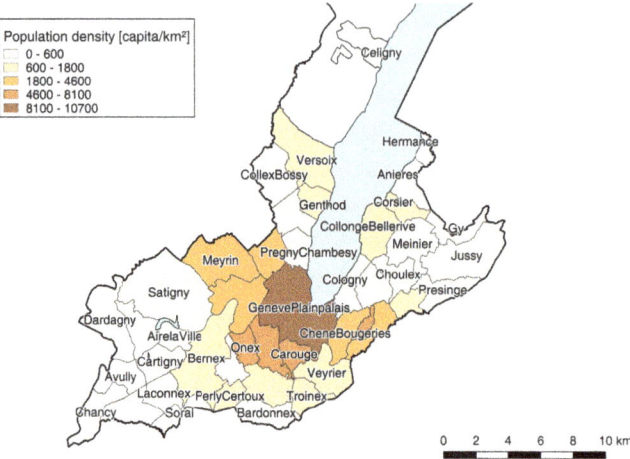

**Figure 6.** Population density of communes in the canton of Geneva.

The building types are distinguished according to the RegBL database [40], as listed in Table 2. The corresponding parameter names in the RegBL report are listed in Table A18 in the Appendix A.4.

**Table 2.** Building types present in the model of the canton of Geneva.

| Building Type | Building Category | Building Class |
|---|---|---|
| Residential SFH | 1021, 1025, 1230 | 1110 |
| Residential MFH | 1025 | 1121 |
|  | 1040 | 1130 |
| Administrative | 1040, 1060 | 1220 |
| Commercial | 1040, 1060 | 1230 |
| Education | 1040, 1060 | 1263 |
| Hospital | 1040, 1060 | 1264 |
| Mixed | 1030 | 1121, 1122 |

The energy reference area (ERA) of the buildings ($A_b^{ERA}$) is computed according to the same database, using the footprint area of the building ($A_b$), the number of floors ($N_b^{floors}$) and a factor of 0.9, an assumption used to account for the inner walls (Equation (16)).

$$A_b^{ERA} \ [m^2] = A_b \ [m^2] \cdot N_b^{floors} \cdot 0.9 \qquad (16)$$

The photo-voltaic rooftop potential is calculated using the rooftop area of the building ($A_b^{roof}$), the average solar irradiation on each roof ($I_b$), a nominal global horizontal irradiation of 1244.334 W/(m²·K) and a factor of 0.75 to account for the part of the roof which cannot be covered with PV panels (e.g., close to the periphery) (Equation (17) [41]).

$$A_{PV,b} \ [m^2] = A_b^{roof} \ [m^2] \cdot \frac{I_b \ [W/(m^2 \cdot K)]}{1244.334 \ [W/(m^2 \cdot K)]} \cdot 0.75 \qquad (17)$$

The number of buildings of each category and renovation stage are considered according to [31] and Figure 7 displays a sample distribution, that of Geneva city center.

*Energies* **2019**, *12*, 2945

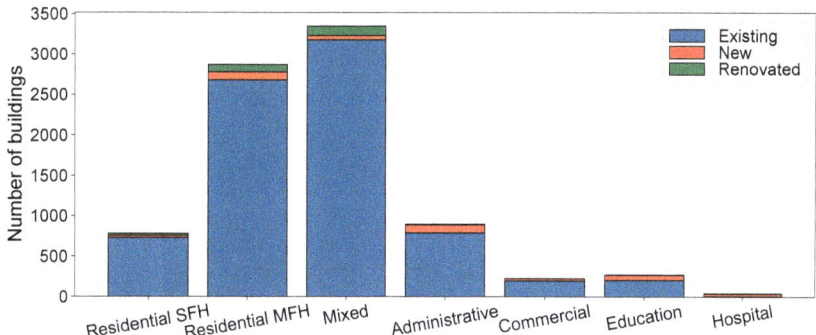

**Figure 7.** Refurbishment level building distribution in Geneva city center.

The demands evaluated are space heating, domestic hot water, air cooling and electricity. The hourly demand profiles are built based on standards and existing heat signature models. The electricity and domestic hot water demand profiles are considered according to the standards of the Swiss society of engineers and architects SIA [42] with a typical day profile repeated throughout the year, while the heating and cooling demands are modeled based on a heating signature profile [31]. These profiles have been calibrated based on statistical data from the energy department of the canton of Geneva [31]. Figure 8, Table 3 and Table A25 display the hourly demand profile of administrative buildings (existing, new and renovated) and their specific yearly demand. The domestic hot water and electricity demand is constant; therefore it is excluded from the hourly variation plots. The corresponding plot/table for all other building categories can be found in Appendix A.5.

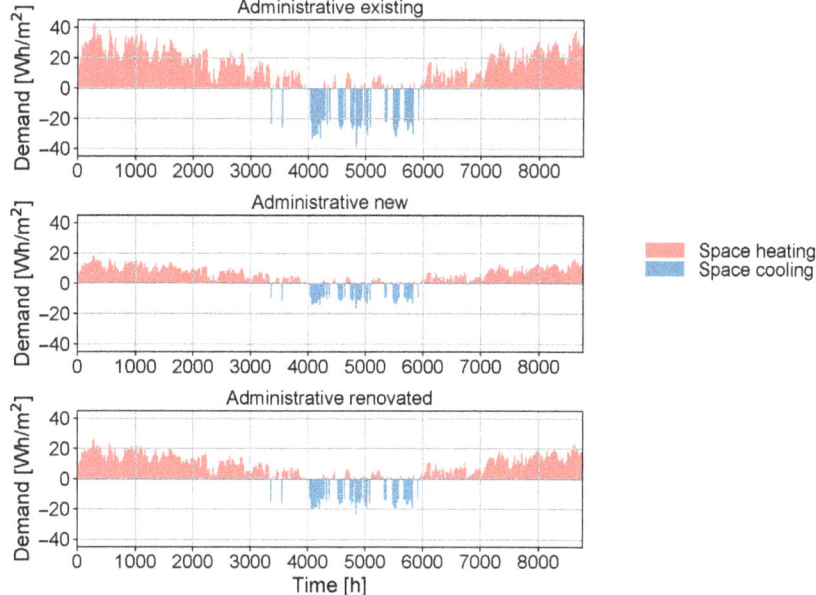

**Figure 8.** Hourly specific energy service demand of administrative buildings.

173

**Table 3.** Yearly specific energy service demand of administrative buildings.

| Building Renovation Stage | Space Heating [kWh/m²] | Air Cooling [kWh/m²] | Dom. Hot Water [kWh/m²] | Electricity [kWh/m²] |
|---|---|---|---|---|
| Existing | 84.2 | 8.9 | 2.9 | 43.2 |
| New | 36.4 | 3.8 | 2.9 | 43.2 |
| Renovated | 51.7 | 5.5 | 2.9 | 43.2 |

*3.4. Time Resolutions: Typical Days Algorithm*

Two time resolutions are used to solve the optimization problem, namely the state-of-the-art monthly averages with two extreme periods, and hourly resolution. Since the computational time for solving the problem increases drastically with the problem size (Figure 9), a k-medoids-based data clustering algorithm is used to reduce the complexity of the problem studied (Figure 10) for hourly resolution. This approach selects the cluster centers based on the smallest sum of distances within each cluster, while the cluster size is selected based on a series of performance indicators [43,44].

Two input parameters are considered for the clustering process, namely the ambient temperature ($T_{ext}$) and the global solar irradiation (GI), since all resources and demands can either be computed using these two parameters, or are assumed constant. Other data such as consumption profiles and their corresponding temperatures of demand are defined based on the computed cluster centers. The k-medoids algorithm is applied between 2 and 25 typical days. A maximum of 12% error in the load duration curve (ELDC) is set and consequently the number of typical days should be greater than five (Figure 9). To select the optimal number of typical days, the Davies-Bouldin (DB) index is used. The DB index is a measure of clustering scheme performance [45]. It accounts for the separation between the clusters—which should be as large as possible—and the within-cluster scatter, which should be as low as possible. The index is defined as the ratio between the cluster separation and the within-cluster distance, where lower values express better cluster separation and the 'tightness' inside the clusters. As observed in Figure 9, the DB index has the lowest value for 8 typical days for the dataset studied here. Therefore, this value is used for further analysis.

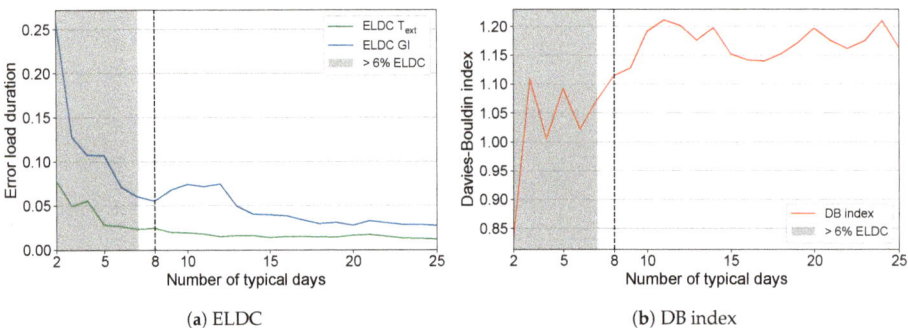

(a) ELDC  (b) DB index

**Figure 9.** Performance indicator evolution using the k-medoids algorithm for selecting the number of typical days.

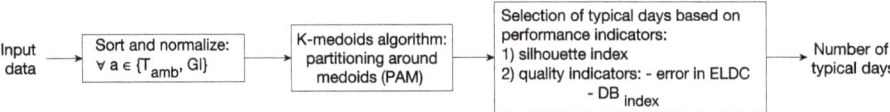

**Figure 10.** Typical days algorithm.

Figure 11 depicts the real profile of the two attributes chosen to cluster the data in grey the computed load duration curve in black. One can see that the load duration curve of both attributes is followed well with the number of typical days chosen.

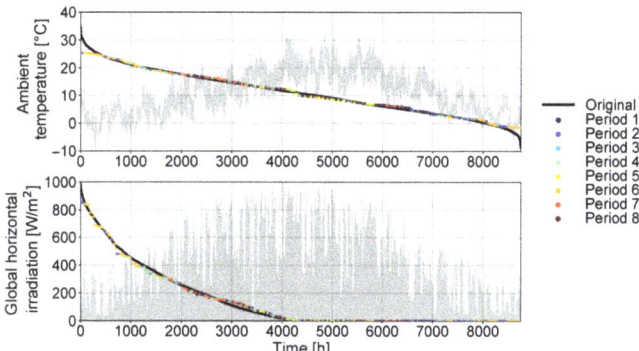

**Figure 11.** Representation of the typical days algorithm.

To clarify contributions of the different time resolutions in the problem formulation, the objective function (Equation (1)) can be assessed in greater detail. For monthly resolution, $p = 1$ (1 year), $t = \{1, 2, ..., 14\}$ represents 12 months and 2 extreme periods, $p_p^{occ} = 1$ represent the occurrence of the year, and $t_t^{op} = \{744, 672, 744, ..., 744, 0, 0\}$ are the number of operating hours in each time step $t$. With hourly resolution, $p = \{1, 2, ..., 10\}$ are the eight typical days and the two extreme hours, $t = \{24, 24, ..., 24, 1, 1\}$ are the number of hours in each time step $t$, $p_p^{occ} = \{54, 46, 17, 49, 52, 68, 49, 30, 1, 1\}$ represents the number of times each operating period appears during the year, and $t_t^{op} = \{1, 1, ..., 1, 0, 0\}$ is the operating time of each time step. For both time resolutions, the operating time of the extreme periods is zero, since they are used only for unit sizing.

### 3.5. Measure of Energy Autonomy

In this work, a urban community is considered energy autonomous when the electricity import from the grid ($E_i$) is zero, or likewise, when the self-sufficiency (SF) factor (Equation (18) [46]) is equal to unity. A solution is considered to be net zero-energy when the power grid export ($E_e$) and import ($E_i$) are equal, which is equivalent to when the self-sufficiency factor (Equation (18)) equals the self-consumption (SC) factor (Equation (21), Figure 12), where $E_g$ represents the electricity generation (e.g., by PV panels, co-generation units).

$$SF = \frac{E_g - E_e}{E_g - E_e + E_i} \tag{18}$$

where the numerator represents the demand:

$$E_g - E_e + E_i = \sum_{p=1}^{P} \sum_{t=1}^{TOP} \left( \dot{M}^+_{el,elheater,p,t} + \dot{M}^+_{el,battery,p,t} + \dot{M}^+_{el,HPs+Ref,p,t} + \dot{M}^+_{el,House,p,t} + \dot{M}^+_{el,CPwinter,p,t} + \dot{M}^+_{el,SOEC,p,t} \right) \tag{19}$$

and the electricity generation is given by:

$$E_g = \sum_{p=1}^{P} \sum_{t=1}^{TOP} \left( \dot{M}^-_{el,PV,p,t} + \dot{M}^-_{el,SOFC,p,t} + \dot{M}^-_{el,Battery,p,t} \right) \tag{20}$$

$$SC = \frac{E_g - E_e}{E_g} \quad (21)$$

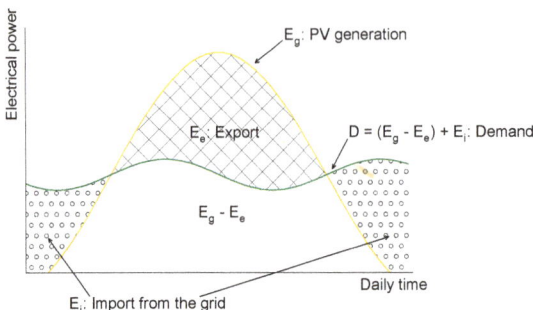

**Figure 12.** Self-sufficiency and self-consumption visual depiction.

## 4. Results and Discussion

### 4.1. Monthly vs. Typical Day Time Resolution

First, the two time resolutions considered (average monthly and typical day hourly) are analyzed for the case study of Geneva city center. Figure 13 depicts the operating-investment cost Pareto front for the two time resolutions, the size of the dots represents the self-sufficiency of the system and the solutions connected to the $CO_2$ network are highlighted in gray. By comparing the two time resolutions, it is observed that for the same investment cost limits, solutions using monthly resolution yield up to 31% lower operating cost, and 18% higher self-sufficiency (for the 8th investment cost limit). This occurs due to the fact that peak shaving is an implicit outcome of data aggregation for the monthly resolution, while peaks must be accounted for explicitly with the hourly resolution and adjustments must be made to buy electricity even when previous electricity sales may have occurred. This results in higher operating cost and lower self-sufficiency by considering scenarios with hourly profiles. This also stresses the importance of considering analysis with enough temporal detail to understand the real system requirements, since grid balancing must be completed on short time scales and thus analysis using average data may lead to problematic scenarios.

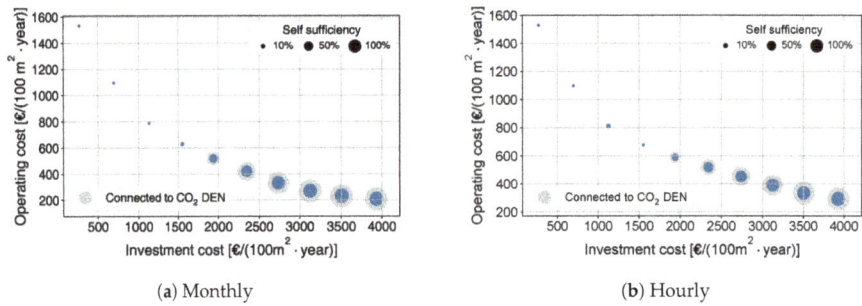

(a) Monthly  (b) Hourly

**Figure 13.** Pareto Geneva city center different time resolutions.

The cost breakdown of the two time resolutions is shown in Figure 14. The first figures on the left show the breakdown of total cost, the biggest contribution being the capex since the system starts optimal solutions require increasing investment to reduce the operating cost and increase the self-sufficiency. A high level of investment is required to supply the peak demand; however, investing approximately 60% of the maximum value yields solutions with self-sufficiency in excess of 60%. The second and third set of figures, the breakdown of investment cost at the building and city levels, show that both time resolutions highlight the same main contributors: heat pumps, SOFCs and PV panels at the building level and PV panels, power-to-gas and the $CO_2$ network pipes at the city level.

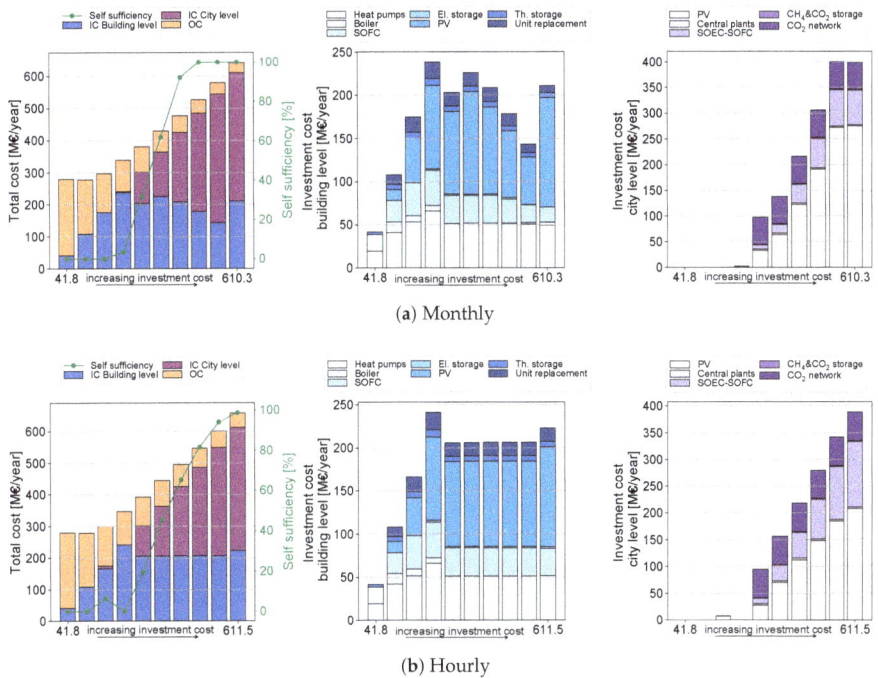

**Figure 14.** Geneva city center cost breakdown for different time resolutions.

As shown in these results, both time resolutions show the same trends and main cost contributors. Therefore, despite the increased accuracy provided by hourly resolution and in the interest of reducing data processing and computation time, the remaining results, at canton level (48 communes), are obtained using a monthly resolution.

### 4.2. $CO_2$ DEN Penetration with Population Density

An $\epsilon$-constraint investment cost optimisation for all 48 communes in the canton of Geneva is performed to study the $CO_2$ DEN penetration depending on the population density. The investment cost of the $CO_2$ network was considered to be explained in Equation (14). Figure 15 depicts the lowest investment cost, 40% of maximum IC, 90% of maximum IC and lowest operating cost scenarios. This figure shows that the scenario with the lowest investment cost does not prompt any of the communes to invest in the $CO_2$ network. With increasing investment cost limits, the communes which are most densely populated, with an investment cost per energy reference area lower than 21.5 k€/100 m$^2$$_{ERA}$, start connecting to the $CO_2$ DEN. Finally, for the minimum operating cost scenario, all communes connect to the low temperature DEN.

The results are also represented using parallel coordinates. Figure 16 shows that higher investment cost limits logically correlate with reduced operating cost and $CO_2$ emissions in the canton. Moreover, higher overall investment cost solutions lead to the largest number of communes connected to the $CO_2$ DEN and the highest self-sufficiency of the canton. Regarding the investment cost at the building level, a mix of high and low investment cost buildings are selected for optimal operating cost, with a moderate investment in PV panels and heat pumps. The solution with the lowest operating cost is selected to explore detailed results, as highlighted in Figure 16. Compared with the current situation (i.e., lowest investment cost solution: mostly boilers supplying heating, no PV market penetration), the best scenario (from an economic standpoint) leads to approximately 90% savings in $CO_2$ emissions and operating cost, with a payback time of 17.5 years.

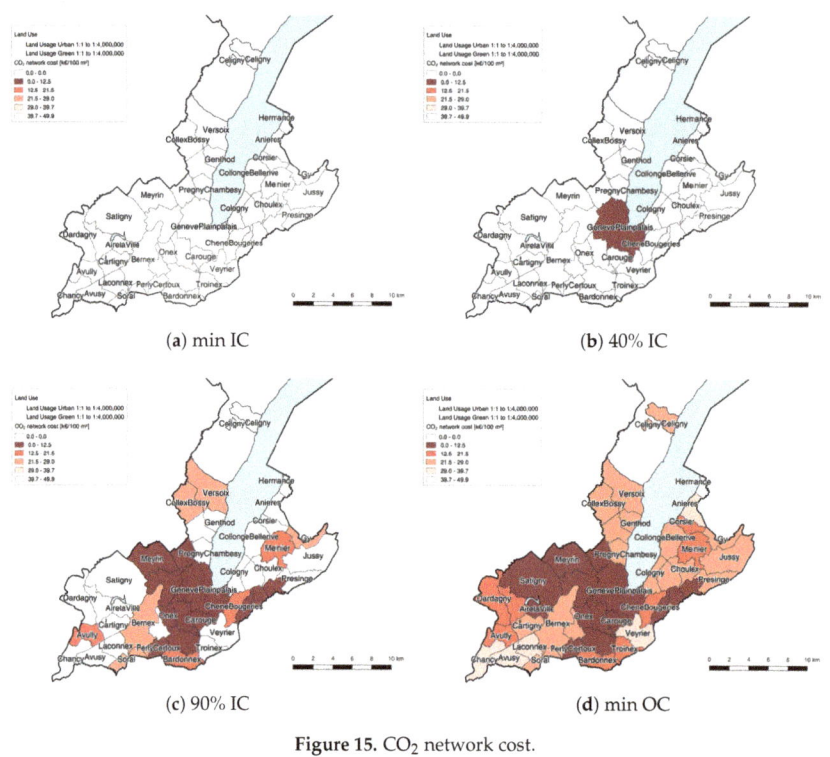

Figure 15. $CO_2$ network cost.

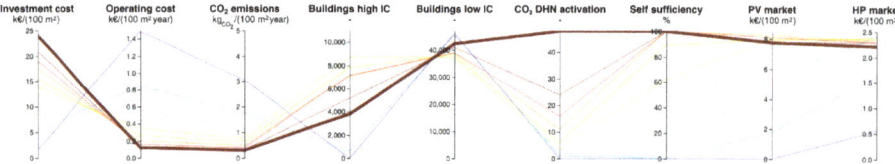

Figure 16. Parallel coordinate representation of the canton solutions.

### 4.3. Detailed Results of Solution with Lowest Operating Cost and Emissions

Figure 17 depicts the details of the solution highlighted above, for each of the 48 communes, sorted by population density. Most of the communes have low population and building densities, and correspondingly low energy flows (i.e., electricity and natural gas import/export). Generally, high population densities are associated with lower district network cost per energy reference area and with high $CO_2$ emissions. However, the environmental impact has a higher correlation with the overall population, i.e., with the total electricity and natural gas import.

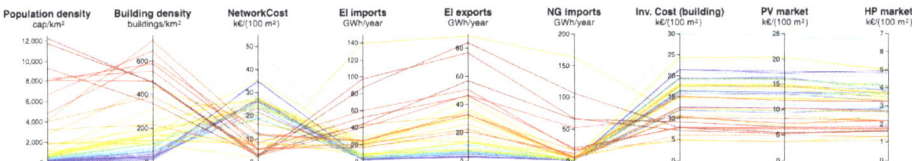

**Figure 17.** Parallel coordinate representation of the communes for the lowest operating cost solution.

Figure 18 is used to detail the energy flows in the communes, by displaying the detailed contributors of electricity and natural gas import/export at the building level for each commune and at the canton level. The results show that:

- the main electricity consumers at the building level are heat pump and refrigeration units ($\approx$35%) and electrical appliances ($\approx$65%);
- the main electricity producers are PV panels, accounting for 91% of the production and SOFC co-generation units supplying the balancing 9%;
- the main natural gas consumers are boilers (47%) and SOFC co-generation units (53%).

At the cantonal level:

- electricity is consumed by the SOEC unit (35%), by the central plants to produce $CO_2$ (9%) vapour, and by net electricity importing buildings (56%);
- electricity is produced by PV panels (77%) and the SOFC co-generation unit (23%);
- natural gas is required for the SOFC unit (18%) and for net natural gas importing buildings (82%);
- natural gas is produced by the methanation unit in the power-to-gas system (18%) and purchased from the grid (82%).

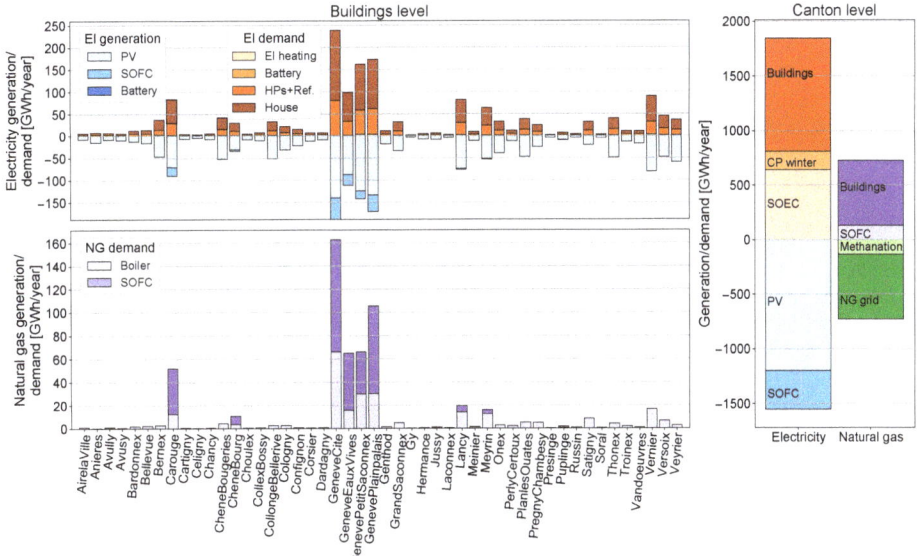

**Figure 18.** Breakdown of electricity/natural gas import/export by commune and at building/canton level.

Figure 19 shows the cost breakdown at the building level, for each commune and at the canton level. Similar to the results shown for Geneva city center, building invesments are principally concentrated in heat pumps and refrigeration units ($\approx$20%), SOFCs ($\approx$3%) and PV panels ($\approx$71%), while the investment cost at the canton level is dominated by the $CO_2$ DEN piping (28%) followed by PV panels (19%) and the the power-to-gas system (9%).

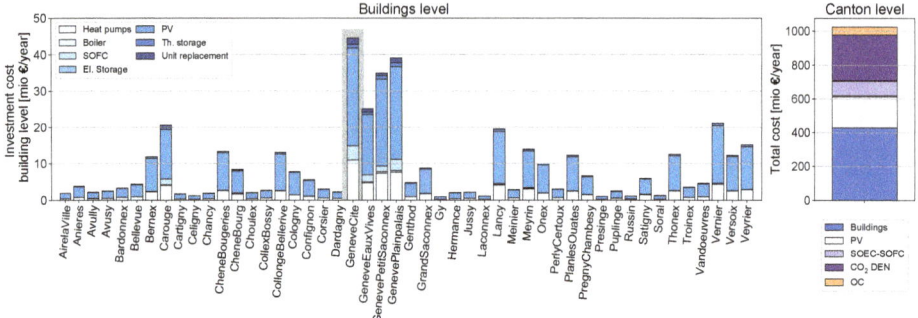

Figure 19. Breakdown of investment and total cost at buildings/canton level.

Also in this case one of the solutions, the one of the commune Génève Cité (highlighted solution in Figures 17 and 19), is selected for additional exploration.

The monthly energy import/export profiles of Génève Cité are shown in Figure 20. As observed in this figure, the electricity consumption of heat pumps is high in winter, when heating is required, and lower in summer, while the electricity demand for electrical appliances is assumed constant over the year. Electricity production from PV panels is higher in summer, corresponding to higher global horizontal irradiation, and the electricity production of SOFC co-generation units is higher in winter, since they provide the electricity requirement of heat pumps and co-generate heat for space heating and domestic hot water demand. Consequently, the natural gas consumption of the SOFC and boiler units are higher in winter, both related to supply of heating services.

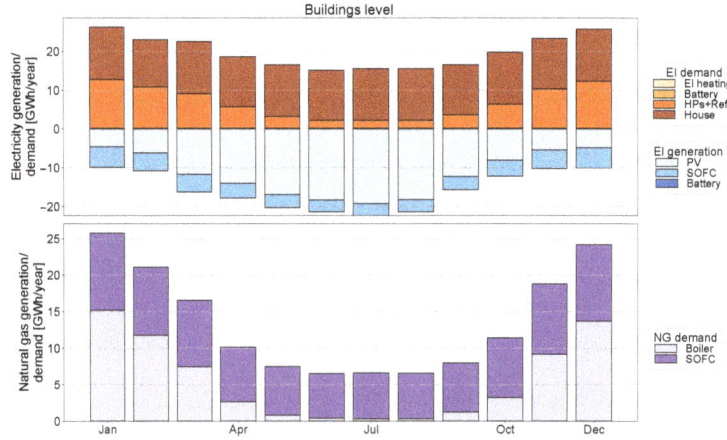

Figure 20. Breakdown of electricity/natural gas import/export by month, for Génève Cité.

## 5. Conclusions

This paper aims at providing a method to systematically integrate multi-energy networks and low carbon resources in cities. The method proposes a double optimisation approach with meta-models to link analysis and optimisation at both building and urban scales. The first optimisation creates a pool of optimal building solutions of different types (residential single- and multi-family houses (SFH, MFH), administrative, commercial, education, hospital and mixed), renovation stages (existing, new and renovated), energy system configurations (existing $H_2O$-based networks or potential low-temperature $CO_2$-based networks) and for different investment cost limits (i.e., budgets). The pool of optimal building meta-models is fed to the optimiser on top, which selects the best mix to optimise energy systems at city/canton level.

Geneva city center is used as a case study to analyse the impact of different temporal resolutions, namely monthly averages and hourly typical days. The results show that implicit peak shaving occurs in the monthly resolution by averaging demands, resulting in lower operating cost and higher self-sufficiency solutions, compared to the hourly resolution. However, the investment cost breakdown proves that the main contributors do not change, irrespective of the time resolution. Therefore, and in view of decreasing data processing and computation time, a monthly resolution is used for the results at canton level.

The second case study, the whole canton of Geneva (48 communes), is analysed to assess $CO_2$ DEN penetration with population density. The results highlight that scenarios with moderate investment limits, only communes with high population density, i.e., a network cost below 21.5 k€/100 $m^2_{ERA}$ connect to the refrigerant-based network, while for the minimum operating cost scenario, all communes are connected to the $CO_2$ DEN. Parallel coordinates are employed to better visualise key performance indicators for the scenarios at the cantonal and communal levels. The energy and cost breakdown results for each commune show that electricity is mostly consumed in heat pumps, refrigeration units and for electrical appliances while being produced by PV panels and SOFC co-generation units, while natural gas is consumed for boilers and SOFCs. Consequently, at the building level, the investment cost is dominated by heat pumps ($\approx$20%), SOFCs ($\approx$3%) and PV panels ($\approx$71%). At the canton level, the electricity importers are the buildings, SOEC unit and central plant, and the electricity exporters are PV panels and the SOFC unit, while the natural gas importers are the building and SOFC unit and the exporters are the methanator and natural gas grid. Consequently, the investment cost at the cantonal level is dominated by PV panels (19%), the power-to-gas system (9%) and $CO_2$ DEN piping (28%).

This work successfully develops an integrated framework, which embeds optimally operating buildings in districts. The framework was validated using the canton of Geneva; however, it is not case specific and can therefore be applied to different urban systems/conditions. This work allows engineers to assess the cost of reaching the Paris agreement targets and reduce the operating cost by approximately 90% in the residential sector, while using low-temperature $CO_2$ district energy networks. The model can also be used to study the integration of other types of large energy systems, e.g., by municipal bodies for future planning of urban energy supply with long planning horizons. Future work includes improving the pool of building meta-models, to cover a wider range of building types and a finer resolution on the building renovation stage and on budget scenarios. A typical day/full hourly resolution is suggested for future work to obtain more precise results and avoid inaccuracies stemming from implicit peak shaving. Further applications of the method in other geographical contexts would create a broader understanding of $CO_2$ DEN penetration and could potentially be extended to a European or global scale to assess feasibility as a multi-energy, fully renewable solution, coupled to long-term energy storage.

**Author Contributions:** Conceptualization, R.S., P.S., I.K., L.G. and F.M.; Formal analysis, R.S., P.S. and I.K.; Investigation, R.S., P.S., I.K., L.G. and F.M.; Methodology, R.S. and P.S.; Project administration, F.M.; Resources, F.M.; Software, R.S. and P.S.; Supervision, I.K. and F.M.; Validation, R.S., P.S. and I.K.; Visualization, R.S., P.S., I.K., L.G. and F.M.; Writing—original draft, R.S.; Writing—review & editing, R.S., P.S., I.K. and L.G.

**Funding:** This project is carried out with the financial support of the Swiss Innovation Agency (Innosuisse-SCCER program).

**Acknowledgments:** This project is carried out with the financial support of the Swiss Innovation Agency (Innosuisse - SCCER program) and of the Swiss Federal Office of Energy in the context of the ERA-NET Co-fund Smart Cities and Communities (ENSCC) project IntegrCiTy, with financial support from Romande Energie, Hoval AG, Holdigaz and Canton of Geneva.

**Conflicts of Interest:** The authors declare no conflict of interest.

## Abbreviations

The following abbreviations are used in this manuscript:

| | |
|---|---|
| DEN | District energy network |
| IDA ICE | IDA indoor climate energy |
| SFH | Single-family house |
| MFH | Multi-family house |
| capex | Capital expenditure |
| SOFC-GT | Solid oxide fuel cell - gas turbine |
| SOEC | Solid oxide electrolysis cell |
| MILP | Mixed integer linear programming |
| PV | Photo-voltaic |
| ERA | Energy reference area |
| DB | Davies - Bouldin (index) |
| ELDC | Error load duration curve |
| GI | Global horizontal irradiation |
| SF | Self-sufficiency |
| SC | Self-consumption |
| BOI | Boiler |
| ELH | Electrical heater |
| AHP | Air-water heat pump |
| COP | Coefficient of performance |
| VAC | Refrigeration cycle |
| SOC | State of charge |
| HST | Heat storage tank |
| HP | Heat pump |
| HE | Heat exchanger |
| BAT | Battery stack |
| DHN | District heating network |

## Appendix A.

*Appendix A.1. Unit Models at Building Level*

Appendix A.1.1. Building

The thermal behaviour of the building is described using a first order dynamic 1R1C model, as illustrated in Figure A1. The entire construction is aggregated into a single capacity $C_b$ while considering a single temperature node $T_b$ [47,48]. Equation (A1) highlights the corresponding energy balance, where $T_b$ denotes the internal temperature, $T^{ext}$ the external temperature, $U_b^{ext} = 1/R_b^{ext}$ the combined thermal transfer coefficient, $\Phi_b^{sun+o}$ the stochastic gains from solar and occupancy sources and $\dot{Q}_b^+$ the heat supplied by the energy system. In the case of partially non-residential dwellings with cooling requirements, a second zone is added to the model and connected through the internal insulation resistance $R_b^{int}$. $T^{min/max}$ in Equation (A2) define the comfort tolerance on the internal temperature.

$$C_b \cdot (T_{b,p,t+1} - T_{b,p,t}) = U_b^{ext} \cdot (T_{p,t}^{ext} - T_{b,p,t}) + \Phi_{b,p,t}^{sun+o} + \dot{Q}_{b,p,t}^+ \quad \forall p \in \mathbf{P},\ t \in \mathbf{TOP} \quad (A1)$$

$$T_{b,p,t}^{min} \leq T_{b,p,t} \leq T_{b,p,t}^{max} \quad \forall p \in \mathbf{P},\ t \in \mathbf{TOP} \quad (A2)$$

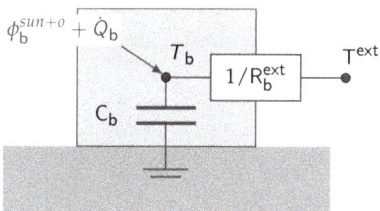

**Figure A1.** 1R1C building model.

### Appendix A.1.2. Boiler (BOI)

The natural gas boiler is described using a static system model formulation (Equations (3)–(7)) and is implemented as an auxiliary heating utility, the sizing dimension being the thermal power output. The main parameter required to model this unit, the thermal efficiency ($\epsilon_{BOI}$), is listed in Table A1.

**Table A1.** Parameter data.

| Parameter | Value | Unit | Ref. |
|---|---|---|---|
| $\epsilon_{BOI}$ | 0.98 | [-] | [44] |

### Appendix A.1.3. Electrical Heater (ELH)

As with the boiler unit, the electrical heater is also described using a static system model formulation (Equations (3)–(7)) and implemented as an auxiliary heating utility, the sizing dimension being the thermal power output. The main parameter required to model this unit, the thermal efficiency ($\epsilon_{ELH}$), is listed in Table A2.

**Table A2.** Parameter data (ELH).

| Parameter | Value | Unit | Ref. |
|---|---|---|---|
| $\epsilon_{ELH}$ | 0.98 | [-] | estimate |

### Appendix A.1.4. Heat Pumps

The air-source heat pump unit (AHP) is described using a static system formulation (Equations (3)–(7)), the sizing dimension being the electrical power input. The conversion efficiency (Equations (A3) and (A4)) is determined using the ideal coefficient of performance (COP) and the second law efficiency $\eta$, which accounts for the irreversibilities in the different cycle components (e.g., compressor). In order to avoid non-linearities coming from the variable supply temperature, the generated heat load is discretised into $n_s = |S_{AHP}|$ streams $s$. When considering different heat sources (e.g., water-source heat pumps) in the problem formulation, a similar model definition can be applied, the solely modification being the source temperature (e.g., $T_{p,t}^{water}$) and the respective second-law efficiency $\eta$. The values of the parameters considered for the air-water heat pump are given in Table A3 and for the corresponding refrigeration cycle (VAC) in Table A4.

$$COP_{AHP,s,p,t} = \frac{T_{AHP,s}^{sink}}{T_{AHP,s}^{sink} - T_{p,t}^{source}} \qquad \forall s \in S_{AHP},\ p \in P,\ t \in TOP \qquad (A3)$$

$$\bar{q}_{AHP,s,p,t} = \eta_{AHP,s,p,t} \cdot COP_{AHP,s,p,t} \qquad \forall s \in S_{AHP},\ p \in P,\ t \in TOP \qquad (A4)$$

**Table A3.** Default parameters values for the AHP second-law efficiency and part-load limit, evaluated from [49].

| | Par. | Tsink [°C] | Tsource [°C] | | | | | | | | |
|---|---|---|---|---|---|---|---|---|---|---|---|
| | | | −20 | −15 | −10 | −7 | −2 | 2 | 7 | 10 | 15 | 20 |
| AHP | $\eta$ | 35 | 0 | 0.464 | 0.458 | 0.458 | 0.469 | 0.462 | 0.435 | 0.416 | 0.37 | 0.307 |
| | | 45 | 0 | 0.445 | 0.463 | 0.464 | 0.46 | 0.446 | 0.439 | 0.436 | 0.43 | 0.396 |
| | | 55 | 0 | 0 | 0 | 0.421 | 0.423 | 0.416 | 0.439 | 0.436 | 0.412 | 0.395 |
| | $\dot{m}^{+,max}_{electricity}$ | 35 | 0 | 0.62 | 0.65 | 0.65 | 0.65 | 0.65 | 0.68 | 0.68 | 0.68 | 0.68 |
| | | 45 | 0 | 0.74 | 0.74 | 0.74 | 0.76 | 0.79 | 0.82 | 0.82 | 0.79 | 0.79 |
| | | 55 | 0 | 0 | 0 | 0.91 | 0.94 | 0.97 | 0.97 | 0.97 | 1 | 1 |

**Table A4.** Default parameters values for the VAC second-law efficiency and part-load limit, evaluated from [49].

| | Par. | Tsink | Tsource [°C] | | | | |
|---|---|---|---|---|---|---|---|
| | | | 20 | 25 | 30 | 35 | 40 | 45 |
| VAC | $\eta$ | 13 | 0.103 | 0.159 | 0.198 | 0.219 | 0.249 | 0.224 |
| | | 15 | 0.076 | 0.14 | 0.181 | 0.243 | 0.243 | 0.224 |
| | | 18 | 0.033 | 0.101 | 0.146 | 0.209 | 0.209 | 0.218 |
| | | 22 | 0 | 0.005 | 0.106 | 0.184 | 0.184 | 0.215 |
| | $\dot{m}^{+,max}_{electricity}$ | 13 | 0.71 | 0.78 | 0.86 | 0.95 | 0.91 | 1 |
| | | 15 | 0.73 | 0.78 | 0.87 | 0.95 | 0.91 | 0.96 |
| | | 18 | 0.73 | 0.8 | 0.89 | 0.96 | 0.93 | 0.89 |
| | | 22 | 0.75 | 0.82 | 0.91 | 1 | 0.95 | 0.8 |

Appendix A.1.5. Storage Units

Battery Stack (BAT)

Stationary batteries are described using a single state dynamic model, the sizing dimension being the electrical energy stored. The model accounts for the system self-discharging rate ($\sigma$) as well as the charging and discharging losses ($\gamma$). To limit any premature degradation of the stack, the minimum (SOC$^{min}_{BAT}$) and maximum (SOC$^{max}_{BAT}$) battery states of charge (SOC) are fixed (Equations (A5) and (A6)). The parameters used to model this unit are listed in Table A5.

$$f_{BAT,p,t} \geq SOC^{min}_{BAT} \cdot f_{BAT} \quad \forall p \in P,\ t \in TOP \tag{A5}$$

$$f_{BAT,p,t} \leq SOC^{max}_{BAT} \cdot f_{BAT} \quad \forall p \in P,\ t \in TOP \tag{A6}$$

**Table A5.** Parameter data (BAT).

| Parameter | Value | Unit | Ref. |
|---|---|---|---|
| $\gamma^{ch}$ | 0.9 | [-] | [39] |
| $\gamma^{dch}$ | 0.9 | [-] | [39] |
| $\sigma$ | 0 | [-] | [39] |
| SOC$^{max}_{BAT}$ | 0.8 | [-] | [50] |
| SOC$^{min}_{BAT}$ | 0.2 | [-] | [50] |

Heat Storage Tanks (HST)

The thermal energy storage tanks are described through a single state, first order dynamic model formulation, the sizing dimension being the unit volume. The minimum state of charge SOC$^{min}$ is set as the current building return temperature $T^{h,r}_{b,p,t}$ during space heating periods, while the maximum operating temperature $T_{HST}^{max}$ is defined as the lowest value between the heat pump operating limit and the nominal supply temperature of the heating system ($T_b^{h,s}$). The required parameters include the tank diameter $D_{HST}$, the specific heat loss rate $U_{HST}$ as well as the charging and discharging efficiencies $\gamma$. The unit is consequently added into the heat cascade formulation through the single charging (cold) and discharging (hot) streams as defined in Equations (A7)–(A10). The parameter values are given in Table A6.

$$\sigma_{HST} = \frac{4 \cdot U_{HST}}{D_{HST}} \cdot \left( T_{HST}^{max} - T_{b,p,t}^{h,r} \right) \quad \forall p \in P, t \in TOP \quad (A7)$$

$$\kappa_{HST} = \frac{4 \cdot U_{HST}}{D_{HST}} \cdot \left( T_{b,p,t}^{h,r} - T^{amb} \right) \quad \forall p \in P, t \in TOP \quad (A8)$$

$$\dot{q}_{HST,s,p,t}^{+} = c_p \cdot \rho \cdot \left( T_{HST}^{max} - T_{b,p,t}^{h,r} \right) \quad \forall p \in P, t \in TOP \quad (A9)$$

$$\dot{q}_{HST,s,p,t}^{-} = c_p \cdot \rho \cdot \left( T_{HST}^{max} - T_{b,p,t}^{h,r} \right) \quad \forall p \in P, t \in TOP \quad (A10)$$

Table A6. Parameter data (HST).

| Parameter | Value | Unit | Ref. |
|---|---|---|---|
| $c_p$ | 4.186 | [kJ/(kg·K)] | estimate |
| $\rho$ | 1000 | [kg/m³] | estimate |
| $\gamma^{ch}$ | 0.99 | [-] | estimate |
| $\gamma^{dch}$ | 0.99 | [-] | estimate |
| $D_{HST}$ | 0.98 | [m] | estimate |
| $U_{HST}$ | 0.0013 | [kW/m²] | [44] |

*Appendix A.2. Unit Models at City Level*

Appendix A.2.1. PV Panels

The PV panels are modeled as described in [47], with $A^{PV}$ the PV area, $\eta^{PV}$ the PV efficiency, $I^{sun}$ the irradiation of the sun, $T^{PV}$ the PV temperature, $U^{glass}$ the thermal transmission coefficient, $T^{amb}$ the ambient temperature, and $f^{glass}$ the factor denoting the portion of the solar irradiation passing through the PV glass:

$$\dot{m}_{PV,electricity} = A^{PV} \cdot \eta^{PV} \cdot I^{sun} \quad (A11a)$$

$$\eta^{PV} = \eta^{PV,ref} - \eta^{PV,var} \cdot \left( T^{PV} - T^{PV,ref} \right) \quad (A11b)$$

$$T^{PV} = \frac{U^{glass} \cdot T^{amb}}{U^{glass} - \eta^{PV,var} \cdot I^{sun}} + \frac{I^{sun} \cdot \left( f^{glass} - \eta^{PV,ref} - \eta^{PV,var} \cdot T^{PV,ref} \right)}{U^{glass} - \eta^{PV,var} \cdot I^{sun}} \quad (A11c)$$

The different parameters assumed are given in Table A7 [47] and the reference stream for $A^{PV} = 100$ m² and $I^{sun} = 100$ W/m² is given in Table A8.

Table A7. Parameters for PV panels.

| Parameter | Value | Unit |
|---|---|---|
| $T^{amb}$ | 288 | K |
| $T^{PV,ref}$ | 298 | K |
| $U^{glass}$ | 29.1 | W/(m²·K) |
| $f^{glass}$ | 0.9 | - |
| $\eta^{PV,ref}$ | 0.14 | - |
| $\eta^{PV,var}$ | 0.001 | 1/K |

Table A8. Streams for PV panel.

| Type | $T^{in}$ [°C] | $T^{out}$ [°C] | $\dot{Q}$ | $\dot{m}^-$ | $\dot{m}^+$ |
|---|---|---|---|---|---|
| Electricity | - | - | - | 1.66 kW | - |

Appendix A.2.2. SOEC-SOFC Co-Generation and Methanation

The co-generation SOFC-GT unit is modeled according to [32] and the co-generation SOEC unit according to [33]. A list of the reference streams in the different units are given in Tables A9–A11.

Table A9. Streams for SOEC unit.

| Type | $T^{in}$ [°C] | $T^{out}$ [°C] | $\dot{Q}$ | $\dot{m}^-$ | $\dot{m}^+$ |
|---|---|---|---|---|---|
| Heat | 91 | 58 | 3.05 kW | - | - |
| Heat | 58 | 27 | 1.66 kW | - | - |
| Electricity | - | - | - | - | 100 kW |
| $H_2O$ | - | - | - | - | 5.98 g/s |
| $H_2$ | - | - | - | 0.67 g/s (94.21 kW) | - |

Table A10. Streams for SOFC-GT unit.

| Type | $T^{in}$ [°C] | $T^{out}$ [°C] | $\dot{Q}$ | $\dot{m}^-$ | $\dot{m}^+$ |
|---|---|---|---|---|---|
| Heat | 648.8 | 260.0 | 16.28 kW | - | - |
| Heat | 109.8 | 35.2 | 9.44 kW | - | - |
| Heat | 35.2 | 30.2 | 1.44 kW | - | - |
| Electricity | - | - | - | 100 kW | - |
| $CH_4$ | - | - | - | - | -2.41 g/s (133.48 kW) |
| $CO_2$ | - | - | - | 6.60 g/s | - |

Table A11. Streams for methanation unit.

| Type | $T^{in}$ [°C] | $T^{out}$ [°C] | $\dot{Q}$ | $\dot{m}^-$ | $\dot{m}^+$ |
|---|---|---|---|---|---|
| Heat | 625.4 | 507.3 | 138.4 kW | - | - |
| Heat | 507.3 | 507.1 | 0.3 kW | - | - |
| Heat | 507.1 | 233.0 | 585.3 kW | - | - |
| Heat | 233.0 | 228.0 | 9.3 kW | - | - |
| Heat | 228.0 | 227.0 | 0.7 kW | - | - |
| Heat | 227.0 | 215.0 | 12.7 kW | - | - |
| Heat | 215.0 | 203.0 | 27.1 kW | - | - |
| Heat | 203.0 | 186.7 | 25.3 kW | - | - |
| Heat | 186.7 | 28.0 | 358.0 kW | - | - |
| Electricity | - | - | - | 100 kW | - |
| $H_2$ | - | - | - | - | 0.2 kg/s (28,349.2 kW) |
| $CO_2$ | - | - | - | - | 1.1 kg/s |
| $CH_4$ | - | - | - | 0.4 kg/s (22,193.6 kW) | - |

The reference flows for the SOEC unit are given for an incoming flow of electricity of 100 kW. The electricity to hydrogen efficiency is computed using the HHV of $H_2$ of 141,746 kJ/kg [51]:

$$\eta = \frac{\dot{m}^-_{H_2} \cdot HHV^{H_2}}{\dot{m}^+_{electricity}} = 94.2\% \quad (A12)$$

The reference flows for the SOFC-GT unit are given for an outgoing flow of electricity of 100 kW. The electrical and thermal efficiencies are calculated using the HHV of $CH_4$ of 55,484 kJ/kg [51]:

$$\eta^{el} = \frac{\dot{m}_{electricity}}{\dot{m}^+_{CH_4} \cdot HHV^{CH_4}} = 74.9\% \quad \text{(A13a)}$$

$$\eta^{th} = \frac{\sum_{k \in K} \dot{Q}_{SOFC}}{\dot{m}^+_{CH_4} \cdot HHV^{CH_4}} = 20.3\% \quad \text{(A13b)}$$

The reference flows for the metahantion unit are given for an incoming flow of electricity of 100 kW.

### Appendix A.2.3. Steam Network

In the steam network, steam is produced at very high pressure and distributed at multiple lower pressure levels. The pressure levels are selected to fit the production profiles of the P2G units. The parameters used to model the steam network are summarized in Table A12.

**Table A12.** Parameters for steam network.

| Type | Header Pressure [bar] | $T^{superheat}$ [°C] | Turbine |
|---|---|---|---|
| Production | 120 | 100 | yes |
| Distribution | 30 | 2 | yes |
| Distribution | 10 | 2 | yes |
| Distribution | 5 | 2 | yes |
| Distribution | 2 | 2 | no |
| Distribution | 1 | 2 | no |
| Distribution | 0.2 | 2 | no |

### Appendix A.2.4. CO$_2$ and CH$_4$ Storage

The storage tanks are modeled using the following equations:

$$SL_{tank,t+1} = SL_{tank,t} + \eta^{ch} \cdot \dot{M}^+_{fuel,t} - \frac{1}{\eta^{dch}} \cdot \dot{M}^-_{fuel,t} \quad \text{(A14a)}$$

$$SL_{tank,t} = f_{tank,t} \quad \forall t \in T \quad \text{(A14b)}$$

where $SL_{tank,t}$ represents the storage level of the tank at time step $t$, $\dot{M}^+_{fuel,t}$ and $\dot{M}^-_{fuel,t}$ the flow rates in and out of the unit at time step $t$, and $\eta^{ch}$, $\eta^{dch}$ the charging and discharging efficiencies. CO$_2$ is stored in liquid form at atmospheric pressure and temperature (i.e., 1 bar, 25 °C). Methane is also stored as a liquid, at the operating pressure of 1 bar and the corresponding temperature required for the liquid state, of −162 °C.

### Appendix A.2.5. Central Plants

The central plant in winter is modeled as a HP using a lake (at a constant temperature of 7.5 °C) as the heat source and CO$_2$ as the refrigerant. A summary of the parameters used for the central plant HP can be observed in table A13.

**Table A13.** Parameters for central plant HP.

| Unit | HP Central Plant |
|---|---|
| $T^{subcool}$ [°C] | 1 |
| $T^{superheat}$ [°C] | 2 |
| $\eta^{comp}$ [−] | 0.8 |
| $dT^{min, evap}$ [°C] | 5.5 |
| $dT^{min, cond}$ [°C] | 1 |

The reference flow of the central plant HP is $\dot{Q}^{cond} = \dot{m}_{CO_2} \cdot L^v_{CO_2}$ and the electricity consumption of the compressor and the heat extracted at the evaporator are calculated solving the thermodynamic cycle. The reference streams of the unit are given in Table A14 for a mass flow of $CO_2$ of 1 kg/s. The COP of the central plant HP is constant throughout the year, at 15.1.

Table A14. Streams for central plant HP (winter).

| Type | $T^{in}$ [°C] | $T^{out}$ [°C] | $\dot{Q}$ | $\dot{m}^-$ | $\dot{m}^+$ |
|---|---|---|---|---|---|
| Heat evaporator | 2 | 4 | $186.4 \cdot \frac{COP-1}{COP}$ kW | - | - |
| Heat condensor | 15 | 13 | 186.4 kW | - | - |
| Electricity | - | - | - | - | $186.4 \cdot \frac{1}{COP}$ kW |
| $CO_2^{vap}$ | - | - | - | 1 kg/s | - |
| $CO_2^{liq}$ | - | - | - | - | 1 kg/s |

The central plant in summer is modeled as a HE with the reference flow $\dot{Q} = \dot{m}_{CO_2} \cdot L^v_{CO_2}$ and a minimum temperature difference $dT_{min} = 5.5°C$. The reference streams of the unit are given in Table A15 for a mass flow of $CO_2$ of 1 kg/s.

Table A15. Streams for central plant HE (summer).

| Type | $T^{in}$ [°C] | $T^{out}$ [°C] | $\dot{Q}$ | $\dot{m}^-$ | $\dot{m}^+$ |
|---|---|---|---|---|---|
| Heat | 7.5 | 9.5 | 186.4 kW | - | - |
| $CO_2^{vap}$ | - | - | - | - | 1 kg/s |
| $CO_2^{liq}$ | - | - | - | 1 kg/s | - |

Appendix A.2.6. Investment Cost of Energy Conversion Technologies

The fixed and variable IC parameters, as well as the reference flows for the different units can be found in Table A16.

Table A16. Parameters for IC.

| Unit | $C^{inv,1}$ [€] | $C^{inv,2}$ [€/kW/€/m²] | Attribute |
|---|---|---|---|
| Boiler | 3990 | 110 | $\dot{Q}$ [kW] |
| Electrical heater | 968 | 13 | $\dot{Q}$ [kW] |
| Heat pumps/Ref cycle | 10,224 | 2232 | $\dot{m}^+_{electricity}$ [kW] |
| Battery stack | 825 | 1290 | $\max(f_{BAT,p,t})$ [kW] |
| Heat storage tank | 1421 | 1945 | V [kW] |
| Domestic hot water tank | 496 | 10,248 | V [kW] |
| PV panels | - | 247 | $A^{PV}$ [m²] |
| SOEC-SOFC | - | 4760 | $\max(\dot{m}^+_{electricity, SOEC}, \dot{m}^+_{electricity, SOFC})$ [kW] |
| HP CP (winter) | 5680 | 1240 | $\dot{m}^+_{electricity}$ [kW] |
| HE CP (summer) | 184 | 197 | $A^{HE}$ [m²] |

*Appendix A.3. Heat Distribution Cost*

The heat distribution cost of the networks is calculated using the formulation of [31]. First, the length of the network ($L^{DHN}$) is calculated based on the number of buildings ($n_b$), the land surface area ($A_l$) and a correlation coefficient (K) [31]:

$$L^{DHN} = 2 \cdot (n_b - 1) \cdot K \cdot \sqrt{\frac{A_l}{n_b}} \quad \text{(A15)}$$

And for each segment (between each two buildings):

$$L_k^{DHN} = \frac{L^{DHN}}{n_b} \tag{A16}$$

Next, the mass flow in the pipes is computed using the maximum heat flow in the pipe $\dot{Q}^{DHN}$ and the specific heat flows $q_{water} = c_{p,\,water}(T_{supply} - T_{return})$, $q_{CO_2} = l_v$:

$$\dot{m}_{max}^{DHN} = \frac{\dot{Q}^{DHN}}{q^{DHN}} \tag{A17}$$

And for each segment ($k$):

$$\dot{m}_k^{DHN} = \frac{\dot{Q}^{DHN} \cdot (n_b - k + 1)}{n_b \cdot q^{DHN}} \tag{A18}$$

Then, the diameter of the pipes ($d^{DHN}$) is calculated using the mass flow $\dot{m}^{DHN}$, the sizing velocity of the fluids (v) [52] and the density of the fluids ($\rho$):

$$d_k^{DHN} = \sqrt{\frac{4 \cdot \dot{m}_k^{DHN}}{\pi \cdot v \cdot \rho}} \tag{A19}$$

Finally, the investment cost ($C^{inv}$) of the networks is computed by summing up the different segments, using the cost coefficients $c_1$ and $c_2$ [52], an interest rate i = 5% and a lifetime lt = 60 years [31]:

$$\tau^{DHN} = \frac{(i+1)^{lt} - 1}{i \cdot (i+1)^{lt}} \tag{A20}$$

$$C^{inv} = \sum_{k=1}^{n-1} \frac{L_k^{DHN}(c_1 \cdot d_k^{DHN} + c_2)}{\tau^{DHN}} \tag{A21}$$

The values of the parameters present in the equations above can be found in Table A17.

Table A17. Network cost parameters.

| Parameter | Unit | Value ($CO_2$ Network) | Value ($H_2O$ Network) |
|---|---|---|---|
| $n_b$ | [-] | 11,903 | 11,903 |
| K | [-] | 0.23 | 0.23 |
| $A_l$ | [m$^2$] | 15,785,286 | 15,785,286 |
| $L^{DHN}$ | [km] | 3630.3 | 3630.3 |
| $q^{DHN}$ | [kJ/kg] | 186.4 | 18.8 |
| $\dot{Q}^{DHN}$ | [MW] | 2938.1 | 2942.7 |
| $\dot{m}_{max}^{DHN}$ | [t/s] | 15.8 | 156.5 |
| v | [m$^2$/s] | 3 (liquid), 6 (vapour) | 3 |
| $\rho$ | [kg/m$^3$] | 837.7 (liquid), 160.9 (vapour) | 1000 |
| $d_{max}^{DHN}$ | [m] | 4 (liquid), 10.4 (vapour) | 33.2 |
| $c_1$ | [€/m$^2$] | 5670 | 5670 |
| $c_2$ | [€] | 613 | 613 |
| i | [-] | 0.06 | 0.06 |
| lt | [-] | 60 | 60 |
| $C^{inv}$ | [M€/y] | 153.5 | 330.6 |

## Appendix A.4. RegBL Database Parameter Names

**Table A18.** RegBL database corresponding parameter notations.

| Parameter Description | Notation (This Paper) | Notation (RegBL) |
|---|---|---|
| Building category | - | GKAT |
| Building class | - | GKLAS |
| Building footprint area | $A_b$ | GAREA |
| Building number of floors | $N^{floors}$ | GASTW |
| Building rooftop area | $A_b^{roof}$ | FLAECHE |
| Building average solar irradiation | $I_b$ | MSTRAHLUNG |

## Appendix A.5. Energy Service Demand of Different Building Categories

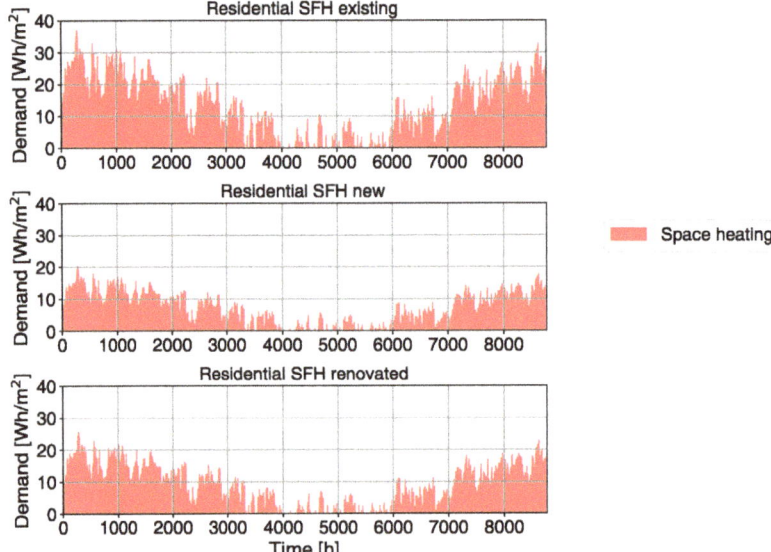

**Figure A2.** Hourly specific energy service demand of residential SFH buildings.

**Table A19.** Yearly specific energy service demand of residential SFH buildings.

| Building Renovation Stage | Space Heating [kWh/m$^2$] | Air Cooling [kWh/m$^2$] | Dom. Hot Water [kWh/m$^2$] | Electricity [kWh/m$^2$] |
|---|---|---|---|---|
| Existing | 80.3 | 0.0 | 13.6 | 18.2 |
| New | 44.0 | 0.0 | 13.6 | 18.2 |
| Renovated | 55.9 | 0.0 | 13.6 | 18.2 |

**Table A20.** Yearly specific energy service demand of residential MFH buildings.

| Building Renovation Stage | Space Heating [kWh/m$^2$] | Air Cooling [kWh/m$^2$] | Dom. Hot Water [kWh/m$^2$] | Electricity [kWh/m$^2$] |
|---|---|---|---|---|
| Existing | 80.3 | 0.0 | 17.8 | 18.4 |
| New | 44.0 | 0.0 | 17.8 | 18.4 |
| Renovated | 55.9 | 0.0 | 17.8 | 18.4 |

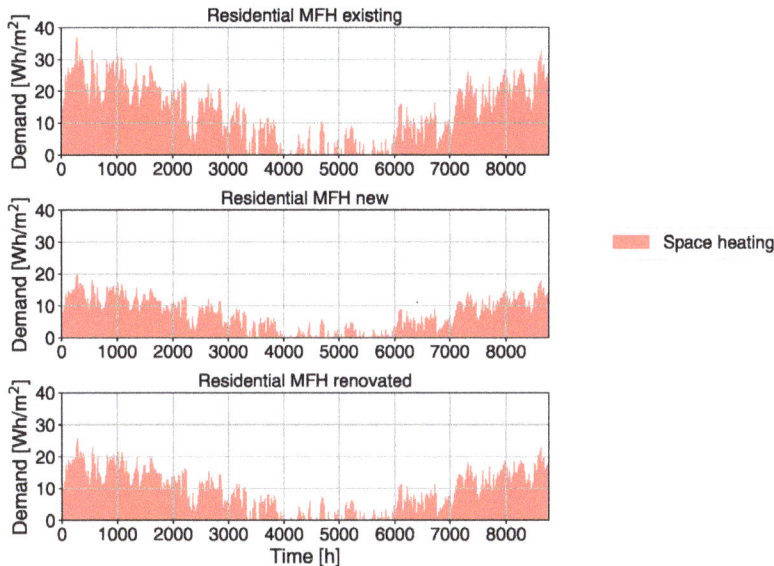

**Figure A3.** Hourly specific energy service demand of residential MFH buildings.

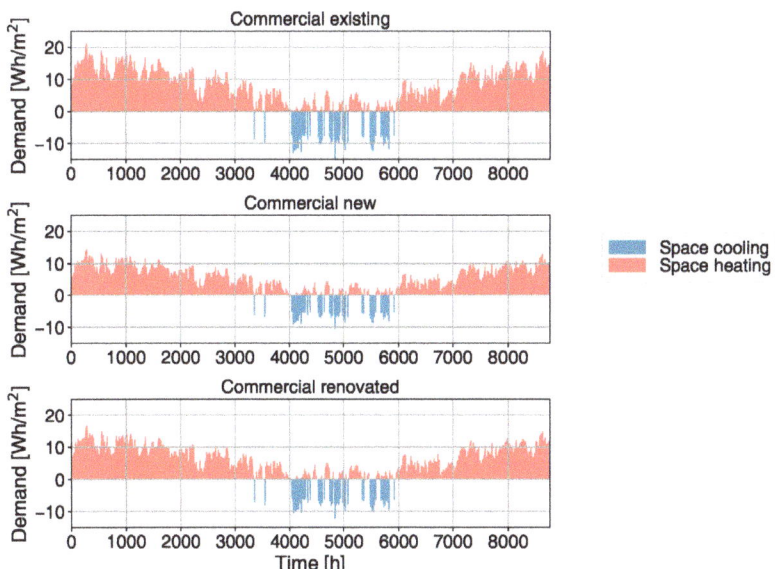

**Figure A4.** Hourly specific energy service demand of commercial buildings.

**Table A21.** Yearly specific energy service demand of commercial buildings.

| Building Renovation Stage | Space Heating [kWh/m$^2$] | Air Cooling [kWh/m$^2$] | Dom. Hot Water [kWh/m$^2$] | Electricity [kWh/m$^2$] |
|---|---|---|---|---|
| Existing | 49.2 | 3.3 | 1.8 | 114.4 |
| New | 33.5 | 2.3 | 1.8 | 114.4 |
| Renovated | 38.4 | 2.7 | 1.8 | 114.4 |

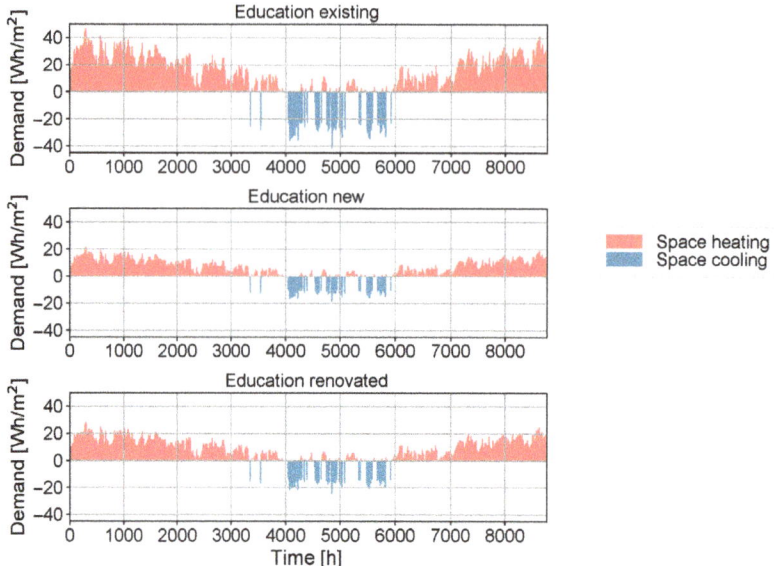

**Figure A5.** Hourly specific energy service demand of education buildings.

**Table A22.** Yearly specific energy service demand of education buildings.

| Building Renovation Stage | Space Heating [kWh/m²] | Air Cooling [kWh/m²] | Dom. Hot Water [kWh/m²] | Electricity [kWh/m²] |
|---|---|---|---|---|
| Existing | 91.8 | 9.7 | 4.5 | 23.8 |
| New | 41.9 | 4.4 | 4.5 | 23.8 |
| Renovated | 55.1 | 5.8 | 4.5 | 23.8 |

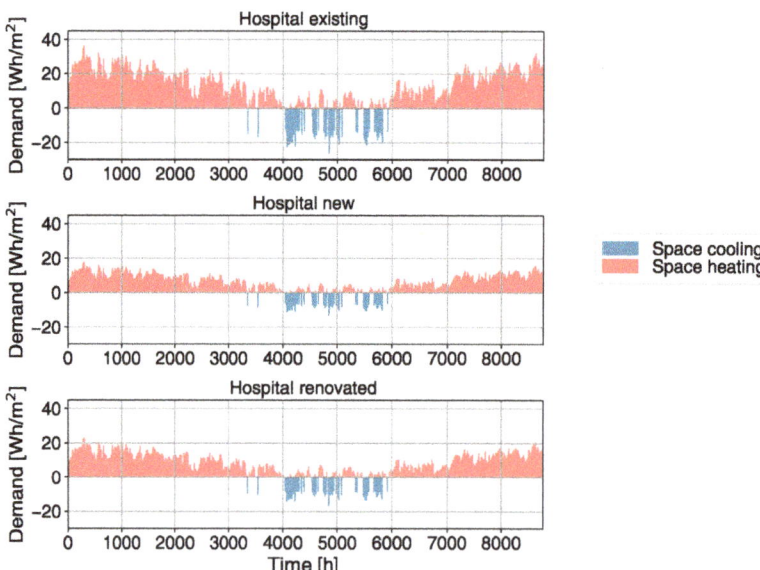

**Figure A6.** Hourly specific energy service demand of hospital buildings.

Table A23. Yearly specific energy service demand of hospital buildings.

| Building Renovation Stage | Space Heating [kWh/m$^2$] | Air Cooling [kWh/m$^2$] | Dom. Hot Water [kWh/m$^2$] | Electricity [kWh/m$^2$] |
|---|---|---|---|---|
| Existing | 83.5 | 5.8 | 34.1 | 34.0 |
| New | 41.2 | 2.8 | 34.1 | 34.0 |
| Renovated | 53.3 | 3.6 | 34.1 | 34.0 |

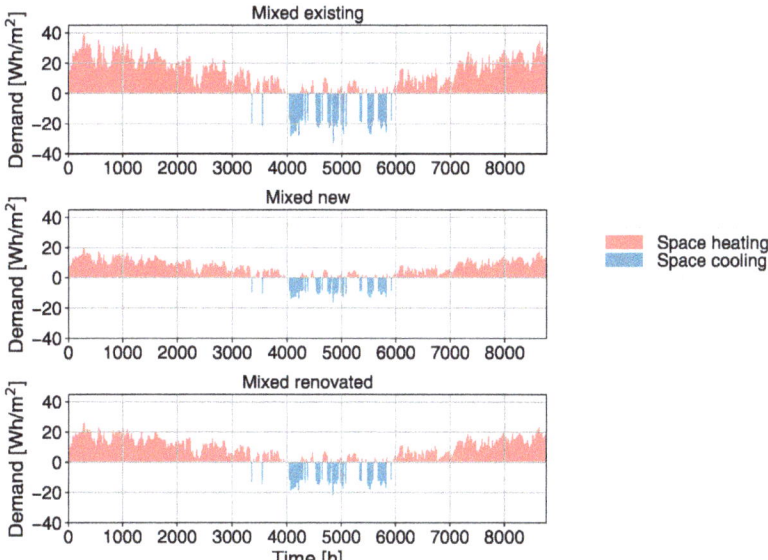

Figure A7. Hourly specific energy service demand of mixed buildings.

Table A24. Yearly specific energy service demand of mixed buildings.

| Building Renovation Stage | Space Heating [kWh/m$^2$] | Air Cooling [kWh/m$^2$] | Dom. Hot Water [kWh/m$^2$] | Electricity [kWh/m$^2$] |
|---|---|---|---|---|
| Existing | 81.6 | 7.5 | 11.9 | 28.4 |
| New | 40.9 | 3.7 | 11.9 | 28.4 |
| Renovated | 54.1 | 4.9 | 11.9 | 28.4 |

Table A25. Heat signature coefficients for all building types and ages.

| Building Type | Building Age | k1 [kW/(m².°C)] | k2 [kW/m²] | $T_{base,h}$ [°C] | $T_{base,c}$ [°C] |
|---|---|---|---|---|---|
| Residential SFH | existing | −1.52 | 23.59 | 15.52 | - |
| | new | −0.83 | 12.91 | 15.55 | - |
| | renovated | −1.06 | 16.43 | 15.5 | - |
| Residential MFH | existing | −1.52 | 23.59 | 15.52 | - |
| | new | −0.83 | 12.91 | 15.55 | - |
| | renovated | −1.06 | 16.43 | 15.5 | - |
| Administrative | existing | −1.87 | 26.51 | 14.18 | 25 |
| | new | −0.8 | 11.41 | 14.26 | 25 |
| | renovated | −1.15 | 16.29 | 14.17 | 25 |
| Commercial | existing | −0.84 | 13.81 | 16.44 | 25 |
| | new | −0.58 | 9.47 | 16.33 | 25 |
| | renovated | −0.67 | 10.89 | 16.25 | 25 |
| Education | existing | −2.03 | 28.84 | 14.21 | 25 |
| | new | −0.93 | 13.19 | 14.18 | 25 |
| | renovated | −1.22 | 17.32 | 14.2 | 25 |
| Hospital | existing | −1.44 | 23.54 | 16.34 | 25 |
| | new | −0.71 | 11.62 | 16.37 | 25 |
| | renovated | −0.91 | 14.96 | 16.44 | 25 |
| Mixed | existing | −1.86 | 27.83 | 14.98 | 25 |
| | new | −0.82 | 12.31 | 15.05 | 25 |
| | renovated | −1.1 | 16.37 | 14.94 | 25 |

*Appendix A.6. Results from Parallel Coordinates*

This section presents the detailed results presented schematically in Figures 16 and 17.

Table A26. Detailed results for Figure 16.

| Investment Cost M€/year | Operating Cost M€/year | CO$_2$ Emissions kt$_{CO_2}$/year | Buildings High IC - | Buildings Low IC - | CO$_2$ Activation - | Self Sufficiency % | PV Market M€/year | HP Market M€/year |
|---|---|---|---|---|---|---|---|---|
| 1.61  | 1.49 | 3.06 | 0      | 46,121 | 0  | 0   | 0.00 | 0.55 |
| 5.53  | 0.84 | 1.77 | 604    | 45,517 | 1  | 0   | 1.78 | 1.62 |
| 9.44  | 0.56 | 1.13 | 10,164 | 35,957 | 2  | 0   | 5.36 | 2.40 |
| 12.8  | 0.44 | 0.92 | 9088   | 37,033 | 3  | 24  | 7.61 | 2.39 |
| 14.03 | 0.35 | 0.76 | 8696   | 37,425 | 4  | 62  | 7.97 | 2.37 |
| 15.50 | 0.26 | 0.62 | 7826   | 38,293 | 6  | 89  | 7.96 | 2.35 |
| 17.56 | 0.20 | 0.50 | 7091   | 39,030 | 11 | 100 | 8.24 | 2.32 |
| 19.05 | 0.16 | 0.41 | 7143   | 38,978 | 16 | 100 | 8.09 | 2.29 |
| 20.86 | 0.13 | 0.34 | 5173   | 40,948 | 24 | 100 | 7.78 | 2.29 |
| 23.96 | 0.12 | 0.31 | 3816   | 42,305 | 48 | 100 | 7.76 | 2.21 |

Table A27. Detailed results for Figure 17.

| Population Density cap/km² | Building Density Buildings/km² | Network Cost k€/(100 m²) | El Imports GWh/year | El Exports GWh/year | NG Imports GWh/year | Investment Cost €/(100 m² year) | PV Market €/(100 m² year) | HP Market €/(100 m² year) |
|---|---|---|---|---|---|---|---|---|
| 40.11 | 10.15 | 27.33 | 2.39 | 3.40 | 0.45 | 21.50 | 17.55 | 4.79 |
| 58.50 | 17.00 | 35.00 | 2.31 | 3.09 | 0.56 | 21.38 | 17.14 | 4.94 |
| 113.57 | 24.25 | 25.83 | 1.76 | 2.96 | 0.33 | 15.65 | 13.06 | 3.23 |
| 116.61 | 31.49 | 23.22 | 1.71 | 2.48 | 0.33 | 16.40 | 13.35 | 3.72 |
| 160.73 | 46.59 | 26.20 | 2.12 | 3.40 | 0.33 | 16.25 | 13.34 | 3.50 |
| 180.15 | 36.63 | 21.53 | 4.70 | 6.01 | 0.89 | 12.60 | 10.29 | 2.77 |
| 187.03 | 31.44 | 34.56 | 3.72 | 5.14 | 0.67 | 19.49 | 16.4 | 4.09 |
| 196.87 | 62.05 | 26.64 | 3.11 | 4.74 | 0.56 | 17.69 | 14.50 | 3.87 |
| 199.46 | 63.62 | 18.84 | 3.17 | 2.97 | 0.89 | 11.77 | 9.24 | 2.89 |
| 214.65 | 57.61 | 29.01 | 3.15 | 6.20 | 0.45 | 19.35 | 16.36 | 3.79 |
| 260.72 | 63.40 | 24.15 | 4.29 | 5.89 | 1.03 | 11.89 | 9.63 | 2.65 |
| 276.87 | 84.11 | 34.66 | 3.96 | 7.11 | 0.67 | 20.66 | 16.96 | 4.50 |
| 291.60 | 76.49 | 39.68 | 4.48 | 11.13 | 0.33 | 19.03 | 15.89 | 3.98 |
| 309.17 | 108.03 | 49.88 | 3.31 | 5.86 | 0.56 | 29.96 | 24.60 | 6.55 |
| 386.72 | 60.20 | 19.02 | 4.56 | 5.63 | 1.19 | 16.54 | 13.34 | 3.62 |
| 396.37 | 108.66 | 12.55 | 4.02 | 5.36 | 1.00 | 10.33 | 8.34 | 2.30 |
| 426.46 | 108.21 | 24.20 | 4.96 | 7.47 | 0.78 | 19.21 | 115.79 | 4.14 |
| 449.19 | 77.06 | 18.32 | 5.62 | 7.31 | 1.33 | 13.30 | 11.00 | 2.77 |
| 451.83 | 69.68 | 12.20 | 4.36 | 6.58 | 1.74 | 13.84 | 11.05 | 2.95 |
| 453.07 | 93.66 | 19.60 | 8.45 | 8.57 | 2.00 | 15.85 | 12.65 | 3.71 |
| 516.24 | 104.00 | 24.50 | 13.50 | 22.37 | 2.80 | 14.61 | 12.30 | 2.90 |
| 522.15 | 108.09 | 24.30 | 9.19 | 11.76 | 2.11 | 15.77 | 12.72 | 3.60 |
| 523.17 | 158.29 | 27.67 | 7.18 | 13.80 | 1.25 | 13.60 | 19.49 | 5.04 |
| 610.19 | 109.39 | 25.75 | 8.43 | 15.36 | 0.94 | 17.77 | 14.90 | 3.61 |
| 612.01 | 153.92 | 18.14 | 7.70 | 9.80 | 1.92 | 15.01 | 12.11 | 3.42 |
| 620.96 | 172.37 | 24.72 | 17.53 | 17.35 | 4.86 | 18.10 | 14.39 | 4.26 |
| 732.32 | 185.79 | 21.03 | 4.15 | 8.62 | 0.58 | 17.98 | 14.97 | 3.76 |
| 741.13 | 239.07 | 27.94 | 6.35 | 13.27 | 0.92 | 19.20 | 15.92 | 4.07 |

## References

1. Anderson, J.E.; Wulfhorst, G.; Lang, W. Energy analysis of the built environment—A review and outlook. *Renew. Sustain. Energy Rev.* **2015**, *44*, 149–158. [CrossRef]
2. *Climate Change 2014: Synthesis Report*; Technical Report; Intergovernmental Panel on Climate Change: Geneva, Switzerland, 2015.
3. International Energy Agency. *World Energy Balances 2018*; Organization for Economic: Paris, France, 2018; OCLC: 1054217453.
4. Sinha, S.; Chandel, S. Review of software tools for hybrid renewable energy systems. *Renew. Sustain. Energy Rev.* **2014**, *32*, 192–205. [CrossRef]
5. Lund, P.D.; Mikkola, J.; YpyÃ, J. Smart energy system design for large clean power schemes in urban areas. *J. Clean. Prod.* **2015**, *103*, 437–445. [CrossRef]
6. Mancarella, P. MES (multi-energy systems): An overview of concepts and evaluation models. *Energy* **2014**, *65*, 1–17. [CrossRef]
7. Niemi, R.; Mikkola, J.; Lund, P. Urban energy systems with smart multi-carrier energy networks and renewable energy generation. *Renew. Energy* **2012**, *48*, 524–536. [CrossRef]
8. Fishbone, L.G.; Abilock, H. Markal, a linear-programming model for energy systems analysis: Technical description of the bnl version. *Energy Res.* **1981**, *5*, 353–375. [CrossRef]
9. Ong, Y.S.; Nair, P.B.; Keane, A.J. Evolutionary optimization of computationally expensive problems via surrogate modeling. *AIAA J.* **2003**, *41*, 687–696. [CrossRef]
10. Eisenhower, B.; O'Neill.; Narayanan, S.; Fonoberov, V.A.; Mezić, I. A methodology for meta-model based optimization in building energy models. *Energy Build.* **2012**, *47*, 292–301. [CrossRef]
11. Ries, R.; Mahdavi, A. Integrated Computational Life-Cycle Assessment of Buildings. *J. Comput. Civil Eng.* **2001**, *15*, 59–66. [CrossRef]
12. Sunikka-Blank, M.; Galvin, R. Introducing the prebound effect: the gap between performance and actual energy consumption. *Build. Res. Inf.* **2012**, *40*, 260–273. [CrossRef]
13. Kofoworola, O.F.; Gheewala, S.H. Life cycle energy assessment of a typical office building in Thailand. *Energy Build.* **2009**, *41*, 1076–1083. [CrossRef]
14. Ochoa, L.; Hendrickson, C.; Matthews, H.S. Economic Input-output Life-cycle Assessment of U.S. Residential Buildings. *J. Infrastruct. Syst.* **2002**, *8*, 132–138. [CrossRef]
15. Junnila, S.; Horvath, A.; Guggemos, A.A. Life-Cycle Assessment of Office Buildings in Europe and the United States. *J. Infrastruct. Syst.* **2006**, *12*, 10–17. [CrossRef]
16. Junnila, S.; Horvath, A. Life-Cycle Environmental Effects of an Office Building. *J. Infrastruct. Syst.* **2003**, *9*, 157–166. [CrossRef]
17. Glaeser, E.L.; Kahn, M.E. The greenness of cities: Carbon dioxide emissions and urban development. *J. Urban Econ.* **2010**, *67*, 404–418. [CrossRef]
18. Kennedy, C.; Steinberger, J.; Gasson, B.; Hansen, Y.; Hillman, T.; HavrÃ¡nek, M.; Pataki, D.; Phdungsilp, A.; Ramaswami, A.; Mendez, G.V. Greenhouse Gas Emissions from Global Cities. *Environ. Sci. Technol.* **2009**, *43*, 7297–7302. [CrossRef] [PubMed]
19. Troy, P.; Holloway, D.; Pullen, S.; Bunker, R. Embodied and Operational Energy Consumption in the City. *Urban Policy Res.* **2003**, *21*, 9–44. [CrossRef]
20. Jones, C.M.; Kammen, D.M. Quantifying Carbon Footprint Reduction Opportunities for U.S. Households and Communities. *Environ. Sci. Technol.* **2011**, *45*, 4088–4095. [CrossRef]
21. Howard, B.; Parshall, L.; Thompson, J.; Hammer, S.; Dickinson, J.; Modi, V. Spatial distribution of urban building energy consumption by end use. *Energy Build.* **2012**, *45*, 141–151. [CrossRef]
22. Gurney, K.R.; Razlivanov, I.; Song, Y.; Zhou, Y.; Benes, B.; Abdul-Massih, M. Quantification of Fossil Fuel $CO_2$ Emissions on the Building/Street Scale for a Large U.S. City. *Environ. Sci. Technol.* **2012**, *46*, 12194–12202. [CrossRef]
23. Keirstead, J.; Jennings, M.; Sivakumar, A. A review of urban energy system models: Approaches, challenges and opportunities. *Renew. Sustain. Energy Rev.* **2012**, *16*, 3847–3866. [CrossRef]
24. Stadler, P.; Girardin, L.; Ashouri, A.; Maréchal, F. Contribution of Model Predictive Control in the Integration of Renewable Energy Sources within the Built Environment. *Front. Energy Res.* **2018**, *6*. [CrossRef]

25. Suciu, R.; Girardin, L.; Maréchal, F. Energy integration of $CO_2$ networks and power to gas for emerging energy autonomous cities in Europe. *Energy* **2018**, *157*, 830–842. [CrossRef]
26. Ruesch, F.; Kolb, M.; Gautschi, T.; Rommel, M. Heat and cold supply for neighborhoods by means of seasonal borehole storage and low temperature energetic cross linking. In Proceedings of the International Conference CISBAT, Lausanne, Switzerland, 4–6 September 2013.
27. De Carli, M.; Galgaro, A.; Pasqualetto, M.; Zarrella, A. Energetic and economic aspects of a heating and cooling district in a mild climate based on closed loop ground source heat pump. *Appl. Thermal Eng.* **2014**, *71*, 895–904. [CrossRef]
28. Bestenlehner, D. Stübler (2014) Energetisches Einsparpotential eines kalten Nahwärmenetzes zur Wärmeversorgung eines Stadtquartiers im Vergleich zu einem konventionellen Nahwärmenetz, 24. In Porceedings of the Symposium Thermische Solarenergie, Bad Staffelstein, Bavaria, Germany, 7–9 May 2014.
29. Kräuchi, P.; Kolb, M. Simulation thermischer Arealvernetzung mit IDA-ICE. In Porceedings of the Fourth German-Austrian IBPSA Conference-Berlin University of the Arts, Berlin, Germany, 26–28 September 2012; pp. 205–211.
30. Molyneaux, A.; Leyland, G.; Favrat, D. Environomic multi-objective optimisation of a district heating network considering centralized and decentralized heat pumps. *Energy* **2010**, *35*, 751–758. [CrossRef]
31. Girardin, L.; Marechal, F.; Dubuis, M.; Calame-Darbellay, N.; Favrat, D. EnerGis: A geographical information-based system for the evaluation of integrated energy conversion systems in urban areas. *Energy* **2010**, *35*, 830–840. [CrossRef]
32. Facchinetti, E.; Favrat, D.; Maréchal, F. Innovative Hybrid Cycle Solid Oxide Fuel Cell-Inverted Gas Turbine with $CO_2$ Separation. *Fuel Cells* **2011**, *11*, 565–572. [CrossRef]
33. Wang, L.; Pérez-Fortes, M.; Madi, H.; Diethelm, S.; Herle, J.V.; Maréchal, F. Optimal design of solid-oxide electrolyzer-based power-to-methane systems: A comprehensive comparison between steam electrolysis and co-electrolysis. *Appl. Energy* **2018**, *211*, 1060–1079. [CrossRef]
34. Marechal, F.; Kalitventzeff, B. Targeting the integration of multi-period utility systems for site scale process integration. *Appl. Thermal Eng.* **2003**, *23*, 1763–1784. [CrossRef]
35. Schütz, T.; Schiffer, L.; Harb, H.; Fuchs, M.; Müller, D. Optimal design of energy conversion units and envelopes for residential building retrofits using a comprehensive MILP model. *Appl. Energy* **2017**, *185*, 1–15. [CrossRef]
36. Weber, C.; Maréchal, F.; Favrat, D.; Kraines, S. Optimization of an SOFC-based decentralized polygeneration system for providing energy services in an office-building in Tōkyō. *Appl. Thermal Eng.* **2006**, *26*, 1409–1419. [CrossRef]
37. Ashouri, A.; Fux, S.S.; Benz, M.J.; Guzzella, L. Optimal design and operation of building services using mixed-integer linear programming techniques. *Energy* **2013**, *59*, 365–376. [CrossRef]
38. Fazlollahi, S.; Mandel, P.; Becker, G.; Maréchal, F. Methods for multi-objective investment and operating optimization of complex energy systems. *Energy* **2012**, *45*, 12–22. [CrossRef]
39. Oldewurtel, F.; Ulbig, A.; Morari, M.; Andersson, G. Building control and storage management with dynamic tariffs for shaping demand response. In Proceedings of the 2011 2nd IEEE PES International Conference and Exhibition on Innovative Smart Grid Technologies, Manchester, UK, 5–7 December 2011; pp. 1–8. [CrossRef]
40. Swiss-Federal-Office-of Statistics, S.F.O. *Catalogue des caractÃšres. Registre fédéral des bÃtiments et des logements. Version 4*; Technical Report; Swiss Federal Statistical Office: NeuchÃtel, Switzerland, 2017. Available online: https://www.housing-stat.ch/fr/accueil.html (accessed on 30 May 2019)
41. Eidgenössisches Departement fÃŒr Umwelt, Verkehr, Energie und.; Kommunikation UVEK. *Solarpotentialanalyse fur Sonnendach.ch Schlussbericht*; Technical Report; Bundesamt für Energie: Zurich, Switzerland, 2016. Available online: https://www.bfe.admin.ch/bfe/de/home/versorgung/statistik-und-geodaten/geoinformation/geodaten/solar/solarenergie-eignung-hausdach.html (accessed on 4 February 2019).
42. SIA. *Données d'utilisation des locaux pour l'énergie et les installations du bÃtiment (SIA 2024)*; Technical Report. Available online: https://shop.sia.ch/normenwerk/architekt/sia_2024/f/2015/D/Product.html (accessed on 27 November 2018)
43. Stadler, P.; Ashouri, A.; Maréchal, F. Model-based optimization of distributed and renewable energy systems in buildings. *Energy Build.* **2016**, *120*, 103–113. [CrossRef]

44. Rager, J.M.F. Urban Energy System Design from the Heat Perspective using mathematical Programming including thermal Storage. EPFL: Lausanne, Switzerland, 2015.
45. Davies, D.L.; Bouldin, D.W. A Cluster Separation Measure. *IEEE Trans. Pattern Anal. Mach. Intell.* **1979**, *PAMI-1*, 224–227. [CrossRef]
46. Luthander, R.; Widén, J.; Nilsson, D.; Palm, J. Photovoltaic self-consumption in buildings: A review. *Appl. Energy* **2015**, *142*, 80–94. [CrossRef]
47. Ashouri, A. Simultaneous Design and Control of Energy Systems. Ph.D. Thesis, Eidgenössische Technische Hochschule ETH Zürich, Zürich, Switzerland, 2014.
48. Fux, S.F.; Ashouri, A.; Benz, M.J.; Guzzella, L. EKF-based self-adaptive thermal model for a passive house. *Energy Build.* **2014**, *68*, 811–817. [CrossRef]
49. Hoval. *Hoval Wärmepumpen Dimensionierungshilfen*; Technical Report; Hoval: Feldmeilen, Switzerland, 2014. Available online: https://www.hoval.ch/de_CH/Effiziente-Wärmepumpen/waermepumpe.html (accessed on 27 May 2019)
50. Koller, M.; Borsche, T.; Ulbig, A.; Andersson, G. Defining a degradation cost function for optimal control of a battery energy storage system. In Prodeedings of the 2013 IEEE Grenoble Conference, Grenoble, France, 16–20 June 2013; pp. 1–6.
51. Shen, V. Standard Reference Simulation Website, NIST Standard Reference Database 173. Available online: https://www.nist.gov/programs-projects/nist-standard-reference-simulation-website (accessed on 20 July 2019).
52. Henchoz, S.; Weber, C.; Maréchal, F.; Favrat, D. Performance and profitability perspectives of a $CO_2$-based district energy network in Geneva's City Centre. *Energy* **2015**, *85*, 221–235. [CrossRef]

© 2019 by the authors. Licensee MDPI, Basel, Switzerland. This article is an open access article distributed under the terms and conditions of the Creative Commons Attribution (CC BY) license (http://creativecommons.org/licenses/by/4.0/).

Article

# Economic Health-Aware LPV-MPC Based on System Reliability Assessment for Water Transport Network

Fatemeh Karimi Pour [1], Vicenç Puig [1,*] and Gabriela Cembrano [1,2]

1. Automatic Control Department, Universitat Politècnica de Catalunya Institut de Robòtica i Informàtica Industrial (CSIC-UPC), C/. Llorens i Artigas 4-6, 08028 Barcelona, Spain
2. Cetaqua, Water Technology Centre, Ctra. d'Esplugues 75, Cornellà de Llobregat, 08940 Barcelona, Spain
* Correspondence: vpuig@iri.upc.edu; Tel.: +34-934-015-752
† This paper is an extended version of our paper published in 2018 IEEE Conference on Control Technology and Applications (CCTA), Copenhagen, 21–24 August 2018; pp. 187–192.

Received: 28 May 2019; Accepted: 29 July 2019; Published: 5 August 2019

**Abstract:** This paper proposes a health-aware control approach for drinking water transport networks. This approach is based on an economic model predictive control (MPC) that considers an additional goal with the aim of extending the components and system reliability. The components and system reliability are incorporated into the MPC model using a Linear Parameter Varying (LPV) modeling approach. The MPC controller uses additionally an economic objective function that determines the optimal filling/emptying sequence of the tanks considering that electricity price varies between day and night and that the demand also follows a 24-h repetitive pattern. The proposed LPV-MPC control approach allows considering the model nonlinearities by embedding them in the parameters. The values of these varying parameters are updated at each iteration taking into account the new values of the scheduling variables. In this way, the optimization problem associated with the MPC problem is solved by means of Quadratic Programming (QP) to avoid the use of nonlinear programming. This iterative approach reduces the computational load compared to the solution of a nonlinear optimization problem. A case study based on the Barcelona water transport network is used for assessing the proposed approach performance.

**Keywords:** drinking water networks; reliability; economic cost; model predictive control; linear parameter varying

---

## 1. Introduction

The management of the urban water cycle (UWC) is a subject of increasing interest taking into account its social, economic, and environmental impacts [1]. Drinking Water Networks (DWNs) are critical infrastructures in urban environments. DWNs are also of vital importance for supporting all kinds of social activities. For the individuals inhabiting a modern city, the water supply service is one of the basic requirements. As the progress of society and the evolution of human civilizations, a growing number of people migrate into cities. Hence, the increasing complexity of the DWNs would generate some complications for the management under multiple objectives, such as economic operations, as well as safety, reliability and sustainability. Moreover, maintaining the quality of the water supplied is another important objective that has already been addressed (e.g., [2]).

DWNs are large-scale systems that have to be flexible and reliable to deal with continuously varying situations, such as unanticipated changes in the demands or faults in some of the elements [3]. According to the literature [4,5], the main goal of the operational control of a DWN is to satisfy the consumer demand, and the operational management of water networks seeks to continuously supply water to the consumer with appropriate quality levels while minimizing production and transport cost, as well as guaranteeing safety levels in tanks and a sustainable source management strategy.

Therefore, the optimal operational management of these systems is a multi-criteria management problem and poses a complicated challenge to the water stakeholder in charge of the operation. In the last decades, MPC has started to attract the attention of both academia and industry due to the possibility of dealing with type of problem including energy optimization and physical load reduction [6]. In general, the MPC approach (using the receding horizon strategy) determines the optimal control action from a sequence of open-loop control actions ahead in a prediction horizon minimizing a set of control objectives and satisfying a set of constraints considering the system dynamic model and physical/operational limits. Furthermore, MPC enables accounting for the multivariable input and output nature, the demand forecasting requirement, and complex multi-objective operational goals of water networks (see, e.g., [7,8]). Generally, standard MPC is formulated as an optimization problem that penalizes the tracking error [9]. Although this method guarantees that the set-point is achieved in a reasonable time, it does not ensure that the evolution among set-points is achieved in an effective manner. However, the common operational goal of many process industries, such as DWNs, is the minimization of economic costs related to energy consumption and water production. To this aim, Economic MPC (EMPC) contributes a systematic approach for optimizing economic performance [10]. The optimization problem behind the EMPC strategy is responsible for finding a family of the optimal set-points taking into account economic profits instead of steering the controlled system to a given set-point [11].

The application of health-aware control strategies based on the system and components reliability allows ensuring the quality of service. To preserve the system reliability, the controller should not only to reduce the operational costs but preserve the actuator health. In this paper, actuator health monitoring is achieved by estimating the reliability of each actuator according to the actuator operational information available.

Recently, system reliability has been considered in the control system design in the context of a Prognosis and Health Management (PHM) framework. Reliability is the ability of a system (or component) to perform its expected function [12]. In this context, reliability allows forecasting the remaining useful life and to anticipate future system faults given the state of its components [13,14].

In recent years, the problem of assessing the lifetime and reliability of the system and its components has received increasing attention. In [15], the actuator lifetime is considered as a controlled parameter to reduce maintenance cost by including an additional objective in the optimal controller. The reliability of a bearing according to its defect growth is estimated by comparing the fatigue crack propagation with the estimation from the diagnostic model in [16]. On the other side, the MPC approach has been proven to be an adequate strategy for implementing health-aware controllers (HAC). In the HAC approach, the online prognostic information of the system is used to adjust the control actions or to develop the mission objective in order to maintain a high level of system health [17]. In [18], an MPC approach is proposed that involves the actuator usage as constraints with the objective to keep the accumulated utilization under a safety level at the end of the mission.

DWN reliability depends on several factors such as the pumps and valves failures rates and quantity/quality of the water, among others [17]. The actuator reliability is usually modeled using an exponential function of the control input [14,19]. On the other hand, the system reliability can be determined by combining each actuator reliability and the interconnection topology. The reliability of water distribution network has already been address in the literature [20]. In [21], the reliability analysis methodologies of water distribution systems are described based on tailor-made "lumped supply–lumped demand" approach and a Monte Carlo framework. In [22], a structure for devising such a proactive risk-based integrity-monitoring approach for the control of urban water distribution networks is proposed.

In previous works, the reliability of actuators in the DWN control has been considered by including a wear index [3,17] or by adding additional constraints to preserve actuator reliability [5]. In those previous methods, the reliability is considered at the actuator level but not at the system level considering the interconnection topology. The main reason is because in this case a set of nonlinear constraints should be considered leading to a nonlinear MPC. Moreover, Economic Nonlinear MPC (ENMPC) is usually computationally expensive and, in general, there is no guarantee that the solution of the optimization problem is the global optimum. Another way of solving the optimization problem in the case of a nonlinear system is translating the nonlinear problem to a quadratic problem by means of linearization approach. In this manner, the system has to be linearized at each iteration, considering the system modeled by an incremental model [23,24]. This approach has been recently improved considering Linear Parameter Varying (LPV) models [25]. LPV models are a class of models with linear structure but including a set of varying parameters that are scheduled online. These models allow representing a nonlinear system into a linear-like system with varying parameters by using the nonlinear embedding approach [26].

This paper proposes a health-aware LPV-MPC controller that uses PHM information provided by the online system reliability evaluation. The reliability model is included into the MPC model. The augmented model including both the reliability and DWN models is transformed into an LPV model. Hence, the control actions obtained satisfy the control objectives/constraints and at the same time preserve the system reliability and lifetime. Finally, the case study considered in this paper to show the effectiveness of the proposed approach is based on a part of a real drinking water transport network of Barcelona. The dynamic model of transport network considered is based on previous works [8,27] where only the flow model is used.

The remainder of the paper is organized as follows. In Section 2, the control-oriented model of DWN from Barcelona is presented. In Section 3, the system reliability modeling is exhibited. In Section 4, the reliability model is integrated into the control algorithm and the economic health-aware controller is presented based on an LPV-MPC approach. In Section 5, results of implementing the proposed control strategy to the DWN network as a case study are compiled. Finally, the conclusion of this work is illustrated and some research lines for future work are introduced in Section 6.

## 2. EMPC of Drinking Water Transport Network

### 2.1. Control-Oriented Modeling

Several control-oriented modeling approaches for DWNs have been proposed in the literature (see, e.g., [28,29]) depending on the layer (transport or distribution) considered. The water transport network is in charge of transporting the water from the sources (typically rivers) to the tanks that supply water to the water distribution network. On the other hand, the water distribution network distributes water to the consumers from the tanks. A suitable description of the dynamic model for the control of the water transport network is based on considering a flow model [8]. The pressure relations are typically relevant for the control of the water distribution network since water should be distributed to the consumers at some pre-established pressure levels [29]. Since this paper is focused on water transport networks, a modeling approach that is based on a flow model is considered that follows the principles introduced by the authors of [8].

Considering the set of compositional elements (as e.g., tanks, valves, pumps, and pipes) and the modeling methodology of each component in the DWN proposed in [8], the control-oriented model of DWN can be described by a linear discrete-time difference-algebraic equations for all time instant $k \in \mathbb{Z}_+$:

$$x(k+1) = Ax(k) + Bu(k) + B_d d_m(k), \tag{1a}$$
$$0 = E_u u(k) + E_d d_m(k), \tag{1b}$$
$$y(k) = Cx(k), \tag{1c}$$

The difference equations (Equation (1a)) model the storage tanks volume dynamics, and the algebraic equations in (Equation (1b)) characterize the network static flow relations (i.e., mass balance at junction nodes). $x(k) \in \mathbb{R}^{n_x}$ is the storage tanks volumes, $u(k) \in \mathbb{R}^{n_u}$ is the actuator (pumps and valves) manipulated flows, $y \in \mathbb{R}^{n_y}$ denotes measured outputs and $d_m(k) \in \mathbb{R}^{n_m}$ is the demanded flow that can be considered as measured (or forecasted) disturbances. $A \in \mathbb{R}^{n_x \times n_x}$, $B \in \mathbb{R}^{n_x \times n_u}$, $B_d \in \mathbb{R}^{n_x \times n_d}$, $E_u \in \mathbb{R}^{n_d \times n_u}$, $E_d \in \mathbb{R}^{n_d \times n_d}$ and $C \in \mathbb{R}^{n_y \times n_x}$ are time-invariant matrices of suitable dimensions dictated by the network topology.

## 2.2. EMPC Formulation of DWN

The application of MPC methods to DWNs allows computing, ahead of time, the optimal actuator set-points to enhance the network performance [1]. This leads to the minimization of a multi-objective convex cost function that includes the following operational goals typically considered in the DWN management:

- Economic: Provide the required amount of water minimizing water production and transport costs.
- Safety: Guarantee the safety levels in each storage tanks that guarantee the water supply under unexpected changes in the demand up to some level.
- Smoothness: Operate actuators in the DWN under smooth control actions to avoid overpressure in pipes and damage in actuators.

### 2.2.1. Minimization of Water Production and Transport Costs

The main control objective of the DWN is to minimize the costs that are related to water production costs and electrical costs associated to pumping. Transferring drinking water from the sources to the tanks through the network includes important electricity costs because of the need of pumping. Hence, the cost function associated to this objective can be formulated as

$$\ell_e(k) \triangleq \alpha(k)^\top W_e u(k), \qquad (2)$$

where $\alpha(k) \triangleq (\alpha_1 + \alpha_2(k)) \in \mathbb{R}^{n_u}$, $\alpha_1 \in \mathbb{R}^{n_u}$ is a fixed water-production cost which is constant and a time-varying water pumping electrical cost is presented by $\alpha_2 \in \mathbb{R}^{n_u}$ that changes at each time instant $k$ according to the dynamic electricity tariff. $W_e$ denotes the weighting term that indicates the prioritization of the economic control objective.

### 2.2.2. Guarantee Safety Management of Water Storage

With the aim of the preserving water supply despite the variation of water demands between two consecutive MPC iterations, a suitable safety volume for each storage tank is needed to be maintained. A possible mathematical expression for this objective can be expressed as follows

$$\ell_s(k) \triangleq \begin{cases} \|x(k) - x_s\|_p, & \text{if } x(k) \leq x_s \\ 0, & \text{otherwise} \end{cases} \qquad (3)$$

where $x_s$ denotes the vector of the safety volumes for all the tanks. To avoid the piecewise linear form of this formulation, this cost function can also be formulated by means of a soft constraint by adding a slack variable $\xi$ that can be expressed as

$$\ell_s(k) \triangleq \xi^\top(k) W_s \xi(k), \qquad (4)$$

where $W_s$ is diagonal positive definite matrix and the following soft constraint is included

$$x(k) \geq x_s - \xi(k). \qquad (5)$$

### 2.2.3. Smoothing of Control Actions

The actuators in the DWN include valves and pumps. Thus, the flow-based control actions determined by the MPC controller should be smooth in order to extend the component lifespan.

To ensure the smoothing effect, the slew rate of the control actions between two consecutive time instants is penalized according to

$$\ell_{\Delta u}(k) \triangleq \Delta u(k)^\top W_{\Delta u} \Delta u(k), \tag{6}$$

where $\ell_{\Delta u}(k)$ corresponds to the penalization of control signal variations $\Delta u(k) \triangleq u(k) - u(k-1)$, and $W_{\Delta u}$ is a diagonal positive definite matrix.

The controller should also operate actuators and tanks inside their bounds and extend the reliability of the system, as presented below. Thus, the MPC optimization problem should be solved considering as constraints the mathematical model of the DWN (Equation (1)) and the operational constraints defined by

$$x(k) \in \mathbb{X} \triangleq \{x(k) \in \mathbb{R}^{n_x} \mid \underline{x} \le x(k) \le \overline{x}\}, \quad \forall k \tag{7a}$$

$$u(k) \in \mathbb{U} \triangleq \{u(k) \in \mathbb{R}^{n_u} \mid \underline{u} \le u(k) \le \overline{u}\}, \quad \forall k \tag{7b}$$

where vectors $\underline{x} \in \mathbb{R}^{n_x}$ and $\overline{x} \in \mathbb{R}^{n_x}$ characterize the minimum and maximum physical state values (tank volumes) of the DWN network, respectively. Similarly, $\underline{u} \in \mathbb{R}^{n_u}$ and $\overline{u} \in \mathbb{R}^{n_u}$ determine the minimum and maximum possible value of manipulated variables, respectively. The EMPC controller design is based on minimizing the following finite horizon cost function

$$J = \sum_{l=0}^{N_p} (\ell_e(l|k) + \ell_s(l|k) + \ell_{\Delta u}(l|k)), \tag{8}$$

where $N_p$ is the prediction horizon. At each time instant, the optimization problem

$$\min_{\mathbf{u}(k), x(k), \xi(k)} J(u(k), x(k), \xi(k)), \tag{9a}$$

subject to:

$$x(l+1|k) = Ax(l|k) + Bu(l|k) + B_d d_m(l|k), \quad l = 0, \cdots, N_p - 1 \tag{9b}$$

$$0 = E_u u(l|k) + E_d d_m(k), \quad l = 0, \cdots, N_p - 1 \tag{9c}$$

$$x(l|k) \ge x_s - \xi(l|k), \quad l = 1, \cdots, N_p \tag{9d}$$

$$u(l|k) \in \mathbb{U}, \quad l = 0, \cdots, N_p - 1 \tag{9e}$$

$$x(l|k), \in \mathbb{X}, \quad l = 1, \cdots, N_p \tag{9f}$$

$$\xi(l|k) > 0, \quad l = 0, \cdots, N_p \tag{9g}$$

$$x(0|k) = x(k), \tag{9h}$$

is solved online, obtaining the optimal sequences $\mathbf{u}^*(k) = \{u(l|k)\}_{l \in \mathbb{Z}_{[0, N_p-1]}}$, $\mathbf{x}^*(k) = \{x(l|k)\}_{l \in \mathbb{Z}_{[1, N_p]}}$ and $\boldsymbol{\xi}^*(k) = \{xi(l|k)\}_{l \in \mathbb{Z}_{[1, N_p]}}$. According to receding horizon philosophy [30], this technique consists of solving the optimization problem in Equation (9a) from the current time instant $k$ to $k + N_p$ using $x(0|k)$ as the initial condition obtained from measurements (or state estimation) at time $k$. According to the philosophy behind the MPC technique [30], only the first value $u^*(0|k)$ from the optimal input sequence $\mathbf{u}^*(k)$ is applied to the system. In this way, the feedback control is included in the controller to make sense a closed loop controller for controlling the system. At time $k + 1$, to compute $u^*(0|k+1)$, the optimization problem in Equation (9a) is solved again from $k + 1$ to $k + 1 + N_p$, updating initial states $x(0|1+k)$ from measurements (or state estimation) at time $k + 1$. The same procedure is repeated for the following time instants. Moreover, the constraint in Equation (9g) is included to force that the slack variable is a positive value.

## 3. Reliability Assessment

### 3.1. Failure Rate and Reliability Concept

As mentioned above, the reliability is the ability of a system (or component) to perform its expected functions and it is defined as follows.

**Definition 1.** *Reliability is characterized as the probability that units, components, types of tools and systems will operate their predesignated role for a specified period of time according to some operating conditions [31].*

More specifically, it is the probability of success in performing a task or reaching a desired property in the process, based on the right component operation. Mathematically, reliability $R(t)$ is the probability that a system will be successful in the interval from time 0 to time $t$:

$$R(t) = P(T > t), \quad t \geq 0 \quad (10)$$

where $T$ is a nonnegative random variable, which signifies time-to-failure. Moreover, the definition of unreliability of a system is presented in the following.

**Definition 2.** *The unreliability of an element (or system) $F(t)$ is determined as the probability that it experiences the first failure or has failed one or more times throughout the time interval 0 to time t.*

Considering the component is regularly in one of the two probable states (failed or operational), the following relation is satisfied

$$F(t) + R(t) = 1. \quad (11)$$

Many different functions have been proposed to describe the reliability as a function of time. Some of the more general reliability functions include the log-normal, exponential and Weibull distributions [32]. In this paper, the exponential distribution is used for modeling the component failure rate. In particular, engineering systems are organized to support varying amounts of loads where they can be expressed in terms of usage rate or occupied period. Reviewing the literature, it has been established that the function load strongly affects the component failure rate [12]. Therefore, it is necessary to consider the load versus failure rate relation when considering system reliability evaluation. In this work, failure rates are determined from actuators under different levels of load according to the applied control input. One of the commonly used relations is based on assuming that actuator fault rates vary with the load by the following exponential law

$$\lambda_i = \lambda_i^0 exp\big(\beta_i u_i(k)\big), \quad i = 1, 2, \ldots, m. \quad (12)$$

where $\lambda_i^0$ indicates the baseline failure rate (nominal failure rate) and $u_i(k)$ is the control action a time $k$ for the $i$th actuator. $\beta_i$ is a constant parameter that depends on the actuator characteristics.

In the useful period of life, the element can be characterized at a given time $t$ by a baseline reliability measure $R_0(t)$. Then, $R_{0,i}(t)$ denotes the reliability of the $i$th actuator obtained under nominal operating conditions

$$R_{0,i}(k) = exp\big(-\lambda_i^0 t\big), \quad i = 1, 2, \ldots, m. \quad (13)$$

Hence, the reliability of the $i$th system component can be computed by applying the exponential function and the baseline reliability level $R_{0,i}$ as follows

$$R_i(k) = R_{0,i} \, exp\left(-\int_0^k \lambda_i(s) \, ds\right), \quad i = 1, 2, \ldots, m \quad (14)$$

In discrete-time, it can be rewritten as

$$R_i(k+1) = R_{0,i}(k)\, exp\left(-T_s \sum_{s=0}^{k+1} \lambda_i(s)\right), \quad i = 1, 2, \ldots, m \tag{15}$$

where $\lambda_i(s)$ is the failure rate obtained from the $i$th component under different levels of load and $T_s$ is the sampling time.

### 3.2. Overall Reliability

The lifetime of a system can be quantified by means of the overall system reliability, denoted as $R_G(k)$. The overall system reliability is computed based on the reliabilities of elementary components (or subsystems). Therefore, $R_G(k)$ depends on the actuators' configuration, which can generally be obtained from series and/or parallel combinations of subsystems (or components) [33]. However, some systems do not follow series, parallel or combination of series and parallel structures. To deal with the more general situation, a graph network model can be used in which it is possible to determine whether the system is working correctly by determining existence of a successful path in the system. A path in a graph network is a set of components that, if operating as expected, the system operation will be successful. A minimal path $P_s$ is a set of components, from which if one were removed would mean the resulting set is no longer a path [33]. Then, the overall system reliability $R_G(k)$ can be computed as

$$R_G(k) = 1 - \prod_{j=1}^{s}\left(1 - \prod_{i \in P_{s,j}} R_i(k)\right), \tag{16}$$

where $j = 1, 2, \ldots, s$ is number of minimal paths.

### 3.3. System Reliability Modeling

For the purpose of integrating the reliability function in the MPC model as a new state variable, a conversion is needed that allows computing reliability in a linear-like form. The proposed transformation is based on applying the logarithm in Equation (16). As stated in Equation (11), Equation (16) can be rewritten as

$$\log(Q_G(k)) = \log\left(\prod_{j=1}^{s}\left(1 - \prod_{i \in p_{s,j}} R_i(k)\right)\right), \tag{17}$$

and by introducing a change of variable

$$z_j(k) = 1 - \prod_{i \in p_{s,j}} R_i(k), \tag{18}$$

Equation (17) leads to

$$\log(Q_G(k)) = \sum_{i \in p_{s,j}}^{s} \log(z_j(k)). \tag{19}$$

According to Equation (18), the $\log(z_j(k))$ can be obtained as

$$\log(z_j(k)) = \frac{\log(z_j(k))}{\log(1 - z_j(k))} \sum_{i \in p_{s,j}} \log R_i(k). \tag{20}$$

Then, by renaming $\beta_j(k) = \frac{\log(z_j(k))}{\log(1-z_j(k))}$ in Equation (20), Equation (17) can be rewritten as

$$\log(Q_G(k)) = \sum_{i \in p_{s,j}}^{s} \beta_j(k) \sum_{i \in p_{s,j}} \log R_i(k). \tag{21}$$

Finally, the system unreliability of system can be estimated from the baseline system unreliability as follows:

$$\log(Q_G(k+1)) = \log(Q_G(k+1)) + \sum_{i \in p_{s,j}}^{s} \beta_j(k) \sum_{i \in p_{s,j}} \log R_i(k). \tag{22}$$

## 4. Economic Health-Aware LPV-MPC

This section presents the incorporation of reliability information in the predictive control law as a new state of the model. As mentioned in Section 2, the reliability of the DWN can be estimated using the control input (actuator commands) information. To include a new objective in the MPC that proposes to extend the system reliability, the reliability model is represented by means of the model in Equation (22). In fact, the new control model of DWN that includes the reliability and dynamic model of DWN is obtained based on the structure shown in Figure 1. Actually, there is a direct relationship between the dynamic model of DWN and its system reliability.

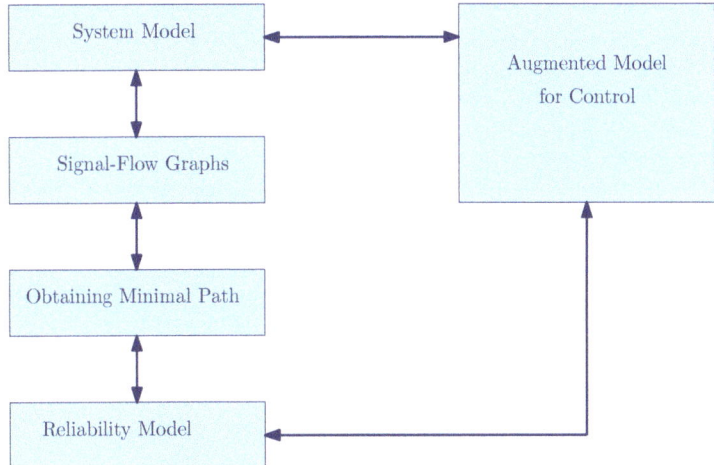

**Figure 1.** Digram of the new proposed control model approach.

Thus, the new MPC model has the following structure

$$\begin{aligned} x_r(k+1) &= A_r x_r(k) + B_r u(k) + B_{r,d} d_m(k), \\ y_r(k) &= C_r x(k), \end{aligned} \tag{23}$$

where the state and output vector are given by $x_r = [x, \log(Q_G), \log(R_1), \ldots, \log(R_i)]^T$ and $y_r = [y, \log(Q_G)]^T$, respectively. The new matrices are defined as

$$A_r = \begin{bmatrix} A & \vdots & 0_{n_x \times n_{j+1}} \\ \hdashline 0_{1 \times n_x} & 1 & \Sigma_{i \in p_{s,j}}^s \beta_j(k) \\ \hdashline 0_{n_i \times n_x} & \vdots & I_{n_i \times n_i} \end{bmatrix}, \quad B_r = \begin{bmatrix} B_{n_u \times n_u} \\ \hdashline 0 \\ \hdashline -\lambda_i \times I_{n_i \times n_i} \end{bmatrix},$$

(24)

$$B_{r,d} = \begin{bmatrix} B_{d,n_u \times n_u} \\ \hdashline 0_{n_{i+1} \times n_{B_d}} \end{bmatrix}, \quad C_r = \begin{bmatrix} C & 0 & 0 & \cdots & 0 \\ 0 & 1 & 0 & \cdots & 0 \end{bmatrix}.$$

Therefore, the new MPC model in Equation (23) can be viewed as an LPV model that has as scheduling variable the control action $u_i(k)$ related to each state and actuator. The new MPC model in Equation (23) cannot be estimated before solving the optimization problem in Equation (9) since the future state sequence is not identified. In fact, $x(l|k)$ depends on the future control inputs $u(k)$ and scheduling parameters, which, for general LPV models, are not expected to be known prior but only to be measurable online at current time $k$. The idea is to obtain a solution to the problem in Equation (9) by solving an online optimization problem as a QP problem. The solution for this problem is to modify the exact LPV-MPC to a linear approximation of the LPV-MPC. This approximation uses an estimation of scheduling variables, $\hat{\theta}$, instead of applying $\theta$. Indeed, the scheduling variables in the prediction horizon are determined and used to update the matrices of the model adopted by the MPC controller. In fact, to solve this problem, the sequence of the control input is used to adjust the system matrices of the model applied in the prediction horizon. Hence, based on the optimal control sequence $\mathbf{u}(k)$, the following sequence of states and predicted parameters can be achieved:

$$\tilde{x}(k) = \begin{bmatrix} x(l+1|k) \\ x(l+2|k) \\ \vdots \\ x(N_p|k) \end{bmatrix} \in \mathbb{R}^{N_p, n_x}, \quad \Theta(k) = \begin{bmatrix} \hat{\theta}(l|k) \\ \hat{\theta}(l+1|k) \\ \vdots \\ \hat{\theta}(N_p-1|k) \end{bmatrix} \in \mathbb{R}^{N_p, n_\theta}.$$

(25)

Therefore, with slight abuse of notation, $f$ can be defined as: $\Theta(k) = f([x^T(k) \quad \tilde{x}^T(k)], u(k))$. The vector $\Theta(k)$ includes parameters from time $k$ to $k + N_p - 1$ whilst the state prediction is accomplished for time $k+1$ to $k + N_p$.

Hence, by using the definitions in Equation (25), the predicted states can be simply formulated as follows

$$\tilde{x}(k) = \mathcal{A}(\Theta(k))x(k) + \mathcal{B}(\Theta(k))u(k) + B_{r,d}d_m(k),$$ (26)

where $\mathcal{A} \in \mathbb{R}^{n_x \times n_x}$ and $\mathcal{B} \in \mathbb{R}^{n_x \times n_u}$ are given by Equations (27) and (28).

$$\mathcal{A}(\Theta(k)) = \begin{bmatrix} I \\ A(\hat{\theta}(k)) \\ A(\hat{\theta}(k+1))A(\hat{\theta}(k)) \\ \vdots \\ A(\hat{\theta}(k+N_p-1))A(\hat{\theta}(k+N_p-2))\ldots A(\hat{\theta}(k)) \end{bmatrix},$$ (27)

and

$$B(\Theta(k)) = \begin{bmatrix} 0 & 0 & 0 & \cdots & 0 \\ B(\hat{\theta}(k)) & 0 & 0 & \cdots & 0 \\ A(\hat{\theta}(k+1))B(\hat{\theta}(k)) & B(\hat{\theta}(k+1)) & 0 & \cdots & 0 \\ \vdots & \vdots & \ddots & \ddots & \vdots \\ A(\hat{\theta}_{k+N_p-1})\cdots A(\hat{\theta}(k+1))B(\hat{\theta}(k)) & A(\hat{\theta}_{k+N_p-1})\cdots A(\hat{\theta}(k+2))B(\hat{\theta}(k+1)) & \cdots & B(\hat{\theta}_{k+N_p-1}) & 0 \end{bmatrix}. \quad (28)$$

By using Equation (26) and augmented block diagonal weighting matrices $\tilde{w}_1 = diag_{N_p}(w_1)$ and $\tilde{w}_2 = diag_{N_p}(w_2)$, the cost function in Equation (8) with new additional objective that aims to maximize the system reliability can be rewritten in vector form as

$$\min_{\mathbf{u}(k),\boldsymbol{\xi}(k),\log Q_G(k)} \sum_{l=0}^{N_p} [\ell_e(l|k) + \ell_s(l|k) + \ell_{\Delta u}(l|k) - \ell_{Rg}(l|k)], \quad (29a)$$

subject to:

$$\tilde{x}(k) = \mathcal{A}(\Theta(k))x(k) + B_{r,d}d_m(k), \mathcal{B}(\Theta(k))u(k), \quad (29b)$$
$$0 = E_u u(l|k) + E_d d_m(k), \quad (29c)$$
$$x(l+1|k) \geq x_s - \xi(l|k) \quad (29d)$$
$$\log Q_G(l+1|k) = \tilde{x}_{nx+1}(l|k) \quad (29e)$$
$$u(l|k) \in \mathbb{U}, \ l = 0, \cdots, N_p - 1 \quad (29f)$$
$$x(l|k), \in \mathbb{X}, \ l = 1, \cdots, N_p \quad (29g)$$
$$\xi(l|k) \geq 0, \ l = 0, \cdots, N_p \quad (29h)$$
$$x(0|k) = x(k), \quad (29i)$$

where $\ell_{Rg}(k) \triangleq \log Q_G^\top w_3 \log Q_G$ is additional objective with the corresponding weight $w_3$ into the EMPC-LPV cost function to maximize the system reliability. Since the predicted states $\Theta(k)$ in Equation (26) are linear in control inputs $u(k)$, the optimization problem can be solved as a QP problem, which is significantly easier than solving a nonlinear optimization problem.

Using this idea, the following iterative approach at each time instant $k$ is applied:

- In the first iteration, the problem in Equation (9) is solved considering that the quasi-LPV model in Equation (23) is instantiated by the LTI model considering that $\theta(0|l) \simeq \theta(1|l) \simeq \theta(2|l) \simeq \ldots \simeq \theta(N_p - 1|l)$ along the prediction horizon $N_p$.
- The parameter varying sequence $\Theta(k)$ is updated using the optimal value of the scheduling variables $\Theta^*(k) = f(\tilde{x}^*(k), \mathbf{u}^*(k))$, where $\tilde{x}^*(k)$ and $\mathbf{u}^*(k)$ are the optimal input and state sequences obtained after the solution of the MPC problem, respectively.
- The parameter varying values for the next iteration $\Theta(k+1)$ are obtained considering $\tilde{x}(k)$ and $\tilde{u}(k)$, i.e., $\Theta_0(k+1) = f(\tilde{x}_1(k), \mathbf{u}_0(k))$.

## 5. Application to the Water Transport Network of Barcelona

In this section, two motivational examples are used to assess the implementation of the proposed economic health-aware LPV-MPC based on system reliability assessment. For both examples, the system under study is a portion extracted from the Barcelona DWN [34]. This network is managed by Aguas de Barcelona (AGBAR), which supplies drinking water for Barcelona and its metropolitan area [35]. The general task of this system is to supply water resources from sources to consumers minimizing the operational costs. The DWN of Barcelona covers a territorial extension of 425 km$^2$, with a total pipe length of 4470 km. Every year, it supplies 237.7 hm$^3$ of drinking water to a population over 2.8 million inhabitants. Regarding the DWN reliability study, sectors, sources, pipelines and tanks are assumed to be perfectly reliable, whereas active elements such as pumps and valves are considered not completely reliable [4]. The results were obtained using a 2.4 GHz and 12.00 Gb RAM Intel(R) Core(TM)i7-5500 CPU. Matlab and Yalmip toolbox were used to perform the simulations.

## 5.1. Water Transport Network (3-Tanks)

In the first example, the proposed study concentrates on a small network based on the Barcelona DWN. Two sources of water and four demand sectors, which represent the district metered area (DMA), are considered (see Figure 2). It is expected that the demand forecast ($d_m$) at each demand sector is known and that every single source can provide this water demand (Figure 3). First, system components must be identified. In this case, there are three pumps, three valves, two sources, three tanks, two intersection nodes, and several pipes.

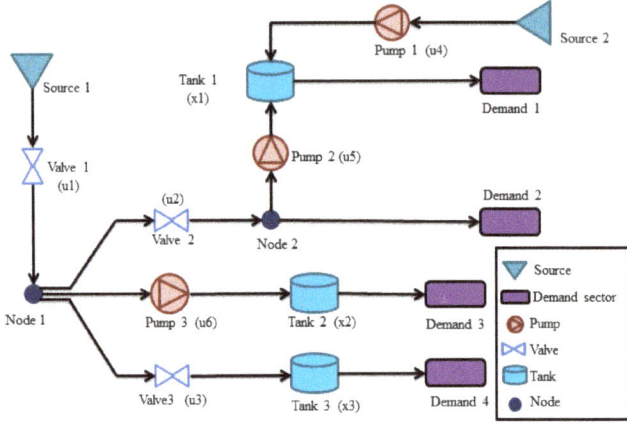

**Figure 2.** Drinking water network diagram (three-tanks).

Afterwards, according the definition of minimal path $P_s$ in Section 3.2, the minimal path sets is determined for the water network, while the $P_s$ is determined based on the relation and the possible connection between each source and demand sector. By considering all the paths from all sources to the demand sector, the combination of all flow paths should follow the functional requirements necessary to satisfy the consumer demands. A minimal path set is composed by those elements which allow a flow path between sources and demand sector, such as pipes, tanks, pumps and valves. Based on this analysis, the following list of each minimal path is presented in Table 1. There are five minimal path sets in the system of Figure 2. The reliability of each minimal path set depends on the reliability of its components. Tanks and pipes are supposed to be perfectly reliable. However, sources are involved in the minimal path sets only for illustrative purposes of the proposed procedure. Table 2 provides the simulation parameters used.

Figure 4 shows the evolution of the valves and pumps commands that were obtained using the new approach of the health-aware LPV-MPC in the three tanks example with and without the reliability-aware objective. As can be seen in Figure 4, the behavior of valve control actions are different from the ones corresponding to the pumps. However, in all of them, the behaviors of control actions in both scenarios are almost the same, thus the reliability-aware objective is not significantly affecting the behavior of the valves and pumps. The comparison of the volume evolution of three tanks based on the health-aware LPV-MPC with and without the reliability-aware objective is presented in Figure 5. The safety volume of each tank is satisfied and hence able to cope with unexpected demands. The system reliability prediction of the DWN, which was obtained when using the proposed controller with and without the reliability-aware objective, is presented in Figure 6. According to these results, it can be observed that, with the use of reliability-aware objective in the MPC, the network reliability is better preserved compared to the case that the reliability is not considered in the MPC objectives. However, the responses of water tanks are similar in both scenarios. The trade-off between the decreasing operating cost and increasing system reliability can be observed in Figure 6. Note that the differences in the amount of the operational cost using the proposed approach is similar to the EMPC controller without reliability objective. Figure 6 shows that the system reliability is increased from 0.9071 to 0.9891 and that is about 9.06% of improvement, while the accumulated cost is increased from 114.6 to 116.7 that is about 1.74% of increment.

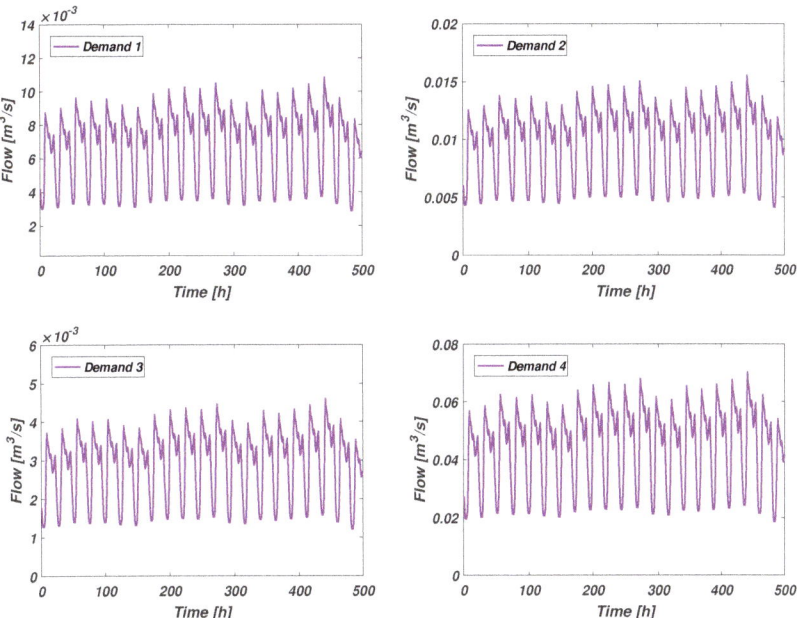

**Figure 3.** Drinking water demand for the three tanks example.

**Table 1.** Success minimal paths of the water transport network of Barcelona (three tanks).

| Path | Component Set |
|---|---|
| $P_1$ | {Source1, Valve1, Valve3, Demand4} |
| $P_2$ | {Source1, Valve1, Pump2, Demand3} |
| $P_3$ | {Source1, Valve1, Valve2, Demand2} |
| $P_4$ | {Source1, Valve1, Valve2, Pump2, Demand1} |
| $P_5$ | {Source2, Pump1, Demand1} |

**Table 2.** Simulation parameters.

| Parameter | Value |
|---|---|
| $Np$ | 24 |
| $T_s$ [h] | 1 |
| $T_m$ [h] | 200 |
| $\alpha_1$ | 0.123  0  0  0.054  0  0 |
| $u_{min}$ [m$^3$/s] | 0  0  0  0  0  0 |
| $u_{max}$ [m$^3$/s] | 1.297  0.05  0.12  0.015  0.0317  0.022 |
| $\lambda_0$ [h$^{-1}$ × h$^{-4}$] | 1.2  3.45  6.3  9.5  1  1 |
| $x_{min}$ [m$^3$] | 0  0  0  0  0  0  0  0  0 |
| $x_{max}$ [m$^3$] | 470  960  3100  1  1  1  1  1  1 |
| $x_0$ [m$^3$] | 0.75  0.62  0.34  0  1  1  1  1  1 |

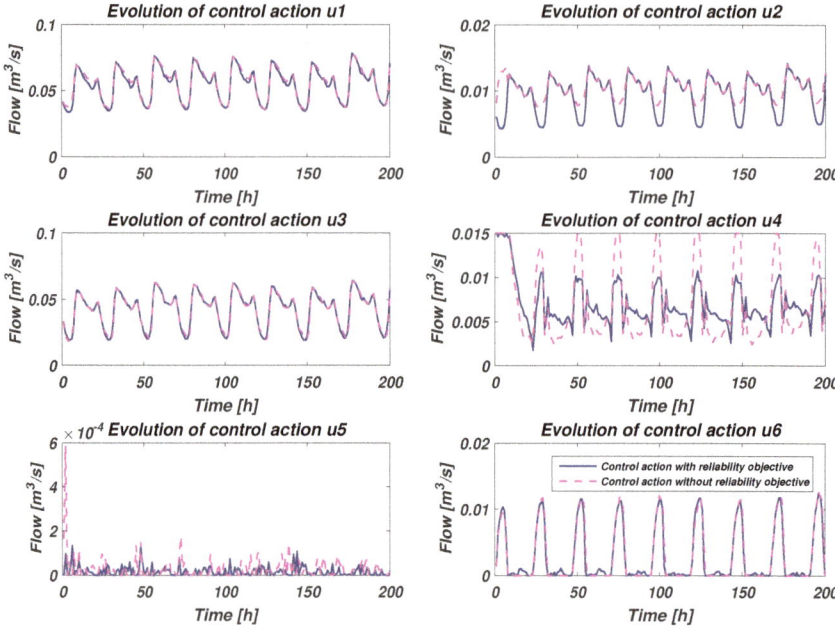

**Figure 4.** Evaluation of the control actions results for three tanks.

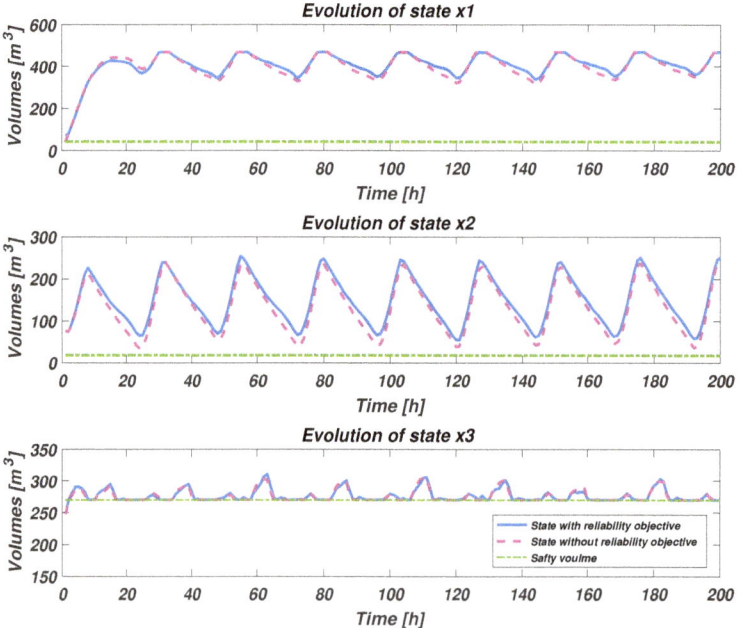

**Figure 5.** Results of the evolutions of storage tanks for three tanks.

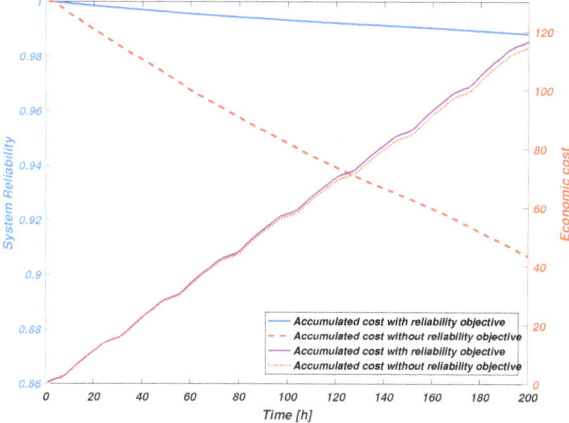

**Figure 6.** Evaluation of system reliability and accumulated economic cost for three tanks.

5.2. Water Transport Network of Barcelona (17-Tanks)

Now, a more complex and realistic example also based on the Barcelona DWN is considered as a case study. This case includes 17 tanks and nine sources, consisting of five underground and four surface sources, which currently provide an inflow of about 2 m$^3$/s. The case study also includes 61 actuators (valves and pumps), 12 nodes and 25 demands. Figure 7 presents the general topology of the network, showing a complex system in terms of its elements and the relationships and connections between them. Figure 8 presents the graph obtained from this network; the nodes correspond to reservoirs or pipe merging/splitting nodes and the arcs correspond to actuators (pumps and valves). The graph of the water network was obtained from the state space representation of the system. This approach is explained with more detail in [36].

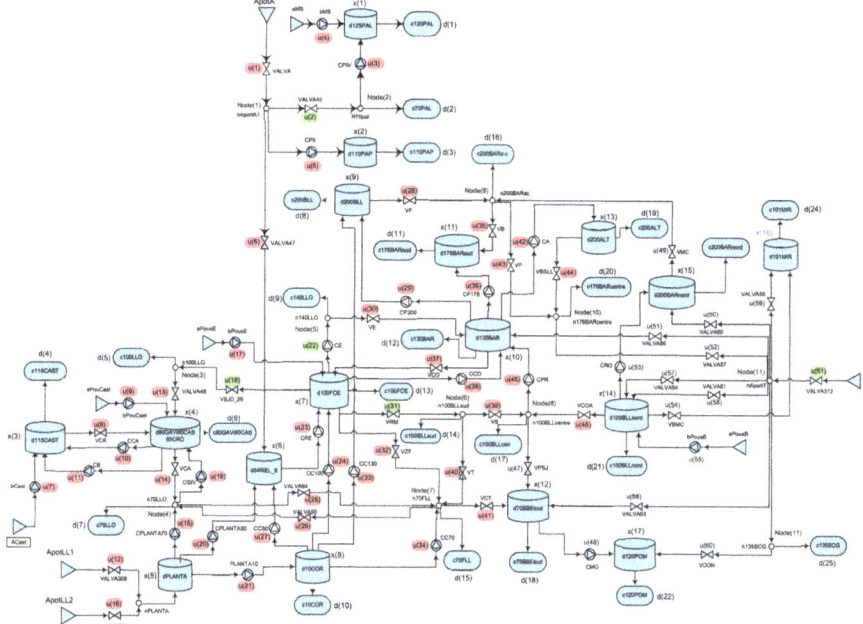

**Figure 7.** Barcelona drinking water network (17 tanks).

As in the previous example, demand sectors, sources, pipelines and tanks are considered perfectly reliable, whereas actuators are not [4]. Moreover, it is expected that the demand forecast ($d_m$) at each demand sectors is known and that every single source can supply the required water demand (see some demand sectors in Figure 9).

The economic reliability-aware LPV-MPC formulation proposed in previous section was applied to the simulation model of the DWN presented in Figure 9.

From the reliability analysis, it could be obtained which states are structurally controllable since the path computation analysis provide all possible paths from a source to a target sectors. Moreover, for each path, an approximate operational cost (according to the electricity cost of each element) and a maximal water flow (according to the physical constraints of the actuators) can also be derived.

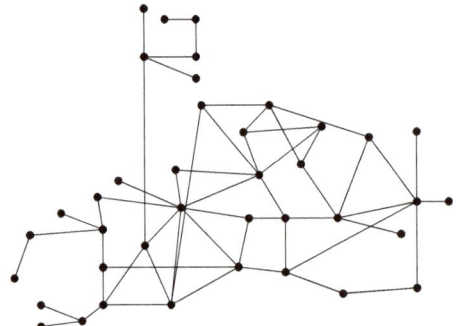

**Figure 8.** Graph of the Barcelona DWN.

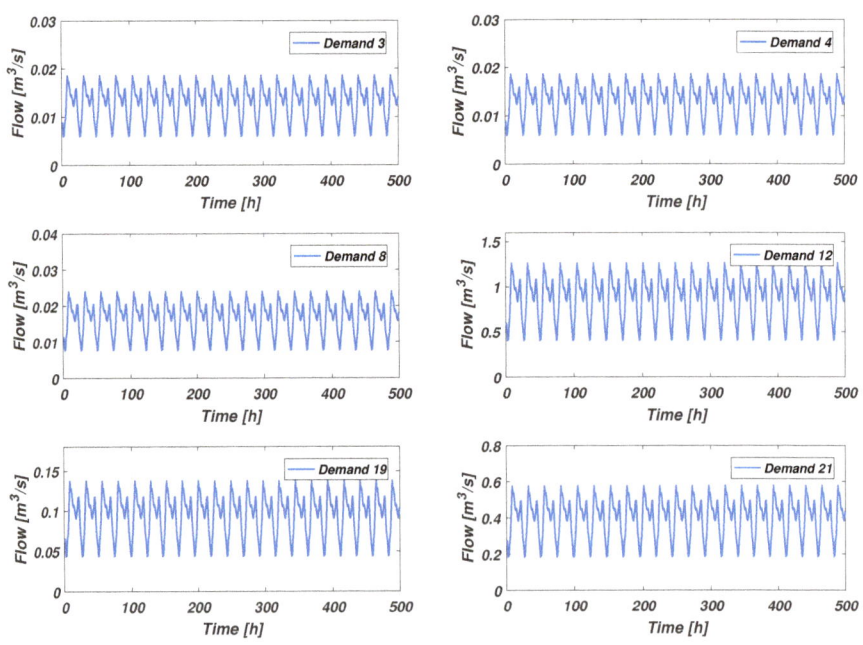

**Figure 9.** Drinking water demand for several sinks.

Tables 3 and 4 show that there exist several critical actuators within the network, considering the topology and the way of network elements are connected, as most actuators (valves or pumps) are the only link between demands and tanks. Consequently, if an actuator fails, then the corresponding demand will not be satisfied. Therefore, if an actuator fails, then the corresponding demand will not be satisfied. Note that the information shown in Tables 3 and 4 is especially significant for AGBAR since it identifies the critical elements in the network for surveillance/correction policies to be implemented in the event of element damage [5]. According to the DWN (Figures 7), Tables 3 and 4 and the above analysis of the success minimal path of the water network, there are 607 minimal path sets in the system of Figure 7. Some samples of success minimal paths are presented in Table 5. The objective of the MPC as explained above is to minimize the multi-objective cost function in Equation (29). The prediction horizon is 24 h because the system and the electrical tariff have periodicity of one day. The sampling time is 1 h. However, in all of them, the behaviors of control actions in both scenarios are almost the same, considering the reliability-aware objective doe not greatly affect the behavior of the valves and pumps.

Table 3. Structural actuators (towards tanks).

| No. | Name | No. | Name | No. | Name | No. | Name |
|---|---|---|---|---|---|---|---|
| $u_1$ | VALVA | $u_{16}$ | VALVA309 | $u_{33}$ | CC130 | $u_{47}$ | VPSJ |
| $u_3$ | CPIV | $u_{17}$ | bPousE | $u_{34}$ | CC70 | $u_{48}$ | CMO |
| $u_4$ | bMS | $u_{19}$ | CGIV | $u_{35}$ | VB | $u_{49}$ | VMC |
| $u_5$ | CPII | $u_{20}$ | CPLANTA50 | $u_{36}$ | CF176 | $u_{50}$ | VALVA60 |
| $u_6$ | VALVA47 | $u_{21}$ | PLANTA10 | $u_{37}$ | VCO | $u_{51}$ | VALVA56 |
| $u_7$ | bCast | $u_{23}$ | CRE | $u_{38}$ | CCO | $u_{52}$ | VALVA57 |
| $u_8$ | VCR | $u_{24}$ | CC100 | $u_{39}$ | VS | $u_{53}$ | CRO |
| $u_9$ | bPouCast | $u_{25}$ | VALVA64 | $u_{40}$ | V | $u_{54}$ | VBMC |
| $u_{10}$ | CCA | $u_{26}$ | VALVA50 | $u_{41}$ | VCT | $u_{55}$ | bPousB |
| $u_{11}$ | CB | $u_{27}$ | CC50 | $u_{42}$ | CA | $u_{56}$ | VALVA53 |
| $u_{12}$ | VALVA308 | $u_{28}$ | VF | $u_{43}$ | VP | $u_{57}$ | VALVA54 |
| $u_{13}$ | VALVA48 | $u_{29}$ | CF200 | $u_{44}$ | VBSLL | $u_{58}$ | VALVA61 |
| $u_{14}$ | VCA | $u_{30}$ | VE | $u_{45}$ | CPR | $u_{59}$ | VALVA55 |
| $u_{15}$ | CPLANTA70 | $u_{32}$ | VZF | $u_{46}$ | VCOA | $u_{60}$ | VCON |

Table 4. Structural actuators (towards demand nodes).

| No. | Name | No. | Name | No. | Name | No. | Name |
|---|---|---|---|---|---|---|---|
| $u_2$ | VALVA45 | $u_{18}$ | VSJD-29 | $u_{22}$ | CE | $u_{31}$ | VRM |
| $u_{61}$ | VALVA312 | | | | | | |

Table 5. Some success minimal paths of the Barcelona DWN.

| Path | Component Set |
|---|---|
| $P_1$ | {aMS, bMS, c125PAL} |
| $P_2$ | {AportA, VALVA, VALVA47, CPIV, c125PAL} |
| $P_3$ | {AportA, VALVA, VALVA45, c70PAL} |
| $P_4$ | {AportA, VALVA, CPII, c110PAP} |
| $P_5$ | {ACast, bCast, c115CAST} |
| ⋮ | ⋮ |
| $P_{607}$ | {AportT, VALVA312, c135SCG} |

Figure 10 shows the comparative evolution of the valves and pumps commands obtained using the new approach with the case without the reliability-aware objective. Note that the plot of control actions 5, 7, 42, and 29 correspond to pump set-points while control actions 51 and 52 correspond to valve set-points (see Figure 7). The behavior of pumps and valves in both scenarios are almost the same even by including the reliability objective inside the cost function of the controller. Figure 11 presents the comparison of volume evolutions of selected storage tanks. Figure 11 shows the proper replenishment planning that the predictive controller dictates according to the cyclic behavior of demands. Notice that the net demand of each tank is properly satisfied along the simulation horizon. The system reliability evolution of the DWN that was obtained from the proposed controller with and without the reliability-aware objective is presented in Figure 12. According to these results, it can be observed that, with the use of reliability-aware objective in the MPC, the reliability of the network is better preserved compared to the case that the reliability is not considered in the MPC design. There is a trade-off between the increasing system reliability and operational cost. However, the operational cost obtained with new proposed approach is almost the same as the EMPC controller that not considers reliability. This figure also shows that the system reliability is improved about 14.73% in the case of the LPV-MPC controller with the reliability objective while keeping the performance and the cost is increased just 2.03%. To have better compare the economics of LPV-MPC with and without the reliability-aware objective, several simulations with different tunings were implemented. Finally, the trade-off curves between the system reliability and economic operational cost for both control schemes are presented in Figure 13. This figure shows that independently of the tuning the economic reliability-aware LPV-MPC control is able to improve and increase the system reliability. The results obtained from the preliminary analysis of Figure 13 are provided in Tables 6 and 7. The results, as shown in these tables, indicate that the system reliability is improved in different tunings while the cost is increased. However, the increased percentage of operational cost is negligible compared to the improvement of reliability obtained.

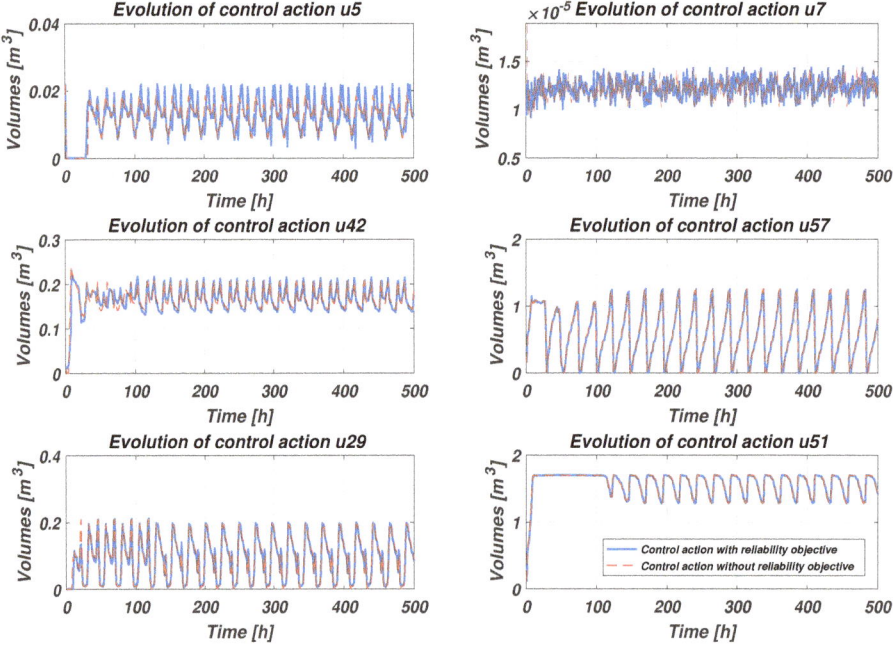

**Figure 10.** Evaluation of the control actions results.

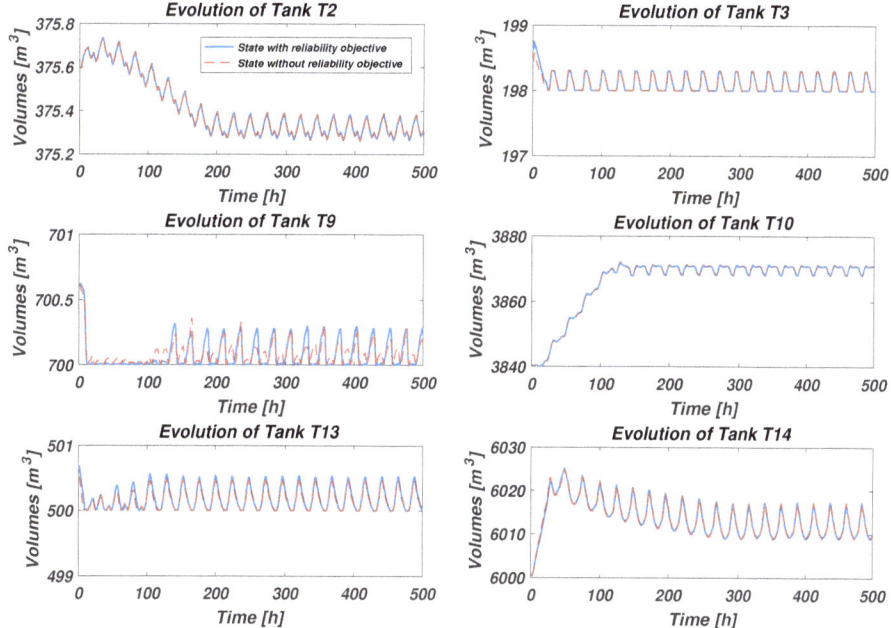

Figure 11. Results of the evolutions of storage tanks.

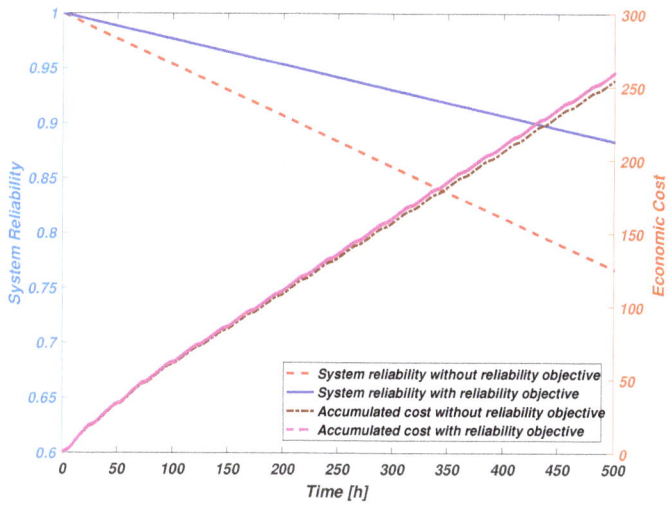

Figure 12. Evaluation of system reliability and accumulated economic cost.

Table 6. Final system reliability.

| Different Weight Tuning Points | 1 | 2 | 3 | 4 | 5 | 6 | 7 |
| --- | --- | --- | --- | --- | --- | --- | --- |
| Economic LPV-MPC with reliability-aware objective | 0.825 | 0.84 | 0.855 | 0.87 | 0.885 | 0.9 | 0.915 |
| Economic LPV-MPC without reliability-aware objective | 0.755 | 0.768 | 0.781 | 0.794 | 0.805 | 0.82 | 0.833 |
| Difference percentage % | 9.27 | 9.38 | 9.48 | 9.57 | 9.94 | 9.76 | 9.84 |

Table 7. Economic cost.

| Different Weight Tuning Points | 1 | 2 | 3 | 4 | 5 | 6 | 7 |
|---|---|---|---|---|---|---|---|
| Economic LPV-MPC with reliability-aware objective | 252.05 | 252.1 | 252.23 | 253.82 | 254.08 | 254.63 | 254.78 |
| Economic LPV-MPC without reliability-aware objective | 252.00 | 252.06 | 252.14 | 253.66 | 254.28 | 254.58 | 254.69 |
| Difference percentage % | 0.02 | 0.02 | 0.04 | 0.08 | 0.16 | 0.02 | 0.04 |

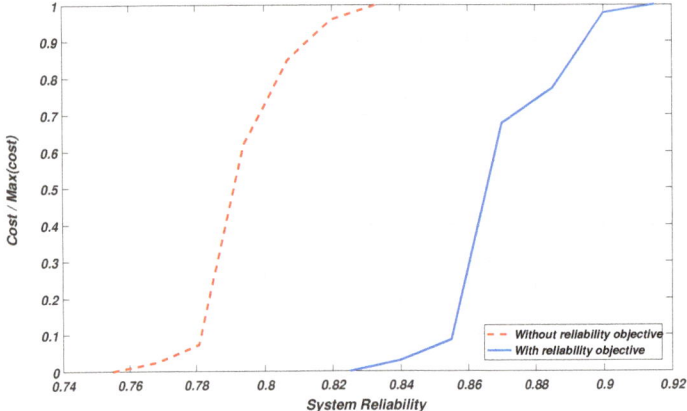

**Figure 13.** Final system reliability vs. economic cost for different weight tuning.

## 6. Conclusions

This paper proposes an economic health-aware LPV-MPC strategy based on the system reliability for water transport networks.

The system reliability is evaluated online considering the control action value. The system reliability is obtained from the reliability of each component and the interconnection topology leading to a nonlinear model. This model is transformed into a linear-like form by means of the LPV framework.

The system reliability is considered during the calculation of the MPC control action by including an extra objective in the cost function and an by augmenting the MPC model with additional states. Then, the by using a LPV-MPC approach, the associated optimization problem can be efficiently solved using quadratic programming. The MPC model is updated at each time iteration instantiating the varying parameters considering the value of the scheduling variables.

The results show that, using the proposed economic health-aware LPV-MPC, the DWN network reliability is maximized with a slight rise in the cost, achieving a good trade-off between both system reliability and cost. In this way, it is possible to maximize the lifetime of elements just by reducing slightly the economic optimality.

Future research will extend the study to water distribution networks by considering the pressure model. Moreover, the problem of re-designing the network by adding some additional paths to overcome the limitations due to critical elements. Moreover, it would be interesting to consider the water quality objective into the proposed approach.

**Author Contributions:** All the authors participate equally in the writing of the paper.

**Funding:** This work has been partially funded by the Spanish State Research Agency (AEI) and the European Regional Development Fund (ERFD) through the projects SCAV (ref. MINECO DPI2017-88403-R) and DEOCS (ref. MINECO DPI2016-76493). This work has been also supported by the AEI through the Maria de Maeztu Seal of Excellence to IRI (MDM-2016-0656).

**Conflicts of Interest:** The authors declare no conflict of interest.

## References

1. Ocampo-Martinez, C.; Puig, V.; Cembrano, G.; Quevedo, J. Application of predictive control strategies to the management of complex networks in the urban water cycle [applications of control]. *IEEE Control Syst.* **2013**, *33*, 15–41.
2. Barbieri, M.; Nigro, A.; Petitta, M. Groundwater mixing in the discharge area of San Vittorino Plain (Central Italy): Geochemical characterization and implication for drinking uses. *Environ. Earth Sci.* **2017**, *76*, 393. [CrossRef]
3. Grosso, J.M.; Ocampo-Martínez, C.; Puig, V. A service reliability model predictive control with dynamic safety stocks and actuators health monitoring for drinking water networks. In Proceedings of the 2012 IEEE 51st Annual Conference on Decision and Control (CDC), Maui, HI, USA, 10–13 December 2012; pp. 4568–4573.
4. Weber, P.; Simon, C.; Theilliol, D.; Puig, V. Fault-Tolerant Control Design for over-actuated System conditioned by Reliability: A Drinking Water Network Application. *IFAC Proc. Vol.* **2012**, *45*, 558–563. [CrossRef]
5. Robles, D.; Puig, V.; Ocampo-Martinez, C.; Garza-Castañón, L.E. Reliable fault-tolerant model predictive control of drinking water transport networks. *Control. Eng. Pract.* **2016**, *55*, 197–211. [CrossRef]
6. Karimi Pour, F.; Puig, V.; Ocampo-Martinez, C. Health-aware Model Predictive Control of Pasteurization Plant. In *Journal of Physics: Conference Series*; IOP Publishing: Bristol, UK, 2017; Volume 783, p. 012030.
7. Ocampo-Martinez, C.; Ingimundarson, A.; Puig, V.; Quevedo, J. Objective prioritization using lexicographic minimizers for MPC of sewer networks. *IEEE Trans. Control. Syst. Technol.* **2008**, *16*, 113–121. [CrossRef]
8. Pascual, J.; Romera, J.; Puig, V.; Cembrano, G.; Creus, R.; Minoves, M. Operational predictive optimal control of Barcelona water transport network. *Control. Eng. Pract.* **2013**, *21*, 1020–1034. [CrossRef]
9. Rawlings, J.B.; Mayne, D.Q. *Model Predictive Control: Theory and Design*; Nob Hill Pub: San Francisco, CA, USA, 2009.
10. Ellis, M.; Durand, H.; Christofides, P.D. A tutorial review of economic model predictive control methods. *J. Process Control.* **2014**, *24*, 1156–1178. [CrossRef]
11. Karimi Pour, F.; Puig, V.; Ocampo-Martinez, C. Economic Predictive Control of a Pasteurization Plant using a Linear Parameter Varying Model. In *Computer Aided Chemical Engineering*; Elsevier: Amsterdam, The Netherlands, 2017; Volume 40, pp. 1573–1578.
12. Karimi Pour, F.; Puig, V.; Cembrano, G. Health-aware LPV-MPC Based on System Reliability Assessment for Drinking Water Networks. In Proceedings of the 2018 IEEE Conference on Control Technology and Applications (CCTA), Copenhagen, Denmark, 21–24 August 2018; pp. 187–192.
13. Edwards, M. *Critical Infrastructure Protection*; IOS Press: Amsterdam, The Netherlands, 2014; Volume 116.
14. Chamseddine, A.; Theilliol, D.; Sadeghzadeh, I.; Zhang, Y.; Weber, P. Optimal reliability design for over-actuated systems based on the MIT rule: Application to an octocopter helicopter testbed. *Reliab. Eng. Syst. Saf.* **2014**, *132*, 196–206. [CrossRef]
15. Gokdere, L.; Chiu, S.L.; Keller, K.J.; Vian, J. Lifetime control of electromechanical actuators. In Proceedings of the 2005 IEEE Aerospace Conference, Big Sky, MT, USA, 5–12 March 2005; pp. 3523–3531.
16. Li, Y.; Kurfess, T.; Liang, S. Stochastic prognostics for rolling element bearings. *Mech. Syst. Signal Process.* **2000**, *14*, 747–762. [CrossRef]
17. Salazar, J.C.; Weber, P.; Sarrate, R.; Theilliol, D.; Nejjari, F. MPC design based on a DBN reliability model: Application to drinking water networks. *IFAC-PapersOnLine* **2015**, *48*, 688–693. [CrossRef]
18. Pereira, E.B.; Galvão, R.K.H.; Yoneyama, T. Model predictive control using prognosis and health monitoring of actuators. In Proceedings of the 2010 IEEE International Symposium on Industrial Electronics (ISIE), Bari, Italy, 4–7 July 2010; pp. 237–243.
19. Karimi Pour, F.; Puig, V.; Cembrano, G. Health-aware LPV-MPC based on a Reliability-based Remaining Useful Life Assessment. *IFAC-PapersOnLine* **2018**, *51*, 1285–1291. [CrossRef]
20. Farmani, R.; Walters, G.; Savic, D. Evolutionary multi-objective optimization of the design and operation of water distribution network: total cost vs. reliability vs. water quality. *J. HydroInform.* **2006**, *8*, 165–179. [CrossRef]
21. Ostfeld, A. Reliability analysis of water distribution systems. *J. Hydroinform.* **2004**, *6*, 281–294. [CrossRef]

22. Christodoulou, S.E. Water network assessment and reliability analysis by use of survival analysis. *Water Resour. Manag.* **2011**, *25*, 1229–1238. [CrossRef]
23. Beal, L.D.; Petersen, D.; Pila, G.; Davis, B.; Warnick, S.; Hedengren, J.D. Economic Benefit from Progressive Integration of Scheduling and Control for Continuous Chemical Processes. *Processes* **2017**, *5*, 84. [CrossRef]
24. Kunz, K.; Huck, S.M.; Summers, T.H. Fast model predictive control of miniature helicopters. In Proceedings of the 2013 European Control Conference (ECC), Zurich, Switzerland, 17–19 July 2013; pp. 1377–1382.
25. Bumroongsri, P.; Kheawhom, S. MPC for LPV systems based on parameter-dependent Lyapunov function with perturbation on control input strategy. *Eng. J.* **2012**, *16*, 61. [CrossRef]
26. Karimi Pour, F.; Puig Cayuela, V.; Ocampo-Martínez, C. Comparative assessment of LPV-based predictive control strategies for a pasteurization plant. In Proceedings of the 4th-2017 International Conference on Control, Decision and Information Technologies, Barcelona, Spain, 5–7 April 2017; pp. 1–6.
27. Cembrano, G.; Quevedo, J.; Salamero, M.; Puig, V.; Figueras, J.; Martı, J. Optimal control of urban drainage systems. A case study. *Control. Eng. Pract.* **2004**, *12*, 1–9. [CrossRef]
28. Mays, L. *Urban Stormwater Management Tools*; McGraw Hill Professional: New York, NY, USA, 2004.
29. Brdys, M.; Ulanicki, B. Operational Control of Water Systems: Structures, Algorithms and Applications. *Automatica* **1996**, *32*, 1619–1620.
30. Maciejowski, J.M. *Predictive Control: With Constraints*; Pearson Education: London, UK, 2002.
31. Gertsbakh, I. *Reliability Theory: With Applications to Preventive Maintenance*; Springer: London, UK, 2013.
32. Jiang, R.; Jardine, A.K. Health state evaluation of an item: A general framework and graphical representation. *Reliab. Eng. Syst. Saf.* **2008**, *93*, 89–99. [CrossRef]
33. Baecher, G.B.; Christian, J.T. *Reliability and Statistics in Geotechnical Engineering*; John Wiley & Sons: Hoboken, NJ, USA, 2005.
34. Ocampo-Martínez, C.; Puig, V.; Cembrano, G.; Creus, R.; Minoves, M. Improving water management efficiency by using optimization-based control strategies: The Barcelona case study. *Water Sci. Technol. Water Supply* **2009**, *9*, 565–575. [CrossRef]
35. Puig, V.; Ocampo-Martinez, C.; De Oca, S.M. Hierarchical temporal multi-layer decentralised MPC strategy for drinking water networks: Application to the barcelona case study. In Proceedings of the 2012 20th Mediterranean Conference on Control & Automation (MED), Barcelona, Spain, 3–6 July 2012; pp. 740–745.
36. Siljak, D.D. *Decentralized Control of Complex Systems*; Courier Corporation: Chelmsford, MA, USA, 2011.

© 2019 by the authors. Licensee MDPI, Basel, Switzerland. This article is an open access article distributed under the terms and conditions of the Creative Commons Attribution (CC BY) license (http://creativecommons.org/licenses/by/4.0/).

Article

# A Multi-Agent Social Gamification Model to Guide Sustainable Urban Photovoltaic Panels Installation Policies

Robert Olszewski [1,*], Piotr Pałka [2,*], Agnieszka Wendland [1,*] and Jacek Kamiński [3,*]

1. Faculty of Geodesy and Cartography, Warsaw University of Technology, 00-661 Warsaw, Poland
2. Faculty of Electronics and Information Technology, Institute of Control and Computation Engineering, Warsaw University of Technology, 00-665 Warsaw, Poland
3. Energy Economics Division, Mineral and Energy Economy Research Institute of the Polish Academy of Sciences, 31-261 Cracow, Poland
* Correspondence: robert.olszewski@pw.edu.pl (R.O.); p.palka@ia.pw.edu.pl (P.P.); agnieszka.wendland@pw.edu.pl (A.W.); kaminski@min-pan.krakow.pl (J.K.)

Received: 1 July 2019; Accepted: 1 August 2019; Published: 6 August 2019

**Abstract:** The paper presents a holistic and quantitative model of social gamification in a smart city, which is likely to stimulate the photovoltaic panels installation. The coupling of multi-agent systems, GIS tools, demographic data, and a spatial knowledge base made it possible to develop and calibrate a computable model of social interaction in a "model smart city," as well as to quantitatively evaluate the deployment of photovoltaic panels. It also enabled the analysis of factors affecting the efficiency of this process, e.g., the photovoltaic potential of solar roofs, the ownership of buildings, the type of building development, the level of social trust, institutional and social incentives, and the development of an information society. The devised model is tested on the city of Warsaw, utilizing spatial and descriptive data provided by city authorities.

**Keywords:** smart city; multi-agent systems; gamification; photovoltaics; renewable energy systems; spatial databases

## 1. Introduction

The contemporary model of urban development led to the urban sprawl and all its adverse effects such as increased car-dependency for the sake of mobility, longer daily trips from home to work, and the depletion of agriculture areas, as well as the degradation of the landscape, the deterioration of air quality, and higher energy consumption [1]. Since the mid-1950s, urbanized land in Europe has expanded by 78%, whereas the population has grown by just 33%, and this ratio continues to decrease [1–3]. Estimated world primary energy demand will rise by almost 60% between 2002 and 2030 [4]. Therefore, creating cities that are low-carbon, resilient, and livable is crucial. Sustainable and low-carbon energy technologies will play an essential role in the energy revolution required to prevent further climate degradation and global warming phenomena [4–6]. Solar power is an integral part of the sustainability of cities and its implementation and execution, through urban planning practices, may play a significant role in improving the energy efficiency of cities [1]. Growing deployment of PVs (photovoltaic panels) can result in social and environmental benefits [7]. Moreover, photovoltaic energy is considered one of the most promising emerging technologies, thanks to its noiselessness, non-toxic emission and relatively simple operation and maintenance [4,5].

The widespread use of renewable energy sources in an urban space, such as PV, requires not only access to highly efficient solar panels, but also financial incentives from municipal authorities as well as, first and foremost, the social engagement of residents. In many agglomerations, the process of molding an information society and the cooperation of smart city residents within the local communities are of crucial importance to the reduction of the environmental impact of the urbanization process, the improvement of the quality of life in the setting itself and the implementation of any sustainable development scenario [8].

This research was motivated by the desire to develop a gamification model, particularly of the so-called non-pecuniary type that uses a mechanism that increases social trust and, consequently, the use of renewable energy. The goal of the paper was to evaluate to what extent gamification could support the diffusion of urban renewable energy systems, especially photovoltaics installation. We proposed a model employing trust-based gamification, in which the municipal authorities propose developments that create a steady increase in social trust, e.g., the construction of playgrounds, parks, transport facilities, and so on, all while encouraging the citizens to install PV.

To meet the goal, a multi-parameter model was developed to stimulate diffusion in the number of PVs using the process of gamification and the methods of social interaction in the so-called smart city. There are several ways of implementing the modelling of this process. This article uses multi-agent system, digital topographic data, and spatially localized demographic information in modelling the process of installing PVs in a city. Thanks to the developed projection variants, the city authorities will be able to take an active role as a stimulator of sustainable development and in the promotion of renewable energy.

The article consists of five chapters. After a short introduction (Section 1), we describe related works, where we describe the state of knowledge regarding the analyzed issue, and we motivate the choice of the methods used (Section 2). Next, we describe the developed work methodology (Section 3). The methodology was divided into the main scenario (Section 3.1), description of the economic model (Section 3.3), the description of the calibration method (Section 3.4), and a description of how the simulations were carried out (Section 3.5). Next, we describe the results obtained (Section 4), where detailed descriptions of results for the city of Warsaw (Section 4.1) and the district of Wawer (Section 4.2) were posted. The work ends with discussion and conclusions (Section 5).

## 2. Related Works

Concerning the use of different energy carriers, tapping into the distributed renewable energy network is an attractive way to maximize urban energy and development sustainability. Solar energy is well-suited to the built environment; in particular, PV is a highly promising urban energy alternative that requires hardly any new infrastructure and offers synergy with building components. The fast-growing PV market and resulting reduction in cost could significantly impact energy, also by reducing $CO_2$ emissions [9,10]. From similar historical observations of earlier energy technologies, Lund investigated the prospects of new energy technologies. A hypothetical fast-track case for PVs, assuming an expansion like that of nuclear energy and oil, would result in it having a 20–25% share of all electricity by 2050 [10].

Investigating the literature regarding motivation for PV installations, one can find works dissolving only economic or technical factors. However, there are also analyses regarding social factors, as well as trust in society. Walker et al. [8] analyzes how interpersonal and social trust is related to the different meanings given to the society in programs and schemes regarding renewable energy, and in the characteristics and consequences that are intended or assumed by taking a community approach.

Nissing and von Blottnitz provided a definition of the concept of sustainable energization, developed by systematically repeating three key elements: The target group, the concept of energy services, and sustainable development [11]. High energy efficiency can be ensured if the urban planning process integrates solar energy in both new and existing urban environments [12]. Amado and Poggi explored the concept of solar urban planning with the goal of developing an operative methodology to achieve the best conditions for zero-energy building [12]. Rylatt et al. described the development of a solar energy planning system, based on the methodology for predicting the solar energy potential of the domestic housing stock with the use of geographical information systems (GIS) [13]. Kammen and

Sunter explored options for establishing sustainable energy systems by reducing energy consumption, particularly in the sectors of buildings and transportation [6]. Thorsten et al. discussed innovative integrated building services engineering systems for energy, water, and organic waste, which could contribute to increased energy efficiency, resource productivity, and urban resilience [14]. Baziliana et al. considered reductions in the underlying costs and market prices of PV systems and their implications for decision-makers [15]. Borri et al. explored the problems, as well as the potential, in the layout and implementation of a multi-agent system supporting an environmental planning process, within the strategic planning perspective [16]. Murakami described saturation phase of building-integrated PVs and coupled multi–agent simulations with electric power flow analysis. The presented agent-based model included two kinds of agents—customer and government agents [17]. Massey et al. described an online map-based gaming platform, which simulates spatial scenarios as MAS (multi-agent system) using human participants as the decision agents [18].

In Poland, the Renewable Energy Sources Act of 20 February 2015 [Ustawa o odnawialnych źródłach energii] has been crucial for the development of renewable energy source (RES) installations and is regarded as conducive to the sustainable use of energy resources meeting the criteria of sustainable development. This act transposes Directive 2009/28/EC into Polish law [19].

The Polish PV market comprises three models of selling electricity to the network along with their corresponding business options: An auction system, a prosumer system, and the so-called "business auto-producer" system (a business prosumer). The auction (large RES investments) and prosumer systems (an energy consumer is also its producer) are official state instruments of support that aim to help satisfy the international commitment for Poland; that is, obtaining a 15% share of energy from renewable energy sources by 2020. On the other hand, market-based principles underlie the "business auto-producer" system (without selling surplus energy) [20]. By the end of May 2018, the total installed capacity in PV was about 300 MW, meaning PVs had a 3.4% share of Polish renewable energy sources. It is estimated, however, that regarding installed capacity, by the end of 2020, solar energy may become one of the leading renewable energy technologies [21].

According to data from the Innogy Polska S.A. press office, which operates the distribution network in Warsaw and the surrounding area, in 2017, the capital city gained 414 PV microgeneration installations (most of them in the final quarter of 2017) with a total capacity of 2317 MW. The number of all installations in Warsaw from September 2008 to 2010 totaled 908 (5.697 MW). According to the analysis of clients who use PV solutions, about 80% of installations are in single-family houses, 15% of clients are various types of enterprises, and 5% are condominiums.

In 2017, the number of subsidies granted to Warsaw residents by the Warsaw City Hall for home-made, renewable energy installations almost doubled, and in the city, the number of applications for projects in renewable energy sources (primarily solar farms but also heat pumps and solar collectors) increases every year [22]. We believe that strengthening this process requires proposing a new approach. The approach proposed in this article establishes an important innovation in a smart city, both in terms of increasing the use of renewable energy and levels of social innovation.

Many papers [23–27] support the idea that solar energy requires long-term policy support. Subsidized feed-in tariffs in PVs increased global demand and allowed for economies of scale - they lowered international PV module costs by 22% with every doubling of capacity. Moreover, due to rapid growth and strong market competition, the prospect for continued cost reductions are still promising [23].

Around the world, various policy instruments and support schemes have been implemented with the aim of stimulating technology diffusion and increasing the deployment of residential and distributed PVs. Solar energy promotion in different countries includes a range of instruments. Chapman et al. [28] review the Australian residential PV policies between 2001 and 2012, in particular the impact of net and gross feed-in-tariffs (FiTs) on electricity prices within the country. From their analysis, they conclude that FiTs have a significant impact on installation rates, being responsible for both a rise in 2008 and a decline post-2008 when they were removed. Similarly, in an empirical analysis focused on PV adoption in California, Schmidt-costa et al. [29] and Drury et al. [30] delve in detail into one of the most successful public policies aimed at PV technology diffusion in the US: The California Solar initiative program. They discuss the impact on PV adoption trends and the policy implications of third-party PV business models in the Los Angeles and Orange counties. Muhammad-Sukki et al. [31] reviewed the financial impact of FiT schemes in Japan and provide a similar financial analysis for Germany, Italy, and the United Kingdom. An alternate policy framework that has been widely implemented around the world in order to promote the deployment of solar PV systems is incentivized self-consumption. In a recent study, Pacudan [32] evaluated the possible options and assessed the impact of the possible implementation of a feed-in tariff and a net metering and billing scheme, including an upfront payment subsided for net metering and a premium payment for exported electricity in Brunei Darussalam. Tilman and Engelmeier highlighted how long-term rents for future energy technologies could be assured even in a developing country suffering severe energy deficits. India's Nawaharlal Nehru National Solar Mission triggers solar investment and manages the necessary subsidies. These are preferential feed-in tariffs, renewable purchase obligations (RPO), the renewable energy certificate (REC), tax holidays, attractive options for the accelerated depreciation of investments, and soft loan schemes [23,24]. In Australia, different amenities are employed depending on the state. For example, the Queensland State Government runs a scheme that pays AUD0.44+ per kW for electricity generated from solar PVs and the states of Victoria and South Australia have introduced minimum five-star energy efficiency requirements for new residential buildings [33]. Based on the examples of China, India, and South Africa, Becker and Fisher compared feed-in and auction-based tariffs and concluded that experiences with feed-in tariffs in Europe were not transferable to emerging countries [25]. However, an approach that applies feed-in tariffs has been effective, for example, in Germany and Spain [33]. The conducted analysis and literature studies indicate that there is a need to create multi-agent systems that take into account social, spatial, and solar data to create predictive models of the increase in the number of solar panels in cities. It should also be examined how different methods of social stimulation (e.g., gamification, direct subsidies, etc.) influence the obtained results.

## 3. Methodology

The conducted research presupposes that to develop a model defining the number of PVs installed in a smart city, one should use a broad spectrum of factors to explain the decision to invest in renewable energy rather than just a single, independent variable (e.g., panel price or the percentage of city subsidies). In addition to those factors that are strictly economic (such as unit price, roof area, subsidies, daylight, and so on), the conducted research assumed that specific characteristics would influence the willingness to install panels. These features would be spatial location, social trust, the development of local communities, relations concerning ownership, policies of the municipal authorities regarding urban residents and businesses, as well as residents' inclination for new technological trends and environmental social responsibility, and the inclination for gamification. Using the studies [3], the authors also chose four social characteristics (age, willingness, trust, wellbeing) that affect the willingness to install solar panels. The values attributed to the social identity parameters of the agents were assigned on the basis of sociological research for specific social groups and city districts.

Figure 1 shows the flowchart of our research methodology. First, we prepared the required data, then we created a model, consisting of a social, economic, and spatial part. The developed model was enriched with specific data from the developed scenarios, and the results were analyzed. A more detailed description of the individual steps is described further.

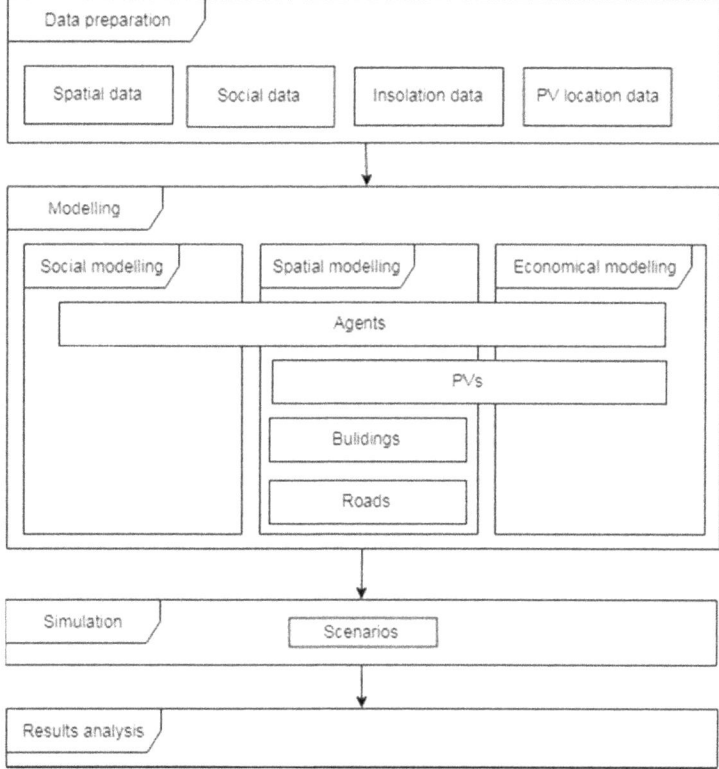

**Figure 1.** Flowchart of methodology used (source: Authors' own work).

*3.1. Scenarios Setting*

We considered 14 various scenarios to stimulate the residents of smart cities to the installation of PVs (Table 1). According to each of the scenarios, the following factors influence the decision to install panels: The location of both residence and work, the manner of moving around the city and the consequent opportunity (or lack of) to view panels on municipal buildings, affluence levels, and the ownership arrangements of property. We assume that individual buildings might belong to the city, companies, or private institutions. Private property may be managed individually (where the owner decides on the possible installation of the panel), in a condominium, or in a housing co-operative. In condominiums, 80% of tenants must consent to install a panel yet in housing co-operatives, the figure is 95%. Adopted values stem from the Polish specificity, yet they have been saved as model parameters and can be easily modified.

**Table 1.** The scenarios considered to stimulate the residents of smart cities to the installation of PVs.

| Number | Name | Features |
|---|---|---|
| 1 | Base case | This model does not entail any benefits; businesses cannot invest in PV. In the basic case, we assume high citizens' wealth. |
| 2 | Installation subsidy | This model assumes that the municipal authorities propose a simple incentive to install PV: a direct installation subsidy of 800–1200 PLN (about 200–300 EUR). |
| 3 | Money-based gamification; the so-called pecuniary-based gamification | This model assumes that the municipal authorities propose a gamification mechanism based on material profits [34]. Everyone who installs PV will receive some material benefits (e.g., free parking, coupons, movie tickets, etc.) worth 200 PLN per month (about 50 EUR). |
| 4 | Trust-based gamification (10% variant); the so-called non-pecuniary-based gamification | This model assumes the so-called non-pecuniary-based gamification established on principles that have a beneficial influence on the level of social trust (social trust increases by 10%). The municipal authorities propose gamification mechanisms that continuously increase this trust, encouraging citizens to install PV, e.g., in building playgrounds and city parks, having road signs powered with PV and so on. |
| 5 | Trust-based gamification (20% variant) | This model assumes the so-called non-pecuniary-based gamification established on principles that it has a beneficial influence on the level of social trust (social trust increases by 20%). |
| 6 | Eco-business | This model assumes that business is active and "eco-friendly," i.e., it does not concentrate on the long-term viability of a PV installation. However, the owner takes into account the cost of the investment. |
| 7 | Small incentives from the City Hall | This model assumes that the municipal authorities have introduced an incentive for residents by installing PVs on one, municipality owned building in each district. |
| 8 | Large incentives from the City Hall | This model assumes that the municipal authorities have decided to introduce an incentive for residents by installing PVs on all large buildings belonging to the city. |
| 9 | High incentives from the City Hall | This model assumes that the municipal authorities have decided to introduce an incentive for residents by installing PVs on 50% of buildings belonging to the city. |
| 10 | Small incentives from the City Hall and active business | This model assumes that the city authorities have decided to introduce an incentive for residents by installing PVs on one, municipality owned building in each district and also that that business is active, i.e., able to invest in PV. |
| 11 | Low wealth | Assumes that low wealth levels of citizens mean that dwellers of single-family homes set aside a monthly amount of 100–1000 PLN (25–250 EUR), while the dwellers of blocks of flats save 10–100 PLN (2.5–25 EUR). |
| 12 | Low wealth, installation subsidy | Assumes low wealth and a simple incentive to install PV from the municipality: A direct installation subsidy |
| 13 | Low wealth, trust-based gamification | Assumes low wealth and trust-based gamification. |
| 14 | Low wealth, money-based gamification | Assumes low wealth and money-based gamification. |

## 3.2. Social Modelling

To validate the devised model, we developed a "model city" comprising seven hexagonal districts with a dominating type of function and structure (Figure 2 and Table 2). The residents inhabit the city, as illustrated by the so-called dot distribution map (Figure 3) and each of 1000 dots, representing a group of residents, was implemented into the system as an agent with specific characteristics. The ArcGIS ESRI (13.0, Esri, Redlands, CA, USA) application was used to develop the source spatial database, enabling the preparation of a set of thematic layers (land cover, buildings, communication routes, districts borders, and the distribution of residents) as SHP (shape) files. These layers were then used to build a multi-agent system in the GAMA environment. GAMA is a modelling and simulation-development environment for building spatially explicit agent-based simulations [35,36]. It is a multiple-application domain platform that uses a high-level, intuitive agent-based language. With GAMA, users can undertake most of the activities that relate to modelling, visualizing, and exploring the simulations by using dedicated tools.

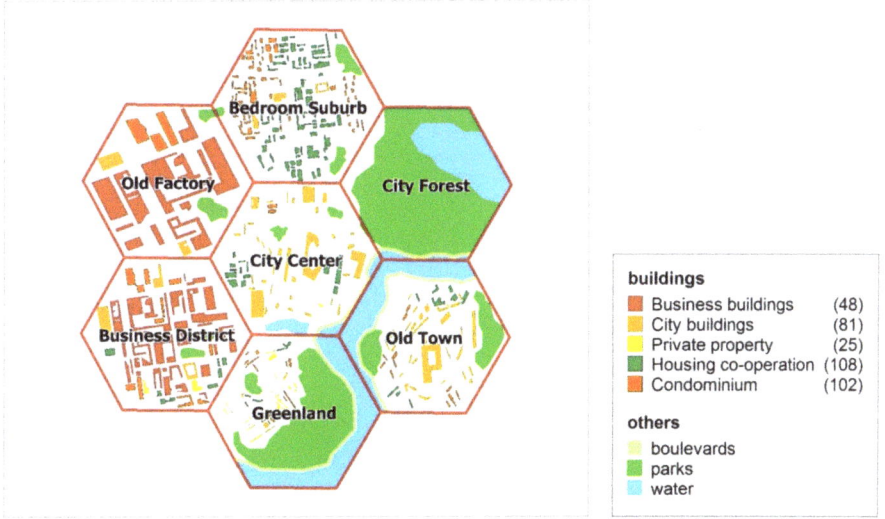

**Figure 2.** Model city (source: Authors' own work).

The "model city" (Figure 2) includes a system of roads on which agents move, a set of buildings (including stand-alone and multi-occupancy residential buildings, factories, office buildings, health clinics, schools, and museums) and green areas, i.e., boulevards, parks, and water reservoirs. The city is also divided into districts that are distinguished by a set of features, which characterize both the district and the people in it, e.g., the office district includes a significant number of office buildings and the people in it are white-collar workers.

An agent models a city resident and has a set of traits that reflect his or her social identity:

- Age, gender, marital status, and number of children.
- The willingness to install PV: A number within the range <0,1>, which changes during an agent's life and depends on both social trust and the number of PV installations that are observed in the environment.
- Trust in other residents/ social trust: A number in the range <0,1>, which determines to what extent a citizen trusts other people. It is not considered to be a pejorative trait—it does not mean naivety, rather faith in the capabilities of others.
- Wealth: A number in the range <0,1> determining the material status of a resident.

Besides the above features, each agent is assigned to a residential place (i.e., a residential building) and a workplace (an office building or a factory). During the simulation, the agents move around the city according to a normal daily rhythm, i.e., in the morning they go to work and return home in the evening. Thereafter, they may go to a museum or a park, visit governmental agencies, or see a doctor. Observing panels installed on other buildings (e.g., on public facilities, private homes, and so on) on the way to work or whilst walking influences the individual agents' willingness to install PVs.

A multi-agent system was used to develop the model, and agents represented both city residents and objects, such as buildings, districts, parks, public utilities, and roads. Each agent possesses the average characteristics of residents, such as wealth, level of social trust, social engagement, the number of children, and so on, and for the model city, each of the 1000 agents represents 1000 residents.

Table 2. Model city districts (source: Authors' own work).

| Number | Name | Number of Agents |
|--------|------|------------------|
| 1 | Greenland | 50 |
| 2 | City Centre | 250 |
| 3 | Bedroom Suburb | 400 |
| 4 | Old Town | 50 |
| 5 | Business District | 100 |
| 6 | Old Factory District | 150 |
| 7 | City Forest | 0 |

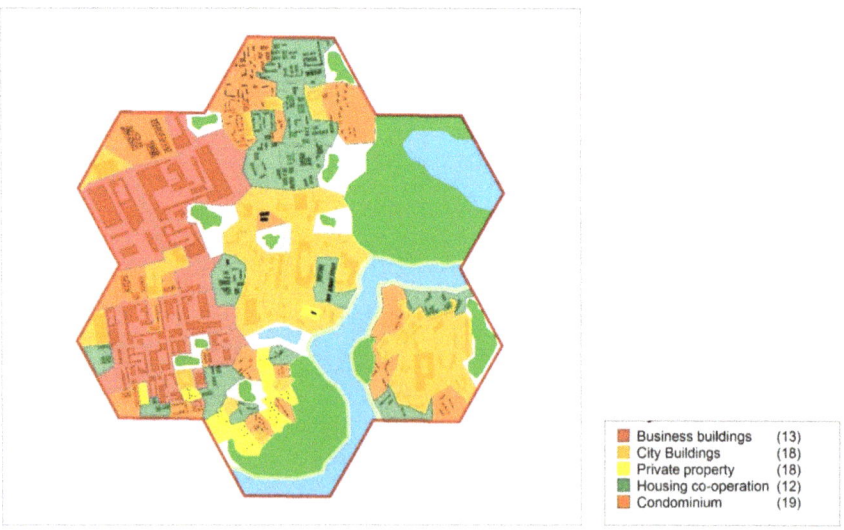

Figure 3. Dot distribution map (each dot represents 1000 inhabitants of the model city) (source: Authors' own work).

In the model, four types of building complex management are considered:

- Private property—single-family houses.
- Private property—retail and office buildings.
- Housing co-operatives.
- Condominiums.
- Public property that is managed by the city hall.

Various guidelines regarding the willingness to install PV characterize each management model, which—particularly for condominiums and housing co-operatives—are created from the aggregation of the preferences of their members.

The decision about whether to build PV in the private properties where there may be several single-family houses, in which it can be located up to several household agents, is made on the basis of voting. When the parameter indicating the willingness to install the panels reaches the threshold value, then an economic analysis is performed, analyzing both the long-term and initial costs of a PV installation. If the installation costs are acceptable (calculated from the wealth agent parameter and the average income parameter) and if the long-term costs after installation are lower than those without, the randomly selected number of household agents chose to install the PV on their buildings. The average income parameter depends on place of living of the agent: Residents of single-family houses have the average income parameter equal to 10,000 PLN (about 2500 EUR), and the residents of condominium and housing co-operatives have this parameter equal to 1000 PLN (about 250 EUR).

For a condominium and a housing co-operative, the total costs of PV installation and long-term costs of installation are considered. Subsequently, household agents belonging to these groups hold a vote, with an agent voting for the installation if the willingness to install panels is greater than the threshold. For condominiums, voting PV through requires more than 80% of votes being in favor, and for housing co-operatives, a proportion of over 95% is required. What differentiates those three types of complexes is a number of inhabitants.

For privately owned retail centers, office buildings, and factories, we assumed that to install panels, the following would be checked:

- For the "eco-business", only whether the investment threshold would not exceed the assumed investment cost threshold for business (10,000 PLN—about 2500 EUR).
- The investment threshold (see above) and long-term analysis of energy costs would be performed.

Public property managed by the city hall includes official buildings, schools, and health clinics. We assumed that PV installation on these buildings would depend on a given scenario:

- The base case.
- The scenario "Small incentives from the City Hall"—with one panel installed in each district.
- The scenario "Large incentives from the City Hall"—with panels installed on large buildings belonging to the city.
- The scenario "High incentives from the City Hall"—with panels installed on 50% of the buildings belonging to the city.

These scenarios differ according to the number of installed panels on buildings managed by the city hall. The presence of PV acts as an incentive for the household agents to install them on their properties.

*3.3. Economic Modeling*

When distributed renewable energy sources, such as PVs, are installed in a given area, then this place may constitute the so-called microgrid and the owners of PV installations the prosumers. For an owner of the microgrid, the legal aspects regulating construction, installation, and cooperation, as well as the connection of the microgrid to the power grid are essential, and so are the aspects that specify payment for the energy transmitted to the network or used within the microgrid.

In Poland, the so-called energy tri-pack, meaning the amendment to the Energy Law and other acts (27 August 2013) contains regulations regarding small RES installations [Dz. U. 2013, item 984, 27 August 2013]. The so-called energy tri-pack promotes improved conditions for microgeneration installation to a network (RES with a joint capacity of up to 40 kW connected to a network of below 110 kV, or with a total thermal power of up to 120 kW). Under the Act, the connection fee is free.

In addition, the procedures leading to the connection of the installation to the network have been simplified. As a result, there are now incentives to install RES. After Billewicz [37], the owner of the microgeneration that generates electricity should consider the fees, which are included in Equations (1)–(4): Excise duty on the production of electricity, income tax on the sold energy, and value added tax (VAT) from resale of energy.

Under the current regulations, electricity generated by means of microgeneration connected to the distribution network must be purchased by the selling energy entity ex officio at the price of 80% of the average electricity sales price from the previous calendar year. Importantly, for an owner of a microgeneration who is a natural person (not an entrepreneur) under the Act on the Freedom of Business Activity, the generation and sale of electricity is not considered an economic activity [38]. In this section, we use the concept of a household agent (representing a natural person) to underline its cognitive capacities.

In the conducted research, we use an economic model in which two variants are compared:

- The installation of PVs on a building or a set of buildings. In this case, we assume that during the period of PV depreciation (20 years), the household agents of a building complex would use less energy from the grid but would have had to bear the costs of the PV installation. Additionally, they would resell the remaining energy produced back to the network.
- No PV installation. In this case, we assume that during the period of PV depreciation (20 years) the household agents would pay for the total amount of energy but would receive interest from a long-term deposit from savings. The initial amount of savings is the costs of PV installation.

Equation (1) describes the annual energy costs with installed PV, it consists of a variable part of the fee and a fixed part of the fee. The variable part of the fee consists of (PG11) price of electricity according to tariff (most popular one: G11), (SVNR) network rate, and quality assurance component of the system fee rate (SFRQAC). The amount of consumed energy (AECPC) is decreased by annual energy production per panel per capita (AEPV), fractioned by energy consumed (HEC). The fixed part of the fee is: (CTR), the network standing charge, and (SF) subscription fee. The costs are increasing. Energy costs are taxed by a value-added tax (VAT). Equation (3) describes the annual energy costs without PV installed. It differs from Equation (1) only with the amount of consumed energy.

Equation (2) describes the annual profit from the resale of energy. The amount of resold energy consists of energy production per panel per capita (AEPV), reduced by $(1 - HEC)$ part of the energy resold to the grid. The amount of energy produced is reduced annually by (DS) drop in cell performance. This energy is sold by the average price of energy on the competitive market (AVEP), reduced by coefficient of change in the energy resale price (ERPCC). In addition, we assume that the price of energy increases year by year (AIEP). Income from energy sales is taxed income tax rate (ITR).

Equation (4) shows annual value on the deposit assuming that the PV is not installed. It comprises the interest rate on the deposit (IRD); it also assumes that the income is taxed by capital gains tax (CGT).

Equations (5) and (6) represent the cost assuming a 20-year amortization. Therefore, Equation (5) shows total energy costs with installed PV, which consists of one-time PV installation costs (CIPV), annual energy costs with installed PV (AECIPV), decreased by annual profit from resale of energy (APER) and profits from gamification (PG). In turn, Equation (6) shows total energy costs without installed PV and consists of annual energy costs without PV installed (AECNPV), decreased by interests income (ADVWPV$^0$ − ADVWPV$^{20}$).

$$AECIPV^Y = (1 + VAT) \\ \times ((CTR + NSC + SF) + (PG11 + SVNR + SFRQAC) \\ \times (AECPC - AEPV \times HEC)) \times (1 + AIEP)^Y \qquad (1)$$

$$APER^Y = AEPV \times (1 - HEC) \times (1 + AIEP)^Y \times (1 - DCP)^Y \times ERPCC \times AVEP \times (1 - ITR) \qquad (2)$$

$$\text{AECNPV}^Y = (1 + VAT) \times ((CTR + NSC + SF) + (PG11 + SVNR + SFRQAC) \times AECPC) \\ \times (1 + AIEP)^Y \qquad (3)$$

$$\text{ADVWPV}^Y = \text{ADVWPV}^{Y-1} + \text{ADVWPV}^{Y-1} \times IRD \times (1 - CGT) \qquad (4)$$

$$\text{TECIPV} = CIPV + \sum_{Y=1}^{20} \left( \text{AECIPV}^Y - APER^Y - PG \right) \qquad (5)$$

$$\text{TECNPV} = \sum_{Y=1}^{20} \text{AECNPV}^Y + \text{ADVWPV}^0 - \text{ADVWPV}^{20} \qquad (6)$$

The parameters for the applied economic model are in Table 3. We also determined that $IINC^0 = CIPV$, where $CIPV$ is the unit cost of a PV installation per capita, meaning that if someone does not install the panels, he or she deposits the money he or she would have spent on the installation into an account with an average annual interest rate.

Table 3. Parameters of applied economic model.

| Description | Abbreviation | Value |
|---|---|---|
| Total Energy Costs with Installed PV | TECIPV | - |
| Total energy costs without PV installed | TECNPV | - |
| Annual energy costs without PV installed | AECNPV | - |
| Annual energy costs with installed PV | AECIPV | - |
| Annual profit from resale of energy | APER | - |
| Costs of PV installation | CIPV | - |
| Profits from gamification | PG | - |
| Annual value on the deposit assuming that the PV is not installed | ADVWPV | - |
| Annual energy consumption per person [39] | AECPC | 800 [kWh] |
| Installation costs 1 kWp PV | IP | 5500 [PLN] (about 1300 EUR) |
| Average PV area for 1 kWp | A | 6.5 [m$^2$] |
| The average price of energy (sale) on the competitive market [40] | AVEP | 0.1637 [PLN/kWh] (about 0.04 EUR/kWh) |
| The coefficient of change in the energy resale price | ERPCC | 80% |
| The interest rate on the deposit [41] | IRD | 1.38%/year |
| Drop in cell performance | DCP | 0.8%/year |
| The annual increase in electricity prices | AIEP | 3%/year |
| The income tax rate | ITR | 18% |
| Capital gains tax | CGT | 19% |
| The Value Added Tax rate | VAT | 23% |
| The price of electricity, tariff G11, Stoen [42,43] | PG11 | 0.3397 [PLN/kWh] (about 0.08 EUR/kWh) |
| The quality assurance component of the system fee rate [4,5] | SFRQAC | 0.0154 [PLN/kWh] (about 0.0039 EUR/kWh) |

Table 3. Cont.

| Description | Abbreviation | Value |
|---|---|---|
| The component of the transitory rate [42,43] | CTR | 0.55 [PLN] (about 0.13 EUR) |
| The variable of the network rate [42,43] | SVNR | 0.1659 [PLN] (about 0.04 EUR) |
| The network standing charge [42,43] | NSC | 6.61 [PLN] (about 1.54 EUR) |
| The subscription fee [42,43] | SF | 2.76 [PLN] (about 0.64 EUR) |
| Annual energy production per panel per capita | AEPV | 800 [kWh] |
| The percentage of energy from PV consumed by the household | HEC | 50% |

In the simulation, such an analysis is carried out each month, because it is assumed that condominium and housing co-operative meetings would occur on a monthly basis.

From Equations (1)–(6), two outputs are obtained:

- The installation costs for PVs per capita; checking whether the cost of installing panels per capita CIPV is acceptable for interested inhabitants.
- A comparison of the long-term costs of installation with that of no PV installation per capita; in this case, checking whether the inequality (7) is true.

$$TECIPV \leq TECNPV \qquad (7)$$

*3.4. Calibration*

Due to the simulations carried out for the model city, it was possible to calibrate the multi-agent system properly, i.e., to determine the value of individual parameters, which then enabled the use of the devised model for a real urban area. In our example, 40,320 iterations correspond to a period of five years during which the municipal authorities, businesses, residents, and condominium and housing co-operatives make decisions, iteratively, regarding PV installation on buildings. The simulation assumes that the step of simulation equals to 15 min, one day equals to 96 simulation steps, and one week takes 672 steps. We use this time structure to reflect the frequency of agents' meetings as they move around the city. However, we assumed that due to the weekly basis of simulation, the representation of the week is crucial, and because of the ease of simulation, we assumed that one week is treated as one month, because we adopted a monthly decision-making process for PV installations. One year lasts 12 months, so 8,064 steps of simulation. The simulation period that was selected was 5 years (40,320 steps of simulation), on one hand to provide enough time for introduce the changes and to assess the changes, and on the second hand to provide reasonable time of computations.

The sensitivity of the model in terms of duration of the simulation step is checked and it turned out that the number of meetings is inversely proportional to the length of the simulation step. On the other hand, the shorter the simulation step, the more time should be spent on simulation, as we wanted to keep the weekly simulation base and the five-year research period. Therefore, 15 min turned out to be a good compromise.

*3.5. Simulations for the Model City*

Using the methodology developed and the multi-agent system, as well as the scenarios presented in Table 1, we conducted preliminary tests of the system on the model data (the so-called model city described in the previous paragraph). The studies made it possible to both adjust the model, that is select and determine critical parameters, and to analyze the influences of location, spatial objects, and demographic parameters on the model city residents. The selection of the most important (critical) parameters was made by using the sensitivity analysis of the method to change individual parameters and to assess the value of the partial correlation coefficient. The assumption was that each agent would represent 1000 residents with averaged characteristics typically seen in a given district of the city.

The results obtained (Figures 4 and 5) indicate that direct subsidies are the most effective form of stimulating the installation of PVs. Compared to the base case (blue color), this variant of stimulation brings about an almost threefold increase in the number of installed panels. It is worth noting that, compared to the base case in which the increase in the number of PVs is significantly affected by installations on private properties and housing communities, in the subsidy model (Figure 4), panels are also installed on buildings managed by the housing co-operatives. The number of PVs installed does not increase for the private properties. This is due to the saturation of the market in the first period (for people who can afford the installation of PVs). Other groups see the positive effects of installing photovoltaic panels in private properties and are encouraged to use such solutions.

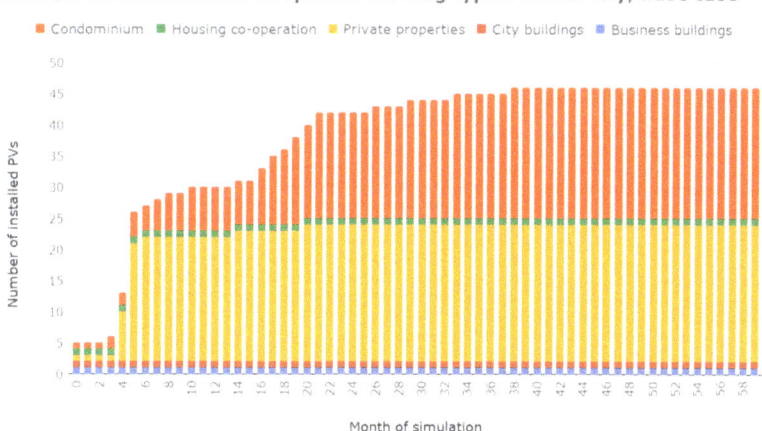

**Figure 4.** The increase in the number of PVs in the model city: The base case. The number of PVs installed on specific building types.

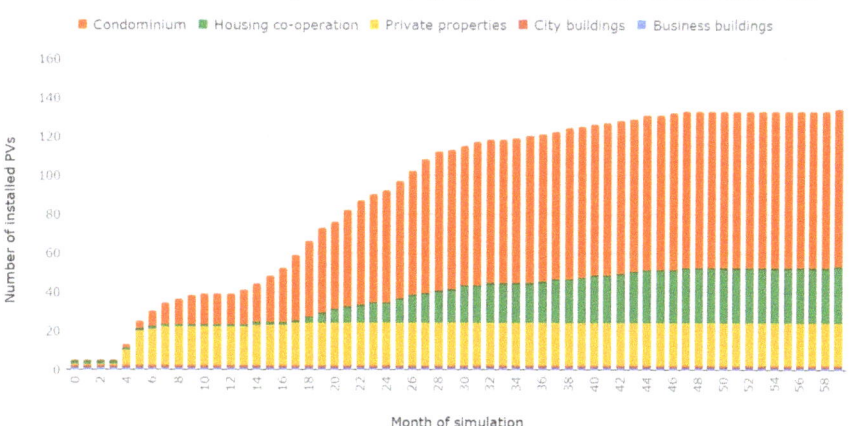

**Figure 5.** The increase in the number of PVs in the model city: Direct subsidy and the number of PVs installed on specific building types.

In the case of subsidies for PV installations, two interesting dependences have emerged (see Figure 6). First, housing co-operatives start to engage: The number of PV systems installed by the housing co-operatives matches the number of those privately installed. However, there is a slower increase in PVs. What is important when analyzing the subsidies (in the amounts of 1,200 PLN and 800 PLN) is that the lower the subsidies, the smaller the share of housing cooperatives in the final number of PVs installed. However, regardless of the level of subsidies, there is a massive increase in PVs installed on buildings belonging to condominiums. The difference in the diffusion of PV in the different building types between the 800 and 1200 PLN surcharge is small—around 15%. A surcharge of 800 PLN is therefore already sufficient to install panels. Due to difficulties in today's economy in Poland, a subsidy of 800 PLN (about 200 EUR) would act as a sufficient incentive for installation. What should be emphasized though is that this method of offering direct subsidies, while effective, does not provide a perfect solution to the problem, both in an economic and a social sense; that is to say, it does not give an incentive that is sufficient to develop a local community.

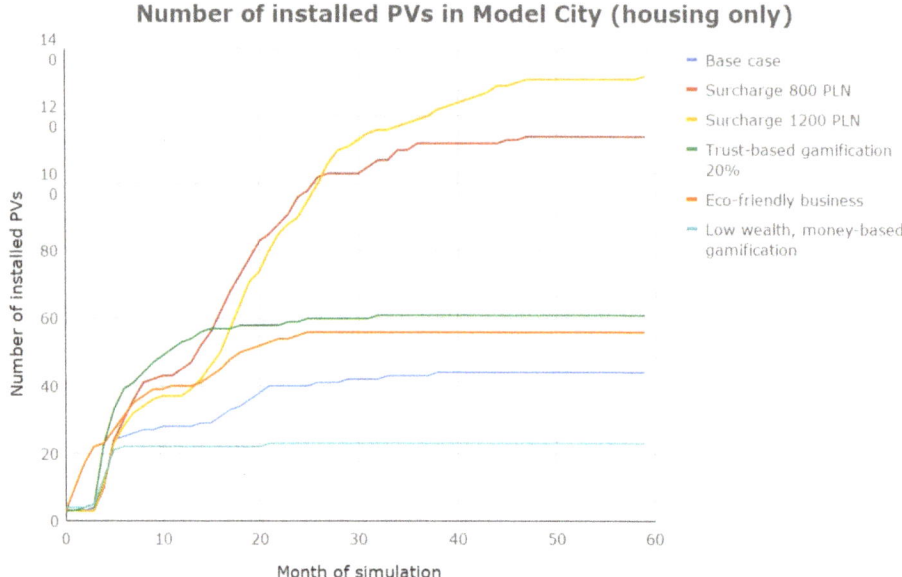

**Figure 6.** Simulation results for selected scenarios for the model city.

Among the other methods of stimulation, trust-based gamification has the highest increase in PV installations, as well as the highest in social trust (Figure 6). In the scenario with a gradual increase in trust of up to 20% (in public institutions as well as mutual trust between residents), there is a rapid increase in the number of new installations during the first 15 months and then a gradual saturation in the subsequent forecasting period. Trust-based gamification is a process that enables the authorities of a smart city to support sustainable development through the installation of PVs. It also enables the development of a local community. In contrast to money-based gamification—a stimulation that generates material profits only to those involved in the installation process (i.e., anyone who installs PV will receive material profits, such as free parking, coupons, cinema tickets. and so on, worth 200 PLN—about 50 EUR)—trust-based gamification contributes to sustainable development.

In the trust-based variant, the city authorities propose gamification mechanisms which create an ongoing increase in social trust, encouraging citizens to install PV. The growth of both interest and trust is achieved, e.g., through the construction of playgrounds, road signs being powered with PV, and so on. In this model, there is a significant increase in the number of panels, especially in those areas of the city where private properties and condominiums dominate. As indicated by the simulations carried out in the southern and eastern parts of the model city (the Greenland and Old Town districts) dominated by private houses, in 60 months, the saturation level achieves nearly 50%. In addition, in the trust-based gamification scenario, the number of installations is also significant in the northern part of the model city where condominiums are common (see Figure 7).

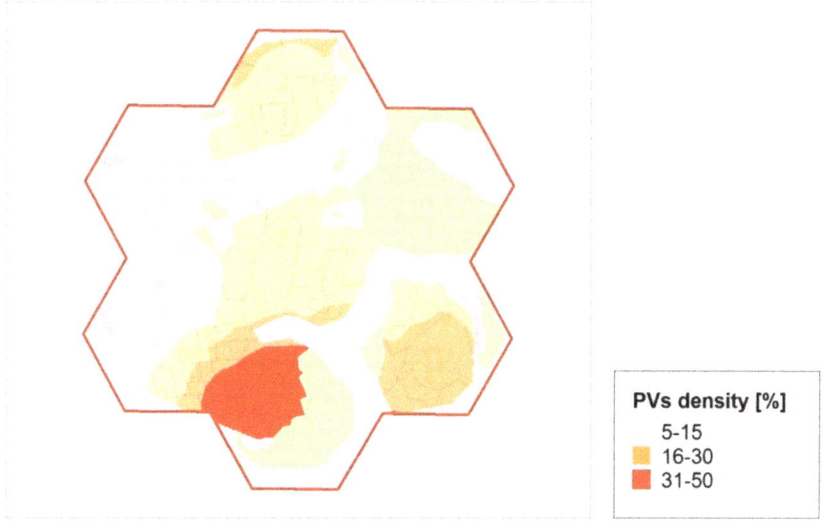

**Figure 7.** The trust-based gamification scenario for a model city.

It should be noted that this variant of stimulation is not sufficiently effectual in districts dominated by white- and blue-collar workers (i.e., the Business and Old Factory Districts). Using other simulation models, e.g., the "eco-business" or "small incentives from the City Hall and active business", leads to a greater interest in PVs in the industrial districts as well. This indicates that the authorities of a smart city need to adopt a hybrid approach, i.e., trust-based gamification for the residents, along with a system of economic incentives and tax reliefs for businesses.

In the case of trust-based gamification, there is a marked increase in the number of installed PVs initially (from 5–20 months in the simulation). This stems from increased levels of social trust being created by these mechanisms. The incentive to install PVs by mounting them on buildings belonging to the city corresponds to this increase. However, the influence is disproportionate to the number of PVs installed to incentivize: The difference in the final number of PVs installed on residential buildings between the large and the small incentives scenarios is five. However, to achieve this, nine more PVs had to be installed on city buildings.

In contrast, when comparing the high-incentives with the large-incentives scenario, the difference in the number of installed PVs is 8, and 30 more PVs were installed on city buildings to achieve this. So that the results were not obscure, it was decided not to activate PV installation on business buildings. A business being active does not affect the number of panels installed on residential buildings. However, including an additional option in which the business is "eco-friendly"—which means that the decision-makers do not consider the long-term costs of the PV installation—increases the number of installed PVs by 12 (compared to the base case).

Simulation of the scenarios listed in Table 1 was carried out for the model city over a period of five years (60 months) to predict an actual increase in the number of panels for the real data. This problem is the subject of the next chapter.

## 4. Results

A model city with defined features was selected as the research area, for which several research scenarios were prepared. We tested the developed methodology for the entire area of Warsaw and the selected district of the city, Wawer. In this area, the most significant development and the biggest number of PV installations in Warsaw can be observed, an initiative undertaken by the residents who install them to save money. At present, those residents who install PVs tend also to be choosing other forms of alternative environmentally friendly activities such as, for example, home wastewater treatment plants.

The base model developed for the model city was calibrated in time according to the number of Warsaw installations provided by the relevant institutions. The assumption for the model city was that each agent represents 1000 residents with averaged characteristics typical for a given district of the city. For the generalized digital geographic data (corresponding to a 1:250,000 scale map) that cover the entire area of Warsaw, we used a model in which a single agent also represented 1000 residents with averaged features. This required using 1750 agents—significantly extending the calculation process. The most complex and long-term calculations were made for the district of Wawer. The study used a topographic source database (equivalent to a 1:10,000 scale map) and detailed demographic data. Each of the 7500 agents represent only 10 residents of the district. Performing these calculations using the GAMA toolset environment that enables multi-agent simulations required about five days of calculation for each of the 10 analyzed models.

The result of choosing the number of agents was, in any case, a compromise between the reality of the model and its computability. The model is quite complex, as it models the everyday life of a citizen moving around the city with a resolution of 15 min, where the whole simulation lasts five years; residents move along the road network in the city according to the plan of the day and week. We analyzed the number of agents for each model and the assumption about the differentiation of the number of agents per simulated citizens (see Table 4). Conclusively, we ensured that the number of agents for each model ranged from 1000 to 7500.

**Table 4.** List of the number of agents and citizens in the models.

| Model | Number of Agents | Number of Citizens | Agents per Citizen |
|---|---|---|---|
| Model city | 1000 | 1,000,000 | 1000 |
| Warsaw city | 1750 | 1,750,000 | 100 |
| Wawer district | 7500 | 75,000 | 10 |

### 4.1. Simulations for the City of Warsaw

A research experiment on actual data, both spatial and demographic, was necessary to verify the developed methodology. A spatial database for the Warsaw area was developed (Figure 8) using the digital general geographical data and demographic data from the census.

Data available on the City Hall of Warsaw's geoportal (http://mapa.um.warszawa.pl/mapaApp1/mapa?service=mapa_oze)—using the so-called RES layer (renewable energy sources) and its so-called solar map—provided information on the PV potential of individual buildings. Due to hardware-based limitations linked to the necessary use of significant computing power of the CPU as the number of agents increased, 1750 agents were employed for the city of Warsaw. Their distribution in individual districts is proportional to the number of residents of a given area (Figure 9).

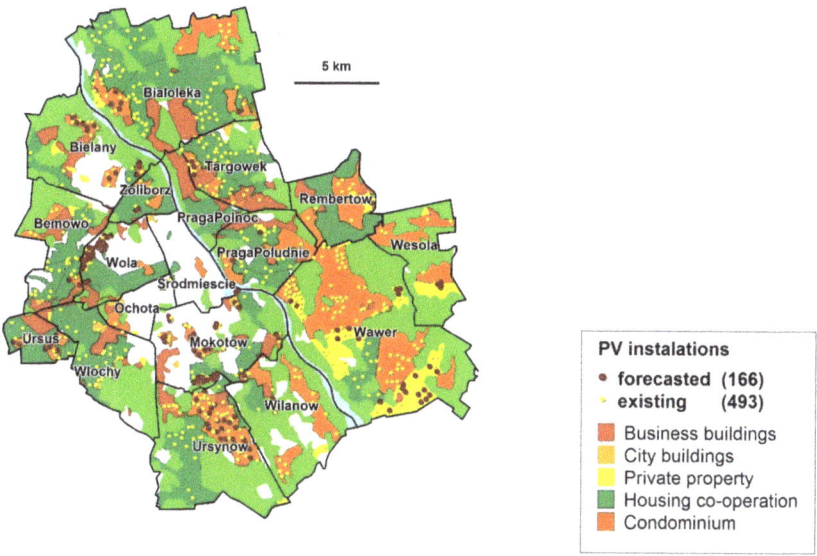

**Figure 8.** Currently existing PVs in Warsaw and panels forecasted in the trust-based gamification scenario.

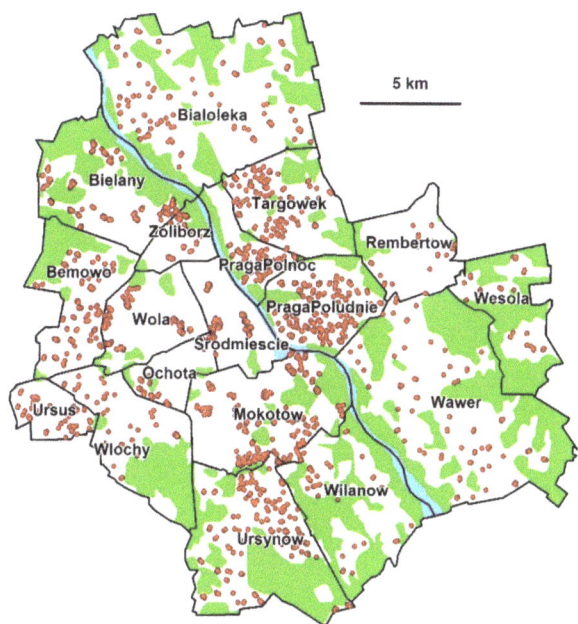

**Figure 9.** Dot density map of the population distribution in Warsaw (each agent (red dot) represents 1000 inhabitants).

Using the principles of cartographic generalization [44], spatial information on development, land cover, and road network was generalized. Averaged demographic characteristics of the population in a given location were assigned to the agents representing the residents of a given district. Information on the number of PVs and their location, in the form of their addresses, was obtained from the City Hall of Warsaw and 'Innogy Polska S.A.' (the electricity supplier). Using the geocoding mechanisms available among GIS tools, all currently installed PVs in Warsaw were added to the spatial database. Besides the spatial location, these data also contain information on the number of modules and the power of the installation, as well as the type of building on which the panel was installed (private, municipal, condominium, housing co-operative, business, etc.).

The methodological starting point, previously devised for the model data for the Warsaw area, proved useful when carrying out a series of development simulations for the PVs. Because of the specific character of Warsaw or, in fact, the specificity of demographic data and Polish economic boundary conditions, the models that had been used before were supplemented with variants in which the citizens' wealth as well as PV installation costs were parameterized. The obtained results indicate (Figures 10 and 11) that for less-wealthy residents, PV costs do not influence the number of installed panels. When residents are affluent, however, the number of installed PVs is significantly higher when the PV installation costs are reduced. However, the most advantageous variant is, again, linked to the trust-based gamification approach. Thus, it is possible to get the most significant number of installed panels with, similarly to the model city, a significant increase in the level of citizens' trust and stimulation of the residents' activity and social development. Spatial distribution (Figures 12 and 13) of the obtained results indicates that the projected increase in the number of panels affects privately owned properties, as well as housing co-operatives and condominiums. By applying this approach, it is also possible to increase installation of PVs on company-owned structures—both factories and office buildings.

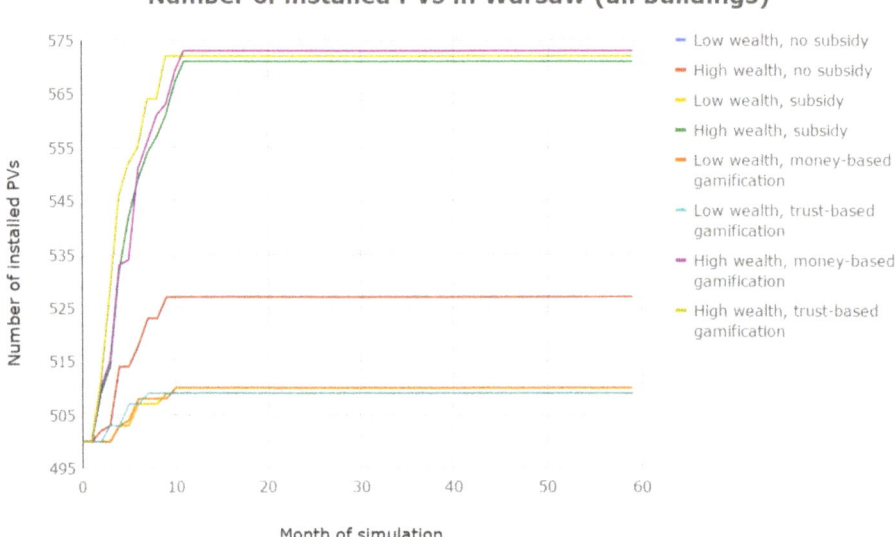

Figure 10. Simulation results for selected scenarios for Warsaw.

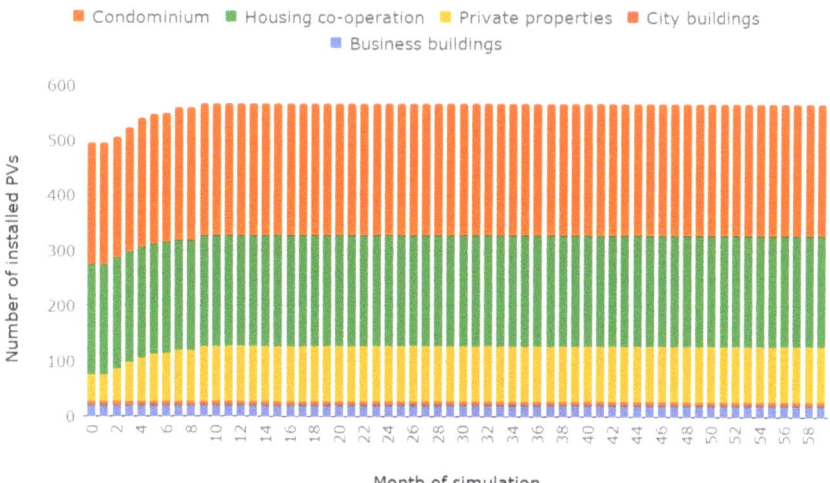

**Figure 11.** The increase in the number of PVs in Warsaw: Trust-based gamification, number of PVs installed on specific building types.

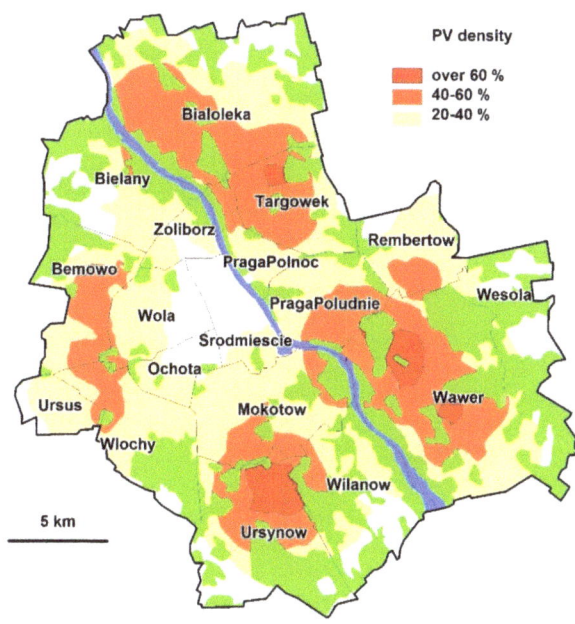

**Figure 12.** Base model simulation for Warsaw (over 60% of PV density—3% of the area of Warsaw, over 40% of PV density—31% of the area of Warsaw).

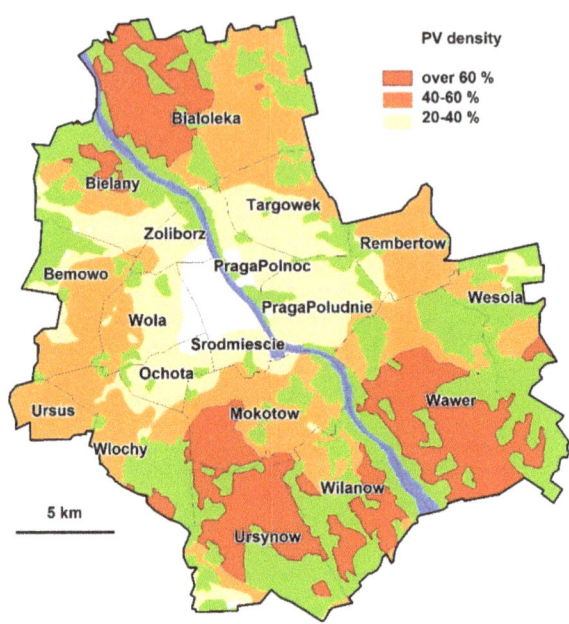

**Figure 13.** Trust-based gamification model simulation for Warsaw (over 60% of PV density—32% of the area of Warsaw, over 40% of PV density—56% of the area of Warsaw).

However, when analyzing the results, one should consider that because of the generalization of source data (both topo- and demographic), the level of generalization of the conducted analysis is also very high. Because each agent represents 1000 residents, any numerical scaling of the obtained results is difficult; many factors considered in the analysis are nonlinear. Since simulation using data made more accurate by two orders of magnitude, was not, due to computation-related reasons, possible for the whole of Warsaw, we decided to carry out this analysis for a selected district of the city.

*4.2. Simulations for Wawer District*

A collection of source data with a much higher level of accuracy was necessary to conduct trial tests for the Wawer district. In the conducted research, the Topographic Objects Database [Baza Danych Obiektów Topograficznych, BDOT 10 k] was used, with 1:10,000 level of detail of the topographical map. In this database, every building is modelled along with precise information regarding the surface and shape of the roof, as well as a detailed road network. It is also possible to allocate the number of residents to individual structures, based on data from the census collected by Poland Statistics [Główny Urząd Statystyczny]. Utilizing the data obtained from the City Hall and the GIS tools, 138 objects with an installed panel located in the Wawer district were assigned a spatial location (Figure 14). The so-called solar map, made available on the official geoportal of the City of Warsaw, provided information on the PVs potential of roofs for all the buildings in Wawer. This form of research made it possible to assign 7500 agents to 75,000 residents of the district. On the one hand, this approach requires considerable computing power (converting one model takes several days); yet on the other, it greatly facilitates the direct interpretation of results.

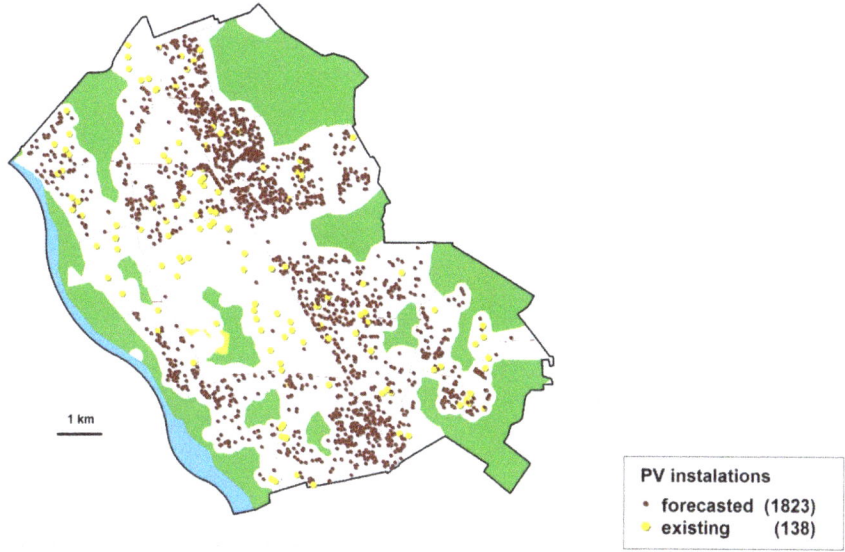

**Figure 14.** Currently existing PVs in Wawer district and panels forecast in the trust-based gamification scenario.

The conducted research has shown that applying the trust-based gamification approach would result in an almost 15-times higher number of PVs being installed in the Wawer district within five years (Figure 15). Moreover, this approach would encourage the engagement of the owners of private properties, as well as those living in condominiums and housing co-operatives (Figure 16). As the spatial distribution of the obtained results indicates (Figures 17 and 18), almost 80% of the Wawer district would be "saturated" with PV installations, far exceeding the 60% level. This is significantly more than using the base variant. What is also crucial is the potential of the generated extra electric energy and levels of avoided greenhouse gas emissions, especially carbon dioxide. The next paragraph deals with this issue in more detail.

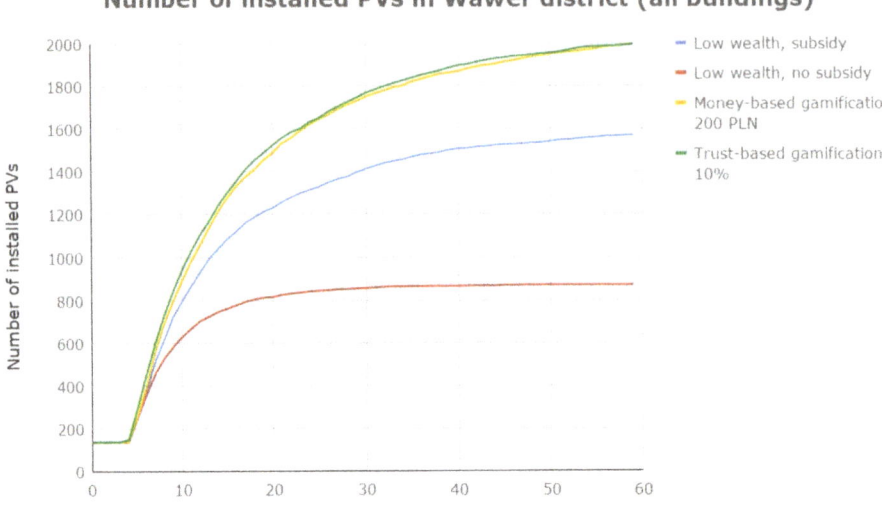

**Figure 15.** Simulation results for selected scenarios for Wawer district.

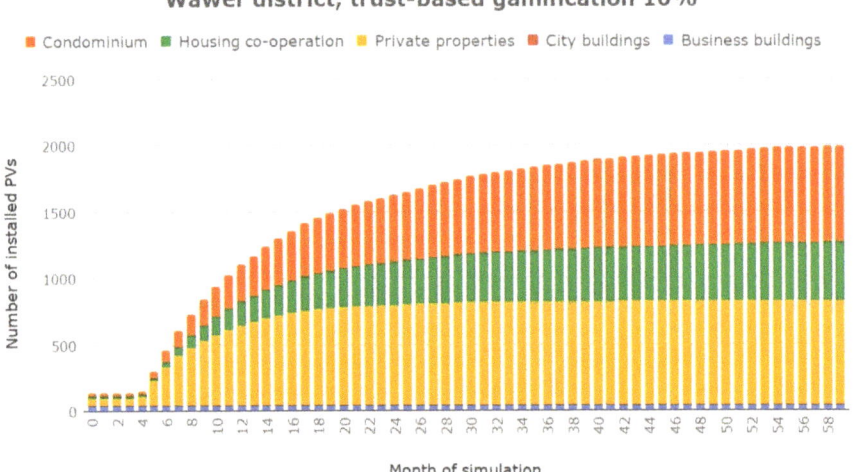

**Figure 16.** The increase in the number of PVs in Wawer district: Trust-based gamification and the number of PVs installed on specific building types.

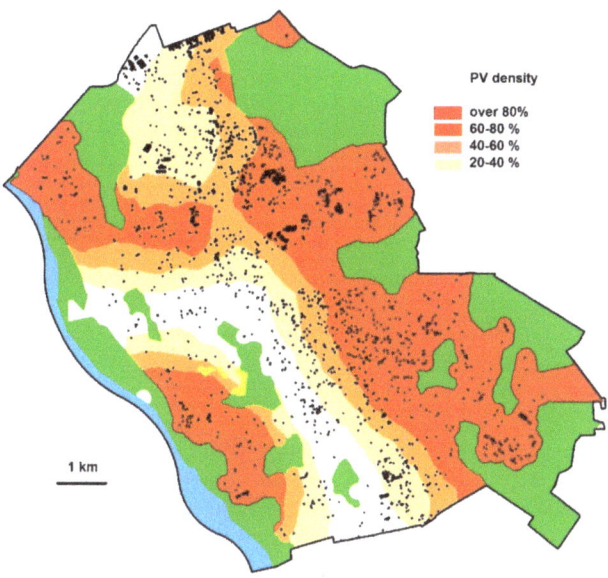

**Figure 17.** The base model simulation for Wawer and dot density map of the population distribution in Wawer (each agent represents 10 inhabitants).

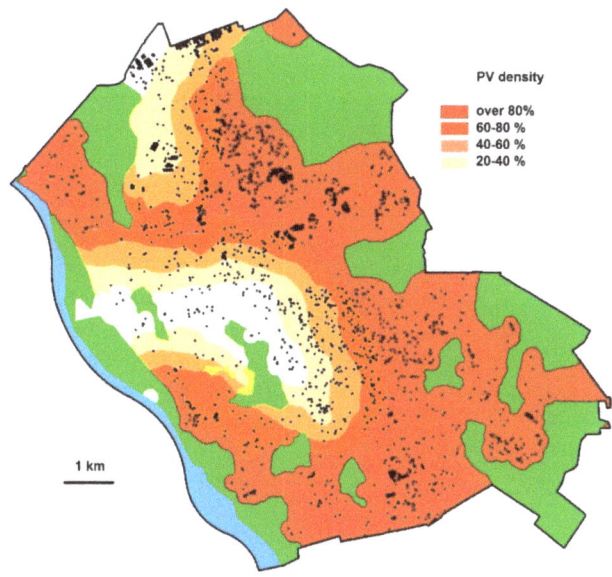

**Figure 18.** The trust-based gamification model simulation for Wawer and dot density map of the population distribution in Wawer (each agent represents 10 inhabitants).

The volume of produced electric energy obtained through PVs in a city and the reduction of $CO_2$ emissions were estimated.

## 5. Discussion and Conclusions

The conducted research has shown that the use of gamification methods allows for much more effective stimulation in the field of PV installations than simple quota subsidies. It is important to compare the results of this methodology with analogous activities carried out in various countries around the world. Several attempts to predict solar PV installations in residential areas have been made. Kwan investigated the influence of different variables on the distribution of residential solar PV in the US with the use of ZIP code level data. Four groups of variables were studied: Environmental (average annual solar radiation data), social (race, age, education, housing density, and the urbanization of the ZIP code), economic (the cost of electricity, average state solar PV incentives, the median household income, and median home value), and political (registered voter party affiliation and city membership of the ICLEI—'Local Governments for Sustainability' organization). We ran a zero-inflated negative binomial regression model to evaluate the influence of selected variables. This allowed us to select (using partial correlation coefficient) the most important parameters for the model. The results of the model indicate that solar insolation, the cost of electricity, and the amount of available financial incentives are important factors influencing the adoption of residential solar PV arrays. The model also shows that strategic government intervention can also be an effective method of promoting solar PV installations [26]. The prediction methodology proposed in the article does not exhaust all analytical possibilities. It is possible to take into account more parameters (e.g., demographic, spatial, etc.), as well as other methods of predicting the increase in the number of solar panels in the city. However, the proposed approach is easy to parameterize and can be extended with additional elements.

With the use of regression analysis, Li and Yi evaluated the relative effectiveness of state and local policy tools in stimulating solar PV installations, with an emphasis on local solar policies. The results of analyzes showed that cities with local financial incentives deploy 69% more solar PV capacities than in cities where such polices do not exist [27].

Cities are now undergoing a rapid and often uncontrolled process of urbanization. They contribute to global climate change by emitting the majority of anthropogenic greenhouse gases. Therefore, it is crucial to study the establishment of sustainable energy systems, which reduce consumption, providing reliable, decentralized, and renewable sources. It is widely acknowledged that access to modern energy plays an essential role in encouraging sustainable urban development, as well as reducing poverty and controlling environmental degradation. In Poland however, there are still no incentives to engage residents to use renewable energy sources (here: PVs), nor are there any incentives to innovate.

The developed methodology has made it possible to conduct a multifaceted simulation of social behaviors targeting the installation of PVs. This, in turn, contributes to sustainable development by limiting the adverse impacts of urbanization on the environment and improving, overall, residents' quality of life. The social impact models and the methods of stimulation by local authorities that we propose indicate that using gamification methods is one of the most effective approaches. It should be said that trust-based gamification is much more valuable (both from an economic and social point of view) than money-based gamification. Not only does it not lead to a side effect, but it also ensures the expanded development of the open information society. According to the conducted research, trust-based gamification not only achieves a high growth rate in the number of panels (up to 15-fold over a five-year period), but also enables the development of social interactions, an increase of trust, and thus, contributes to social engagement in a smart city. Using the model proposed in this article will also mean it is possible to estimate both the power of energy generated from renewable sources and the amounts of $CO_2$ emissions that have been curbed.

Spatial databases and quantitative models originating from applied social sciences made it possible to perform a multifaceted study of the methods of social stimulation aimed at implementing renewable energy sources to reduce the impact on the environment and to improve the quality of life in an urban environment. Taking a quantitative approach, this issue requires the development and analysis of the effectiveness of the computational applications connected to the utilization of the process of social gamification for the sustainable development of urban solutions, including the use of renewable energy sources. On the other hand, using a multi-agent system in which every agent represents a resident, characterized by a set of social attributes, makes it possible to simulate and analyze the social processes occurring in a city.

Based on the methodology developed, simulations were carried out for both Warsaw and the selected district of the city. The analyses performed for the Wawer district, resting on detailed topographical and demographic data, particularly enabled reliable estimation of the projected increase in the number of panels, energy production, and avoided $CO_2$ emissions. This was possible due to the use of multi-agent systems, in which each agent represents only a few people and simulates their actions in a real urban environment, represented in the digital spatial database. Designed models and tested scenarios are critical to determine the real impact of adopting solutions to reduce the consumption of fossil primary energy carriers and, consequently, reduce the emission of atmospheric pollutants, especially $CO_2$. The emission projections developed for "Polityka energetyczna Polski do 2040 roku" [The energy policy of Poland until 2040] were used to determine the amount of avoided $CO_2$ emissions. The results of the calculations revealed that to reduce $CO_2$ emissions, the trust-based gamification scenario was the most effective. Implementing the instruments of environmental policy proposed in this scenario would lead to a reduction in $CO_2$ emissions for one of the districts of Warsaw.

In the research, the average values of the number of modules and the power of PVs were considered. However, actual roof surfaces, as well as real PV potential, and the number of residents were also considered. In further research, we plan to consider the different parameter sizes of PV installations, as well as a higher number of factors that could influence the decision-making of agents who represent the residents. The solution found is generic and parametrizable in many ways. Subsequent research will also aim to verify the developed methodology for other agglomerations with different urban, demographic, and economic conditions, as well as with a much more diversified PV potential.

Future work and further investigations will be focused on an in-depth analysis of various variants of stimulation of the panel installation process, with particular emphasis on trust-based gamification, as well as taking into account other demographic and social parameters of residents. The authors intend also to verify the model developed for Polish spatial, solar, demographic, cultural, and economic conditions on the example of other cities on different continents.

**Author Contributions:** Conceptualization, R.O. and P.P.; methodology, R.O. and P.P.; software, P.P.; validation, A.W. and J.K.; formal analysis, R.O. and J.K.; investigation, R.O., P.P. and A.W.; resources, A.W.; data curation, P.P. and A.W.; writing—original draft preparation, R.O., P.P. and A.W.; writing—review and editing, A.W.; visualization, R.O., A.W., P.P. and J.K.; supervision, R.O.; project administration, A.W.; funding acquisition, R.O.

**Funding:** This research received no external funding.

**Conflicts of Interest:** The authors declare no conflicts of interest.

### References and Note

1. Amado, M.; Poggi, F. Solar urban planning: A parametric approach. *Energy Procedia* **2014**, *48*, 1539–1548. [CrossRef]
2. European Environment Agency (EEA). *Urban Sprawl in Europe—The Ignored Challenge*; EEA Report No 10/2006; EEA: Copenhagen, Denmark, 2006.
3. European Environment Agency (EEA). *Urban Sprawl in Europe*; Joint EEA-FOEN Report No 11/2016; EEA: Copenhagen, Denmark, 2016.
4. Lupangu, C.; Bansal, R.C. A review of technical issues on the development of solar photovoltaic systems. *Renew. Sustain. Energy Rev.* **2017**, *73*, 950–965. [CrossRef]
5. Moosavian, S.M.; Rahim, N.A.; Sevaraj, J.; Solangi, K.H. Energy policy to promote photovoltaic generation. *Renew. Sustain. Energy Rev.* **2013**, *25*, 44–58. [CrossRef]

6. Kammen, D.; Sunter, D.A. City-integrated renewable energy for urban sustainability. *Sci. Urban Planet* **2016**, *352*, 922–928. [CrossRef]
7. Pereira da Silva, P.; Dantas, G.; Pereira, G.I.; Camara, L.; De Castro, N.J. Photovoltaic distributed generation—An international review on diffusion, support policies, and electricity sector regulatory adaptation. *Renew. Sustain. Energy Rev.* **2019**, *103*, 30–39. [CrossRef]
8. Walker, G.; Devine-Wright, P.; Hunter, S.; High, H.; Evans, B. Trust and community: Exploring the meanings, contexts and dynamics of community renewable energy. *Energy Policy* **2019**, *38*, 2655–2663. [CrossRef]
9. Niemi, R.; Mikkola, J.; Lund, P.D. Urban energy systems with smart multi-carrier energy networks and renewable energy generation. *Renew. Energy* **2012**, *48*, 524–536. [CrossRef]
10. Lund, P.D. Fast market penetration of energy technologies in retrospect with applications clean energy futures. *Appl. Energy* **2010**, *87*, 3575–3583. [CrossRef]
11. Nissing, C.; von Blottnitz, H. Renewable energy for sustainable urban development: Redefining the concept of energization. *Energy Policy* **2010**, *38*, 2179–2187. [CrossRef]
12. Amado, M.; Poggi, F. Towards solar urban planning: A new step for better energy performance. *Energy Procedia* **2012**, *30*, 1261–1273. [CrossRef]
13. Rylall, M.; Gadsden, S.; Lamas, K. GIS-based decision support for solar energy planning in urban environments. *Computers Environ. Urban Syst.* **2001**, *25*, 579–603.
14. Thorsten, S.; Joong-Won, L.; Tae-Goo, L. Sustainable Urban (re-)Development with Building Integrated Energy, Water and Waste Systems. *Sustainability* **2013**, *5*, 1114–1127. [CrossRef]
15. Baziliana, M.; Onyejia, I.; Liebreichd, M.; MacGille, I.; Chased, J.; Shahf, J.; Gieleng, D.; Arenth, D.; Landfeari, D.; Zhengrongj, S. Re-considering the economics of photovoltaic power. *Renew. Energy* **2013**, *53*, 329–338. [CrossRef]
16. Borri, D.; Camarda, D.; De Liddo, A. Multi-agent Environmental Planning: A Forum-based Case Study in Italy. *Plan. Pract. Res.* **2008**, *23*, 211–228. [CrossRef]
17. Murakami, T. Saturation-phase prediction of building-integrated photovoltaics by using agent-based simulations. In Proceedings of the IEEE International Symposium on Industrial Electronics, Bari, Italy, 4–7 July 2010; pp. 2469–2474.
18. Massey, D.; Ahlqvist, O.; Vatev, K.; Rush, J. A Massively Multi-user Online Game Framework for Agent-Based Spatial Simulation. In *CyberGIS for Geospatial Discovery and Innovation*; Wang, S., Goodchild, M.F., Eds.; Springer: Dordrecht, The Netherlands, 2019; Volume 118, pp. 213–224.
19. Założenia do planu zaopatrzenia w ciepło, energię elektryczną i paliwa gazowe dla m.st. Warszawy, Warszawa, 2018. Available online: https://infrastruktura.um.warszawa.pl/sites/infrastruktura.um.warszawa.pl/files/1-_robert_szkopek.pdf (accessed on 5 July 2019).
20. Instytut Energetyki Odnawialnej, Rynek fotowoltaiki w Polsce, 2018, Warszawa. Available online: https://ieo.pl/pl/projekty/raport-rynek-fotowoltaiki-w-polsce-2018 (accessed on 5 July 2019).
21. Raport OZE Polityka Energetyczna m.st. Warszawy w roku. Personal Communication, 2017.
22. Moc mikroinstalacji fotowoltaicznych w Warszawie wzrosła o 2 MW. Available online: http://gramwzielone.pl/energia-sloneczna/30669/moc-mikroinstalacji-fotowoltaicznych-w-warszawie-wzrosla-o-2-mw (accessed on 5 April 2018).
23. Tilman, A.; Engelmeier, T. Boosting solar investment with limited subsidies: Rent management and policy learning in India. *Energy Policy* **2013**, *59*, 866–874.
24. Sahoo, S.K. Renewable and sustainable energy reviews solar photovoltaic energy progress in India: A review. *Renew. Sustain. Energy Rev.* **2016**, *59*, 927–939. [CrossRef]
25. Becker, B.; Fisher, D. Promoting renewable electricity generation in emerging economies. *Energy Policy* **2013**, *56*, 446–455. [CrossRef]
26. Kwan, C.L. Influence of local environmental, social, economic and political variables on the spatial distribution of residential solar PV arrays across the United States. *Energy Policy* **2012**, *47*, 332–344. [CrossRef]
27. Li, H.; Yi, H. Multilevel governance and deployment of solar PV panels in U.S. cities. *Energy Policy* **2014**, *69*, 19–227. [CrossRef]
28. Chapman, A.J.; McLellan, B.; Tezuka, T. Residential solar PV policy: An analysis of impacts, successes and failures in the Australian case. *Renew. Energy* **2016**, *86*, 1265–1279. [CrossRef]

29. Schmidt-costa, J.R.; Uriona-maldonado, M.; Possamai, O. Product-service systems in solar PV deployment programs: What can we learn from the California Solar Initiative? *Resour. Conserv. Recycl.* **2019**, *140*, 145–157. [CrossRef]
30. Drury, E.; Miller, M.; Macal, C.M.; Graziano, D.J.; Heimiller, D.; Ozik, J.; Perry, T.D. The transformation of southern California's residential photovoltaics market through third-party ownership. *Energy Policy* **2012**, *42*, 681–690. [CrossRef]
31. Muhammad-Sukki, F.; Abu-Bakar, S.H.; Munir, A.B.; Mohd Yasin, S.H.; Ramirez-Iniguez, R.; McMeekin, S.G.; Stewart, B.G.; Sarmah, N.; Mallick, T.K.; Rahim, R.A.; et al. Feed-in tariff for solar photovoltaic: The rise of Japan. *Renew. Energy* **2014**, *68*, 636–643. [CrossRef]
32. Pacudan, R. Feed-in tariff vs incentivized self-consumption: Options for residential solar PV policy in Brunei Darussalam. *Renew. Energy* **2018**, *122*, 362–374. [CrossRef]
33. Kuwahata, R.; Monroy, C.R. Market stimulation of renewable-based power generation in Australia. *Renew. Sustain. Energy Rev.* **2011**, *15*, 534–543. [CrossRef]
34. Deterding, S. Gamification: Designing for motivation. *Interactions* **2012**, *19*, 14–17. [CrossRef]
35. Grignard, A.; Taillandier, P.; Gaudou, B.; An Vo, G.; Huynh, N.Q.; Drogoul, A. GAMA 1.6: Advancing the art of complex agent-based modeling and simulation. In Proceedings of the International Conference on Principles and Practice of Multi-Agent Systems, Dunedin, New Zealand, 1–6 December 2013; Springer: Berlin/Heidelberg, Germany, 2013.
36. Grignard, A.; Drogoul, A. Agent-Based Visualization: A Real-Time Visualization Tool Applied Both to Data and Simulation Outputs. In Proceedings of the AAAI-17 Workshop on Human-Machine Collaborative Learning, San Francisco, CA, USA, 4–5 February 2017. Technical Report.
37. Billewicz, K. Mikrogeneracja—Aspekty Nieuwzględnione w Polskiej Legislacji. In *Zarządzanie Energią i Teleinformatyka*; Kaprint: Lublin, Poland, 2014; pp. 79–92.
38. Dąbrowski, J.; Hutnik, E.; Włóka, A.; Zieliński, M. Analiza wykorzystania instalacji fotowoltaicznej typu on-grid do produkcji energii elektrycznej w budynku mieszkalnym. *Rynek Energii* **2014**, *1*, 53–59.
39. Roczne zużycie energii w domu. 2014. Available online: https://optimalenergy.pl/aktualnosci/cena-pradu/roczne-zuzycie-energii-elektrycznej-w-domu/ (accessed on 10 October 2018).
40. Średnie ceny sprzedaży energii elektrycznej na rynku konkurencyjnym za IV kwartał 2017 i za rok 2017. 2018. Available online: https://www.ure.gov.pl/pl/urzad/informacje-ogolne/aktualnosci/7480,Srednie-ceny-sprzedazy-energii-elektrycznej-na-rynku-konkurencyjnym-za-IV-kwarta.html (accessed on 10 October 2018).
41. Średnie oprocentowanie lokat: Styczeń. 2018. Available online: https://www.comperia.pl/artykuly/o_lokatach_bankowych/32104-srednie-oprocentowanie-lokat-styczen-2018.html (accessed on 10 October 2018).
42. RWE/Innogy—ceny dystrybucji prądu. Available online: https://enerad.pl/rynek-energii/dodatkowo/osd-dystrybutorzy/rwe-ceny-dystrybucji-pradu/-dystrybucja (accessed on 10 October 2018).
43. Available online: http://www.innogy.pl/pl/dla-domu/oferta/najprostsza-dla-ciebie (accessed on 10 October 2018).
44. Mackaness, W.; Ruas, A.; Sarjakoski, T. *Generalisation of Geographic Information: Cartographic Modelling and Application*; Elsevier: Amsterdam, The Netherlands, 2007.

© 2019 by the authors. Licensee MDPI, Basel, Switzerland. This article is an open access article distributed under the terms and conditions of the Creative Commons Attribution (CC BY) license (http://creativecommons.org/licenses/by/4.0/).

*Case Report*

# Aligning Urban Policy with Climate Action in the Global South: Are Brazilian Cities Considering Climate Emergency in Local Planning Practice?

**Debora Sotto [1], Arlindo Philippi Jr. [1], Tan Yigitcanlar [2,\*] and Md Kamruzzaman [3]**

1 School of Public Health, University of São Paulo, Av. Dr. Arnaldo, 715 São Paulo, SP 01246-904, Brazil
2 School of Civil Engineering and Built Environment, Queensland University of Technology, 2 George Street, Brisbane QLD 4000, Australia
3 Faculty of Art, Design and Architecture, Monash University, 900 Dandenong Road, Caulfield East, VIC 3145, Australia
\* Correspondence: tan.yigitcanlar@qut.edu.au; Tel.: +61-7-3138-2428

Received: 6 August 2019; Accepted: 3 September 2019; Published: 5 September 2019

**Abstract:** Climate change is the biggest global threat of our time. As a signatory nation of the Paris Agreement, Brazil has made a climate action commitment, and expressed its nationally determined contribution to reduce its greenhouse gas emissions by 37%. The Brazilian population is highly urban, and Brazilian cities are mostly responsible for greenhouse gas emissions, and the worst effects of global warming are experienced in cities. Hence, the fulfillment of the Brazilian climate commitments depends on the active engagement of municipalities. Nevertheless, the Brazilian national government does not monitor local climate actions, and it is not clear how local urban policy is aligned with climate action. In order to bridge this gap, this study tackles the question of: "Are, and if yes how, cities considering the climate emergency in their local planning mechanisms?" This question is investigated by placing five major Brazilian cities under the microscope. The methodological approach includes literature review and applied qualitative analysis to scrutinize how climate issues and actions are factored in urban planning regulations to verify if and to what extent local policies contribute to the fulfillment of the Brazilian nationally determined contribution, and sustainable development goals. The results disclose that investigated cities have adequately incorporated climatic issues in their urban planning mechanisms. However, policy concentrates more on adaptation rather than mitigation, and policy implementation yet to be realized.

**Keywords:** climate change; climate emergency; climate crisis; global warming; sustainable urban development; sustainable development goals; smart cities; disasters; urban health; urban policy

## 1. Introduction

The Anthropocene, a geological era of human domination on Earth's resources, has generated many complex problems that threaten the well-being and existence of many species—including the human kind [1,2]. The single most important problem we face today, in the Anthropocene age, is climate change or more correctly the climate emergency or crisis [3,4]. The impacts, just to name a few, include more frequent and extreme heat episodes, rising seas and increased coastal flooding, longer and more damaging wildfire seasons, more destructive hurricanes, costly and growing health impacts, heavier precipitation and flooding, destruction of marine ecosystems, more severe droughts in some areas, increased pressure on groundwater supplies, melting ice cap, disruptions to food supplies, plant and animal range shifts, climate refugees, and changing seasons [5–7].

During the last few years, various solutions have been put forward to combat the problem, particularly targeting the unsustainable urban development and cleaner production [8–10]. These include adopting new development paradigms to make cities more sustainable, resilient, and smarter—e.g., smart city and smart urbanism movements [10,11]—and incentivizing green technology and cleaner industrial development [12–14]. However, the solutions targeting more sustainable urbanization and industrialization with limited emissions have not found wider application grounds across the globe, and also existing initiatives were not efficient enough to make a significant difference [15,16].

There have been international and national platforms and initiative to intervene with the status quo, and also provide possible directions and solutions to combat the climate change problem—more correctly the factors causing it. For example, as nation signatories of the United Nations' Framework Convention on Climate Change (UNFCCC) and to the Paris Agreement [17] many countries have made the commitment, expressed in their nationally determined contribution (NDC), to reduce its greenhouse gas emissions (GHG) [18]. However, as most of the world population is urban, compliance with NDC depends on the engagement of local governments—i.e., municipalities [19]. Furthermore, it is required for cities to take firm action on climate adaptation and resilience promotion, as the worst effects of global warming—floods, droughts, landslides, epidemics—are experienced by the population at the local level, deeply affecting public health and quality of life, especially among the most vulnerable social sectors [20].

Engagement of cities across the world in tackling climate change, autonomously of their respective nation-states, is a relatively recent phenomenon [21–23], with significant and positive impacts on the latest accomplishments of international climate negotiations. Despite the lack of a legal status under the international law, cities have intensified their actions in international relations through the construction of networks, such as Local Governments for Sustainability (ICLEI), United Cities and Local Governments (UCLG), 40 Cities for Climate, and Compact of Mayors (C40 Group), active in sustainable urban development, and climate change. As a result, cities have been identified as key actors for environmental protection and sustainable development—firstly participating in UNFCCC Conferences as 'observant organizations,' and from the 2010 Conference of the Parties (COP-16) onward as "governmental actors".

The final declaration of the United Nations Conference on Sustainable Development (Rio +20), "The Future We Want" [24], has reaffirmed the strategic role of cities in promoting sustainable development. The Rio +20 declaration also set out the commitment to formulate "sustainable development goals" (SDGs), replacing the "millennium development goals" (MDGs). These SDGs were issued in September 2015 [25] a set of 17 main goals, broken down into 169 targets and monitored by correspondent indicators, to be pursued by all nations, regardless of their developmental stage.

The sustainable development goal (SDG) no. 13 (SDG-13) refers specifically to "climate action", expressing the global commitment to take urgent action on climate change and its impacts within the UNFCCC framework. SDG Target 13.B fosters capacity building, promotion for productive climate change-related planning and management in local communities in the least developed countries, as well as women, youth, and vulnerable minorities. SDG-11, on the other hand, expresses the commitment to make cities and human settlements safe, resilient, and sustainable. Among the foreseen axes of action, Target 11.B establishes the goal to substantially increase, by 2020, the number of cities and human settlements adopting and implementing integrated policies and plans towards inclusion, resource efficiency, mitigation and adaptation to climate change and resilience to disasters, in line with the Sendai Framework for Disaster Risk Reduction (2015–2030)—the first major international agreement of the post-2015 development agenda.

Both SDG-11 and 13 strongly connect to strategic goals and actions established by the 2016 "new urban agenda" (NUA), approved in the city of Quito (Ecuador) [26]. NUA expressly envisions cities and human settlements that adopt and implement disaster risk reduction and management, reduce vulnerability, build resilience and responsiveness to natural and human-made hazards, and promote mitigation and adaptation to climate change. By doing so, NUA highlights the mandatory connection existing between SDG-13 (action against global climate change), SDG-11 (sustainable cities and communities), and SDG-3 (health and well-being).

Not only SDG-3, health and well-being, is indispensable for the quality of life in cities, but it is also a central component in the construction of urban resilience. In this sense, Target 3.D commits the nation-states to strengthen the capacity of all countries, and also local governments, for early warning, reduction, and management of these risks, considering the increasing national and global risks associated with extreme climatic events.

It is important to stress that the most vulnerable social segments, which already have the worst housing and sanitation conditions and the worst levels of access to essential services, including health, are also the most exposed to the adverse effects of extreme weather events and other shocks [27–29]. Hence the commitment expressed by Target 1.E, deployed from SDG-1, eradicating poverty, to build the resilience of the poor and those in a vulnerable situation and to reduce their exposure and vulnerability to extreme events, shocks, and disasters. In an increasingly urban world, building the resilience of the poorest and most vulnerable is one of the significant challenges to be faced by urban planners and managers.

Against this backdrop, the aim of this study is to examine how cities, and particularly cities in the global south (an emerging term used by the World Bank to refer to countries located in Asia, Africa, Latin America (including Brazil), and the Caribbean and considered to have low- and middle-income levels [30]), are considering the climate emergency in their local planning mechanisms. The study concentrates on Brazil as the case study context, and places five major Brazilian cities—i.e., Manaus, Salvador, Goiânia, São Paulo, and Curitiba—under the microscope to address the research aim. The methodology of the investigation contains undertaking a literature review, collecting public policy data, and conducting a quantitative analysis. The paper is structured as follows: After this introduction section, Section 2 presents materials and methods. Section 3 discloses the results. Section 4 discusses the findings, and Section 5 concludes the paper.

## 2. Materials and Methods

### 2.1. Case Study

The study selects Brazil as a relevant country context from the global south—with vast land, large population, pristine natural assets, and a big appetite for economic growth—to explore how well local urban policy is aligned with climate action [31]. The research is developed as a multiple case study [32]. It involved literature review, data collection in digital repositories, and the qualitative analysis of the data using a content analysis software—i.e., NVivo [33]. As it would not be possible to study all 5570 Brazilian municipalities, in order to cover the different realities and main biomes of Brazilian territory, five major Brazilian cities were targeted to be selected for the analysis at the ratio of one city per each region of the country—i.e., North, Northeast, Midwest, Southeast, and South.

The case study selection process obeyed the following criteria: (a) Population above 500,000 people (the Brazilian Institute of Geography and Statistics (IBGE) classifies Brazilian cities in three categories: small (less than 100,000 inhabitants); medium (more than 100,000 inhabitants and less than 500,000 inhabitants) and large (more than 500,000 inhabitants. IBGE's methodology does not divide large cities in any other additional categories. Large cities in Brazil are more typically "urban" than medium and small cities, which is why this investigation opted to select only cities with more than 500,000 inhabitants); (b) Local environmental bodies and councils, according to 1981 National Environmental Policy [34]; (c) Existence of a Master Plan, approved accordingly to 2001 Statute of the City [35]; (d) City's participation in at least one of the following international city networks, selected for their outstanding performance in the field of climate change: Local Governments for Sustainability (ICLEI), Mercociudades, as representative of the Communities and United Local Governments in South America (UCLG), and C40 Group, and; (e) Availability of data and information pertinent to municipal management and planning, accessible through the Internet. Based on these criteria (Table 1), the following cities were selected as the object of the case study: Goiânia (Central-West Region), Salvador (Northeast Region), Manaus (North Region), São Paulo (Southeast Region) (in Brazil, all municipalities with more than 500,000 inhabitants must follow the same institutional framework as cities with a population of one million, two million or 12 million people. Therefore, the fact that São Paulo is significantly larger than the other selected cities does not distort or affect the results of the proposed investigation.), and Curitiba (Southern Region) (Brasília, capital city of the nation, was excluded from the selection process because, as the seat of the Federal District, it presents unique institutional characteristics, distinct from the other Brazilian municipalities). As seen in Table 1, despite listed similarities, the case cities are in different sizes and characteristics, thus a one-on-one comparison would not be a correct approach [36]. Here the purpose is to capture the general trends related to climate policy in Brazilian cities. Hence, having cities with different attributes is valid approach to address the research question [37].

Table 1. Case study cities' salient characteristics.

| City | Population | Environmental Body and Council | Master Plan in Accordance with the Statute of the City | City's Networks | Data Available on the Internet |
|---|---|---|---|---|---|
| Goiânia State of Goiás Midwest Region | 1,466,105 | Yes | 2007 | ICLEI UCLG | Yes |
| Salvador State of Bahia Northeast Region | 2,953,986 | Yes | 2016 | C40 ICLEI UCLG | Yes |
| Manaus State of Amazonas North Region | 2,130,264 | Yes | 2014 | ICLEI | Yes |
| São Paulo State of São Paulo Southwest Region | 12,106,920 | Yes | 2014 | C40 ICLEI UCLG | Yes |
| Curitiba State of Paraná South Region | 1,908,359 | Yes | 2015 | C40 ICLEI UCLG | Yes |

*2.2. Methodology and Data*

Assuming that an effective climate policy at any level must take as its starting point the findings of GHG emissions inventories, we firstly collected the inventories drawn up by the surveyed cities, in order to compare their findings with Emission Inventories edited by the respective states and 2010 and 2015 national emissions inventories issued by the Brazilian government. The policy document search was organized according to four areas, or climate action axis: (a) Mitigation; (b) Adaptation; (c) Resilience as defined by the IPCC [38], and; (d) Cooperation (cooperation viewed as decentralized cooperation, or city's diplomacy, comprehending agreements celebrated between cities and any other international actors, such as other cities, other countries, international organizations, as well as actions developed within the context of city's international networks). Table 2 presents local policies, plans, programs, and projects, classified according to the four climate action axes of mitigation, adaptation, resilience, and cooperation.

The documents, compiled by the researchers in the first quarter of 2018 in official digital data bases (In Brazil, due to national law provisions, all federative entities, municipalities included, must provide free access to all public interest information. In compliance, the union (federation), states and most of the large municipalities have made their legislation (laws and regulations) available online, in official digital databases), and organized by city and topic and listed in Table 3. Compiled documents were processed with NVivo, firstly sorted out in cases, each case corresponding to one of the investigated cities as well as to the nation. The word search tool was used to find references to the four axes of climate action—i.e., adaptation, cooperation, mitigation, and resilience—in the texts of the compiled documents. The word search tool was also used to locate references, in the local documents, to national regulations related to climate change—e.g., National Emissions Inventory, 2009 National Climate Change Policy (PNMC), 2016 National Adaptation Plan (PNA), Brazil's NDC, 2013 Sectoral Health Plan for Climate Change Mitigation and Adaptation, and 2012 National Protection and Civil Defense Policy (PNPDEC). All policy documents were in Portuguese language. Analyses were conducted in Portuguese, and the results were translated into English.

Table 2. Local policies, plans, programs, and projects.

| Mitigation | Adaptation | Resilience | Cooperation |
| --- | --- | --- | --- |
| GHG emissions inventories | GHG emissions inventories | GHG emissions inventories | City networks |
| Master plan | Master plan | Master plan | Master plan |
| Land use regulations | Land use regulations | Land use regulations | Decentralized cooperation |
| Housing local plan | Housing local plan | Housing local plan | Decentralized cooperation |
| Mobility local plan | Mobility local plan | Mobility local plan | Decentralized cooperation |
| Sanitation local plan | Sanitation local plan | Sanitation local plan | Decentralized cooperation |
| Waste management plan | Waste management plan | Waste management plan | Decentralized cooperation |
| Drainage local plan | Drainage local plan | Drainage local plan | Decentralized cooperation |
| Public health local plan | Public health local plan | Public health local plan | Decentralized cooperation |
| Climate mitigation plan | Climate adaptation plan | Disasters management plan | Decentralized cooperation |

We have used the AND, OR, NEAR, (*), REQUIRED, FUZZY and (") operators in at least five successive rounds of word search queries in order to obtain more precise results, avoiding redundancies, ambiguities, and false positive results (NVivo, as qualitative analysis tool, allows the election of objective criteria and comparison standards for data contained in large masses of text. The obtained results present a satisfactory degree of precision for the proposed analysis, but not, of course, absolute precision, as despite the efforts to reduce the investigated topics to expressions as accurate and straightforward as possible, some inconsistencies have persisted, mostly due to redundancies and ambiguities typical of the Portuguese language). The references obtained in the word search rounds were grouped in four nodules, corresponding to each of the climate action axes: Adaptation, Cooperation, Mitigation, and Resilience, as well as a fifth node, devoted to the identification of possible connections between public health and other local policies (Table 4). Lastly, several rounds of queries were performed, crossing the evidence compiled in each node with the cases, thus obtaining results that are further discussed in order to offer some insights.

Table 3. Compiled policy documents organized by city and topic.

| Topic | Type | CURITIBA | GOIÂNIA | MANAUS | SALVADOR | SÃO PAULO |
|---|---|---|---|---|---|---|
| Climate | Emissions Inventory | 2013 Emissions Inventory—3rd Edition | 2016 Emissions Inventory—2nd Edition | | 2013 Emissions Inventory—1st Edition | 2009 Emissions Inventory—2nd Edition |
| | Plan | 2009 Climate Action Plan | 2011 Sustainable Goiânia | The Card of Amazonia | Coastal Management Plan | 2009 Climate Change Municipal Policy |
| | Legislation/Report | 2016 Curitiba Strategic Actions: Climate and Resilience | 2016 Local Policy on the Prevention, Reduction and Compensation of $CO_2$ and Other Vehicular Emissions | 2011 Environmental Code (proxy to a local climate plan) | 2015 Sustainable Development and Innovation Incentives Municipal Program | |
| | Report | 2014 Curitiba's Environmental and Socio-Economic Vulnerabilities Evaluation | | | 2015 Environmental and Sustainable Development Municipal Policy | |
| Urban Planning | Plan | 2015 Master Plan | 2007 Master Plan | 2014 Master Plan | 2016 Master Plan | 2014 Master Plan |
| Land Use Regulations | Legislation | 2000 Land Use Regulations | 1994 Urban Zoning | 2014 Land Use Legislation | 2016 Land Use Regulations | 2016 Zoning and Land Use Regulations |
| | | | 2008 Environmental Zoning | 2014 Urban Perimeter Legislation | | |
| | | | 2008 Building Code | 2014 Building Code | 2017 Building Code | 2017 Building Code |
| Housing | Plan | Social Housing Municipal Plan | Provisions on Social Housing | | 2008–2015 Municipal Housing Plan | Municipal Housing Plan |
| Mobility | Legislation/Plan | 2014–2017 Urban Mobility Plan | Urban Mobility Plan | 2015 Urban Mobility Plan | 2018 Sustainable Urban Mobility Municipal Plan | 2015 Urban Mobility Plan |
| | | | | | | 2001 Municipal Transportation System |
| Sanitation | Plan | 2017 Sanitation Municipal Plan | Municipal Sanitation Plan | 2014 Municipal Sanitation Plan | 2010–2011 Municipal Sanitation Plan | 2010–2030 Municipal Sanitation Plan |
| | | 2017 Water Distribution Municipal Plan | | | | |
| | | 2017 Sewage Municipal Plan | | | | |

Table 3. Cont.

| Topic | Type | City | | | | | |
|---|---|---|---|---|---|---|---|
| | | CURITIBA | GOIÂNIA | MANAUS | SALVADOR | SÃO PAULO |
| Waste Management | Plan | 2017 Waste Management Municipal Plan | 2014 Waste Management Local Plan | 2010 Waste Management Municipal Plan | 2012 Waste Management Municipal Plan | 2014 Waste Management Municipal Plan |
| Drainage | Plan | 2017 Urban Drainage Plan | 2014 Drainage Regulations | | | 2012 Drainage and Rainwater Management Manual |
| Disasters Management | Legislation/ Plan | Municipal Civil Defense System | 1998/2010 Municipal Civil Defense System | 2014–2017 Municipal Public Health Plan | 2015 Rain Contingency Plan | 2006 Municipal Civil Defense System |
| | | | | | 2017 Urban Arborization Plan | |
| | | | | | 2013 Civil Defense System | |
| Public Health | Plan | 2018–2021 Municipal Public Health Plan | 2014/2017 Municipal Public Health Plan | 2014–2017 Municipal Public Health Plan | 2018–2021 Municipal Public Health Plan | 2014–2017 Municipal Public Health Plan |
| International Cooperation | Legislation/ Platform/ Agreement | Participation in city's networks: ICLEI, CGLU, C40; International Affairs Secretariat reports; Nazca platform | Participation in Cities Networks: ICLEI, CGLU, UBERLAC, Sustainable and Emergent Cities Initiative; International Affairs Municipal Advisory reports; Nazca Platform; Vulnerability Assessment (UNFCCC Adaptation Platform) | Participation in City's Networks: ICLEI; Nazca Platform; Municipality's reports | Participation in City's Networks: ICLEI, CGLU, C40; Nazca Platform; Decentralized cooperation agreements | Participation in City's Networks: ICLEI, CGLU, C40; International Affairs Secretariat Reports |
| | | | | | | Twin-Cities Agreement Regulations |

Table 4. NVivo codebook.

| Node | Expressions |
|---|---|
| Climate change | "mudança climática" OR "mudanças climáticas" OR "mudança do clima" OR "mudanças do clima" OR "alteração do clima" OR "alterações do clima" OR "alteração climática" OR "alterações climáticas" |
| Adaptation | Adaptaç* |
| Risk | Risco* |
| Vulnerability | Vulnerabilidade OR vulnerave* |
| Water | Água OR "recursos hídricos" OR permeabilidade OR drenagem |
| International agreements | "acordo internacional" OR "acordos internacionais" OR "cooperação internacional" OR "cooperação descentralizada"OR "relações internacionais" OR "internacionalização" |
| International financing | "financiamento internacional" OR "financiamentos internacionais" |
| International organizations | "organização internacional" OR "organizações internacionais" Or "agência internacional" OR "agências internacionais" OR "ONU" OR "HABITAT" OR "Banco Mundial" Or "Banco Interamericano de Desenvolvimento" OR "BID" OR "World Bank" OR "Organização Mundial da Saúde" |
| International networks | "rede internacional" OR "redes internacionais" Or "ICLEI" OR "CGLU" OR "C40" |
| NDC | Ndc OR indc |
| PNA | PNA OR "Plano Nacional de Adaptação" OR "Plano Nacional de Adaptação à Mudança do Clima" |
| PNMC | "política nacional sobre mudanças do clima" OR "política nacional sobre mudança do clima"OR "política nacional de mudanças climáticas" OR "política nacional de mudança do clima" OR "PNMC" |
| Global warming | "aquecimento global" |
| Greenhouse effect | "efeito estufa" |
| Carbon emissions | "emissão de carbono" OR "emissão de gases" OR "emissões" |
| Energy | energia |
| Mitigation | Mitig* |
| Pollution | poluição |
| Resilience | Resiliência OR resiliente* |
| PNPDEC | "política nacional de proteção e defesa civil" OR PNPDEC OR "Sistema Nacional de Proteção e Defesa Civil" OR SINPDEC OR "Conselho Nacional de Proteção e Defesa Civil" OR CONPDEC |
| Disaster | Desastre* OR enchente* OR escorregamento* OR seca* OR INCÊNDIO* |
| Extreme events | "evento extremo" OR "eventos extremos" OR "evento climático extremo" OR "eventos climáticos extremos" OR evento*+extremo* |
| Resilience | Resiliência OR resiliente* |
| Sectoral public health plan for climate change mitigation and adaptation | "plano sectorial da saúde para mitigação e adaptação à mudança do clima" OR "PSMC-Saúde" OR "plano setorial da saúde para mitigação adaptação às mudanças climáticas" OR "plano sectorial da saúde para mitigação e adaptação à mudança climática" |
| Health | saúde |

255

## 3. Results

### 3.1. National, State, and Local GHG Emissions

The Brazilian government, through the Ministry of Communications, Technology, Innovation, and Science (MCTIC) has been inventorying national GHG emissions since 1991. The 3rd National Inventory, published in 2016, with data up to 2010, indicates that the economic sectors responsible for the majority of the national emissions are "Agriculture", with 33% and "Energy", with 29%—e.g., Brazil is world's biggest ethanol producer [39]—followed closely by "Land Use and Forests", with 27% of the total emissions. "Industrial Processes" and "Waste Treatment" respectively account for 7% and 4% of the national emissions (Figure 1). The estimated total emissions for 2010, calculated by the IPCC 2006 methodology (GWP-SAR), was 1,364,197 GtCO2e [40].

Data from the National Emission Register System (SIRENE) compiled in preparation for the 4th National Inventory indicates slightly higher emissions for the year 2015: 1,368,000 GtCO2e in total, but still well below the emission limit set in Brazil's NDC of 2,068,000 GtCO2e for 2020. The "Energy" sector, with 33% of emissions, seems to have surpassed the "Agriculture" sector, with 31% of emissions, followed by the "Land Use" sector, with 24%, "Industrial Processes" with 7% and "Waste Treatment", with 5% (Figure 1).

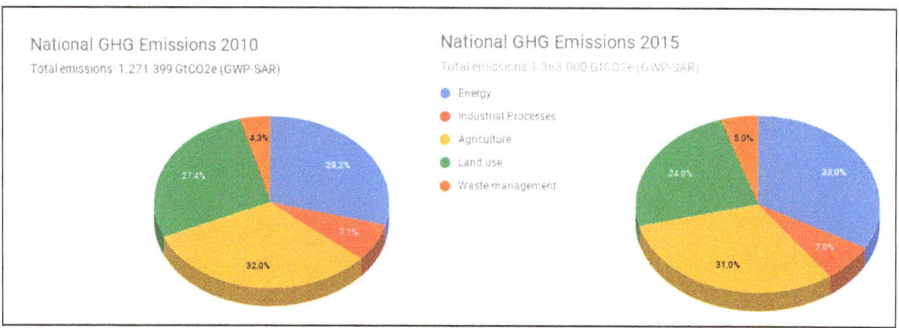

**Figure 1.** National GHG emissions in 2010 and 2015 [40].

The evolution of the national emissions shows that, between 1990 and 2015, the total emissions line follows the same path as the "Land Use" emissions line, so that the emissions peaks observed in 1995 and 2005 correspond precisely to the "Land Use" high points, observed due to the increase in deforestation in the same periods (Figure 2), especially in the Amazon region (the Amazon, the lungs of the planet, cannot be recovered once it is gone. The August 2019 bush fires blazing in Amazon are part of a larger deforestation crisis, accelerated by President Jair Bolsonaro's policies. According to an article published on 28 August 2019 in New York Times, hundreds of government workers on the front lines of enforcing Brazil's environmental laws signed an open letter warning that their work has been hampered by President Jair Bolsonaro, contributing to a rise in deforestation and the fires sweeping through the Amazon. One of the reasons for such heightened reaction is that the Amazon rainforest fires that have been blazing out of control in Brazil for weeks could have far-reaching effects on environmental and public health—e.g., air quality, weather patterns, loss of biodiversity deforestation, diseases, mental health) [41–43].

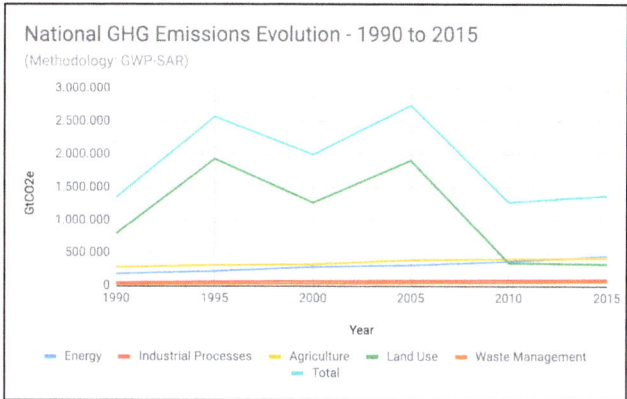

**Figure 2.** National-level GHG emissions (1990–2015) [40].

The contribution of the "Land Use" and "Agriculture" sectors is comparatively higher than the other sectors', coherently with the fact that, although 85% of the Brazilian population lives in cities, only 0.63% of the total 8.156 million km$^2$ of national territory correspond to urban areas [44]. Therefore, at the national level, deforestation control is decisive to the fulfillment of the mitigation targets established on Brazil's NDC. The examination of the inventories of the states of Amazonas, Bahia, Goiânia, Paraná, and São Paulo, where the five cities investigated in this study are located, also corroborates the significant weight of the "Land Use" and "Agriculture Sectors" to the fulfillment of Brazil's NDC mitigation targets (Figure 3).

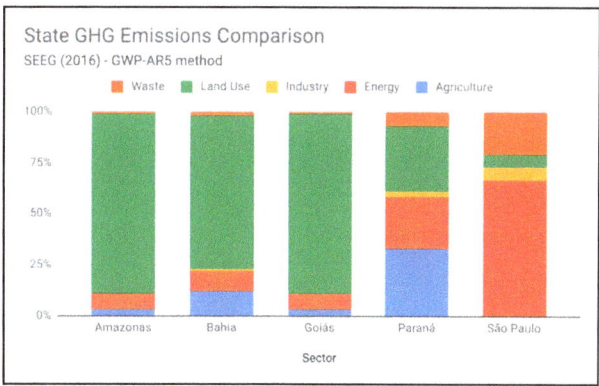

**Figure 3.** State-level GHG emissions [45].

According to the SEEG GHG inventories (Amazonas and Goiás did not issue official GHG inventories. To fill in this gap, this study has used the SEEG platform emissions estimates, issued by the Climate Observatory according to the IPCC guidelines [46]), the states that are closest to the national emissions framework are Paraná and Goiás. In both, the "Agriculture" sector is responsible for most of the emissions (33% in Paraná and 54% in Goiás), followed by the "Land Use" sector (32% in Paraná and 28% in Goiás). The "Industry" sector has practically no contribution to emissions in Goiás, while in Paraná it accounts for 3% of emissions. Coherently, the "Energy" sector accounts for 25% in Paraná, while in Goiás it corresponds to 13% of the emissions.

The states closest to the national emissions framework are Paraná and Goiás. In both states, the "Agriculture" sector is responsible for most of the emissions (33% in Paraná and 54% in Goiás), followed by the "Land Use" sector (32% in Paraná and 28% in Goiás). The "Industry" sector has practically no contribution to emissions in Goiás, while in Paraná it accounts for 3% of emissions. Coherently, the "Energy" sector accounts for 25% in Paraná, while in Goiás it corresponds to 13% of the emissions.

In the states of Bahia and Amazonas, the "Land Use" sector is the main responsible for most of the emissions, 75% in Bahia and 88% in Amazonas. Nevertheless, there are some significant differences between the two states inventories: in Bahia, the "Agriculture" and "Energy" sectors have higher weight than in Amazonas, accounting respectively for 12% and 10% of total emissions, in comparison to only 3% and 8% in Amazonas. The "Industry" sector practically does not contribute to the emissions of the state of Amazonas, while in the State of Bahia, industrial emissions represent only 1% of the total.

In comparison to the national emissions framework, the most discrepant inventory is the state of São Paulo's, the most urbanized and industrialized State in the country. In São Paulo's inventory, the "Energy" sector alone is responsible for 53% of the emissions. Although the "Agriculture" sector comes in second place, with 20% of the State's emissions, the other sectors, "Energy", "Industry" and "Waste", more closely related to urban activities, account for 75% of the total carbon emitted in the State.

About the five investigated cities, it is essential to highlight that all of them have issued legal provisions determining the edition of local emissions inventories. Insofar, Curitiba, Goiânia, Salvador and São Paulo have all issued local emissions inventories, in compliance with municipal regulations. Manaus 2014 Master Plan foresees the implementation of a local vehicle and industrial emissions control program, as the city—located in the heart of the Amazon Forest—has an estimated fleet of 713,000 motor vehicles [47] and about 500 industries installed in the Industrial Zone (Zona Franca), operating mainly in the segments of electronics, two-wheeled vehicles, and chemicals [48]. Regardless, the city has not yet issued an emissions inventory.

A study published by Nobre et al. [49] indicates that the Amazon Forest can undergo irreversible changes due to the continuous loss of biodiversity and deforestation. According to the authors, either a temperature increases above 4 degrees Celsius or deforestation exceeding 40% of the forest area would be sufficient to lead to the definitive *savanization* of the region, and in the last 60 years, the temperature in the Amazon has increased by about 1 degree Celsius, with deforestation of about 20% of the total area of tropical forest. The Amazonas emissions inventory published in the SEEG platform, in turn, indicates that 89% of total emissions from the state—where the city of Manaus stands—result from changes in land use and 3% from agricultural activities. Following a downward trend in emissions observed between 2003 and 2009, the survey of the SEEG platform indicates that emissions from the state of Amazonas have been increasing since 2014.

A study published by Liu et al. [50] indicates that pollution from the city of Manaus, more specifically, nitrogen oxides produced by automotive vehicles and industries, significantly interferes with the climate and the ecosystem equilibrium in the Amazon Forest by changing isoprene oxidation, thus interfering with the formation of rains in the region. In view of this situation, the absence of an official emissions inventory for the city of Manaus is hugely worrying, especially considering the significant weight that the Amazon Forest preservation represents for the country's (if not the world's) climate balance as a whole and the fulfillment of the goals assumed by the Brazilian government in its NDC.

As for the other surveyed cities, Curitiba, Salvador, and Goiânia have all issued emissions inventories using the Global Protocol for Community-Scale GHG Emission Inventories (GPC), a method developed specifically for cities by the World Resources Institute (WRI), the C40 Group and ICLEI, based on parameters signed by the Intergovernmental Panel on Climate Change (IPCC). Only the city of São Paulo has applied the IPCC 2006 method (the same used by the Federal Union in the National Inventory). There are no significant incompatibilities between the two methods, thus allowing some comparison between the four local inventories (Figure 4).

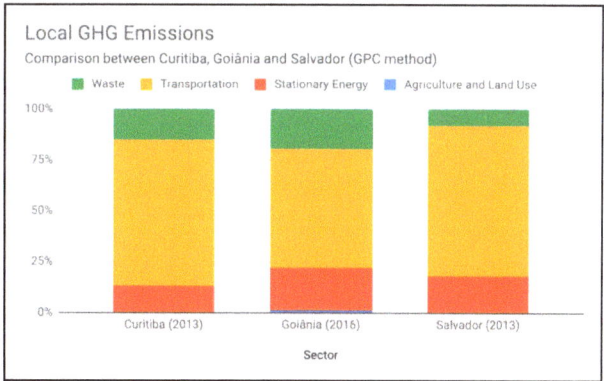

**Figure 4.** Local-level GHG emissions [51–53].

Considering the three cities that used the GPC method, the sector that occupies the first place in all three inventories is the "Transportation" sector: 78% of emissions in Salvador, 73% in Curitiba and 58.47% in Goiânia. In Salvador and Goiânia, the second place in emissions is occupied by the "Stationary Energy" sector, with 20.58% in Goiânia and 18% in Salvador. In Curitiba, the "Stationary Energy" sector occupies the third place, with 12%, just behind the "Waste" sector, with 15% of total emissions. In Salvador and Goiânia, the "Waste" sector also has critical participation, occupying the third place, respectively with 8% and 9.55% of the emissions. Among the three cities, only the city of Goiânia recorded some contribution to the "Agriculture and Land Use" sector, with 1.4% of total emissions (Figure 5).

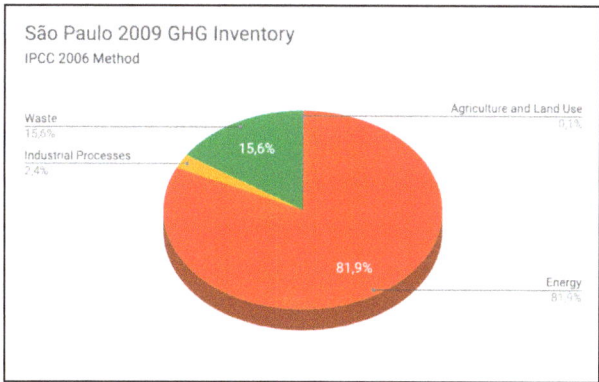

**Figure 5.** São Paulo's 2009 GHG inventory [54].

São Paulo's emissions inventory, prepared by the IPCC 2006 method, has computed emissions in four sectors: "Agriculture, Forestry and other Soil Use" (AFOLU), "Energy", "Industrial Processes and Product Usage" (IPPU) and "Waste" The first place, accounting for 82% of the emissions, was the "Energy" sector, covering mostly the fuel combustion emissions (99.7%) and in a tiny part fugitive energy (0.3%). Secondly, the "Waste" sector accounted for 16% of emissions, followed by the "IPPU" sector, with about 2% of emissions. The emissions of the "AFOLU" sector were practically irrelevant, which is not surprising considering that most of São Paulo's territory is urban.

Considering that "Energy" emissions are a product of energy consumption in transportation and buildings, it is possible to conclude that the emissions distribution in the city of São Paulo does not differ from the pattern observed in the other three cities, where "Transportation" and "Stationary Energy" are responsible for more than 75% of all three cities' emissions. Another noteworthy observation, concerning decentralized cooperation, lies in the fact that the majority of the official emissions inventories had the technical support of an international actor, as shown in Table 5.

Table 5. State- and city-level GHG inventories * [51–57].

| Entity | Edition | Year | Method | Source | Technical Support |
|---|---|---|---|---|---|
| Manaus | | | | No official emissions inventory | |
| Amazonas | | | | No official emissions inventory | |
| Salvador | 1st | 2013 | GPC | Prefeitura De Salvador (2016) | ICLEI WRI Brazil |
| Bahia | 1st | 2008 | IPCC 2006 | Governo Do Estado Da Bahia (2010) | None |
| Goiânia | 2nd | 2016 | GPC | Abreu (2017) | Inter-American Development Bank |
| Goiás | | | | No official emissions inventory | |
| São Paulo | 2nd | 2009 | IPCC 2006 | Instituto Ekos Brasil, Geoklock Consultoria e Engenharia Ambiental (2013) | World Bank |
| Estado de São Paulo | 1st | 2008 | IPCC 2006 | Governo Do Estado De São Paulo (2011) | British Embassy |
| Curitiba | 3rd | 2013 | GPC | Prefeitura De Curitiba (2016) | ICLEI |
| Paraná | 1st | 2012 | IPCC 2006 | Governo Do Estado Do Paraná (2015) | None |

* During the study, the City of Goiânia had not officially published the results of its emissions inventory. Request for access to the contents of the inventory based on the right of information, made at the Transparency Portal of the City of Goiânia, was not met.

The comparison between the local, state, and national inventories allows for some additional considerations. Firstly, the main connection point between the national inventory with the state and municipal inventories is the "Waste" sector, which highlights the importance of synergies and coordination between the National, State and Municipal Sanitation and Waste Management Policies in the country.

Secondly, the main connection point between state and municipal inventories is the "Energy" sector, which in cities is strongly related to the high energy consumption by both the "Transport"—mostly, fossil fuels—and the "Stationary Energy" sectors—hydropower and thermoelectric energy. This points out to the need not only to structure state and municipal policies focused on energy efficiency and sustainability but also to coordinate these policies, as well as the resulting plans, programs and projects (according to the 1988 Brazilian Constitution, the Union and the Federated states, not the municipalities, are in charge of all energy sources regulation, production and distribution. Therefore, municipality actions in this field are restricted mostly to energy consumption). This also highlights the strategic importance of Urban Mobility and Public Transport Plans, in their various modalities, including mobility on foot and by bicycle, to emissions control at both the state and the local level.

Finally, the two main connection point between the national and state inventories, respectively, the "Agriculture" and "Land Use" sectors, present very little weight as emission sources in the local inventories. This circumstance, however, does not authorize to discard these sectors as focal points of public policies also at the local level, especially considering the environmental services provided by the environmental protection areas and the preventive role to environmental degradation that can be played by sustainable agriculture in the urban fringes of large cities. The comparison between national, state, and local inventories indicates that cities can contribute to Brazil's NDC more in terms of climate adaptation and resilience than in emissions reduction. Nevertheless, mitigation actions at the local level are also of critical importance, as they contribute to air pollution control, microclimate regulation, and the promotion of public health and quality of life.

3.2. Local Climate Plans

Starting in 2009, when the National Climate Change Policy (PNMC) was published, all investigated cities, except for Manaus, have issued their own climate plans. The cities of Curitiba and São Paulo edited their climate plans in the same year of 2009, contemplating Mitigation, Adaptation, and

Resilience actions. Goiânia, in turn, edited its Sustainable Goiânia Plan in 2011, within the framework of the Inter-American Development Bank's Emerging and Sustainable Cities Platform (IDB). Goiânia's Plan is broader than a strict climate plan but contemplates in-depth climate actions. The city of Salvador, in a similar way, incorporated climate actions into two regulations issued in 2015, one approving the Municipal Program for Sustainable Development and Incentives for Innovation and another approving the Environmental and Sustainable Development Policy of the Municipality. For the city of Manaus, its 2011 Environmental Code has been used in this study as a proxy of "climate plan," since it also contemplates, indirectly, some local climate actions. Table 6 and Figure 6 present the qualitative analysis results of climate plans.

Table 6. Climate action references in local climate plans.

| Climate Plan | Adaptation | Cooperation | Mitigation | Resilience |
|---|---|---|---|---|
| Curitiba (2009) | 3 | 1 | 7 | 0 |
| Goiânia (2011) | 63 | 15 | 46 | 26 |
| Manaus (2011) | 35 | 1 | 45 | 0 |
| Salvador (2015) | 28 | 4 | 57 | 2 |
| São Paulo (2009) | 23 | 1 | 99 | 7 |
| Total | 152 | 22 | 254 | 35 |

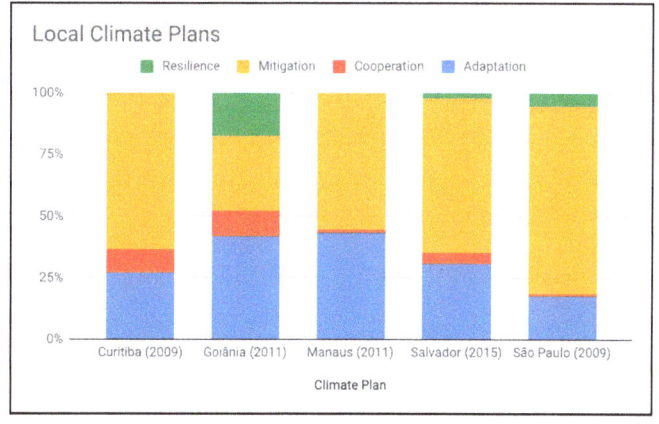

Figure 6. Climate action references in local climate plans.

In four of the five local climate plans, except Goiania's, Mitigation was the most referenced climate axis, what is a consequence of the investigated cities' concern with the consumption of fossil fuels and sustainable waste management. In the case of São Paulo's, more than 75% of the references relate to Mitigation.

This finding is consistent with the fact that combating air pollution, mainly particulate matter produced by motor vehicles, is a critical challenge for the city of São Paulo [58]. According to Saldiva [59], the reduction of air pollution in the city of São Paulo to the levels recommended by the World Health Organization (WHO) could bring about an increase in the life expectancy of the population of about three years. Thus, alternatives to fossil fuels in urban mobility is a crucial issue in São Paulo's climate strategy. Besides, in these four investigated cities adaptation was the second most mentioned climate action axis. Goiânia's Climate Plan presented a higher number of references to Resilience in comparison to the other cities' Climate Plans. In all cities, the cooperation axis was the least mentioned. Considering the total number of references observed in the five Climate Plans, Mitigation obtained 54% of the references, Adaptation, 32.8%, followed by references to Resilience, with 7.6%, and Cooperation, with only 4.8% (Figure 7).

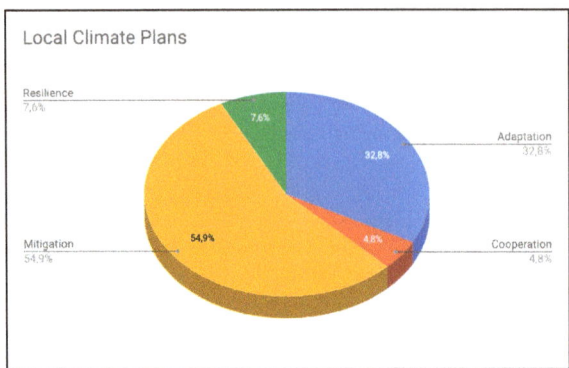

Figure 7. Climate action references in local climate plans.

## 3.3. City Master Plans

In Brazil, according to the 1988 Brazilian Constitution [60], whereas environmental protection is a competence shared by all the federated entities, urban planning and urban management are, mostly, local or municipal issues. Municipalities with a minimum population of 20,000 inhabitants are hence obligated to edit their Master Plans, not only to set their essential territorial planning guidelines but also to coordinate and integrate land use, environmental protection and sectoral policies regulations at the local level. Hence, Master Plans should present at least some reference to climate action amongst their provisions. All the five investigated cities have issued Master Plans following the guidelines and minimum requirements set by the 2001 Statute of the City, Brazil's national law on urban planning and management (Table 7).

Table 7. Climate action references in master plans.

| Master Plan | Adaptation | Cooperation | Mitigation | Resilience |
|---|---|---|---|---|
| Curitiba (2015) | 44 | 1 | 22 | 9 |
| Goiânia (2007) | 59 | 0 | 10 | 7 |
| Manaus (2014) | 42 | 0 | 3 | 0 |
| Salvador (2016) | 143 | 0 | 63 | 45 |
| São Paulo (2014) | 253 | 0 | 62 | 42 |
| Total | 541 | 1 | 160 | 253 |

All of the investigated Master Plans contain references to Adaptation, Mitigation, and Resilience actions, except for Manaus, which did not show results for Resilience actions. Furthermore, only Curitiba's Master Plan has presented one reference to decentralized cooperation actions. All five Master Plans have shown a higher number of references to Adaptation actions, in the first place, and to Mitigation actions, in the second place. Taking the sum of all the Master Plans' references to climate action, the most mentioned axis was Adaptation, with 67.2%% of the references, followed by Mitigation, with 19.9%, Resilience, with 12.8% and Cooperation, with 0.1% (Figure 8).

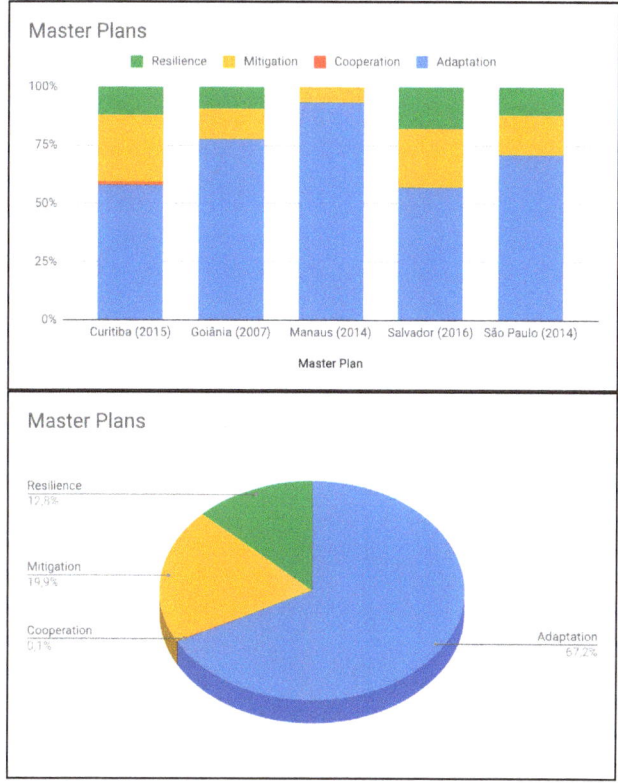

**Figure 8.** Climate action references in master plans.

*3.4. Land Use Regulations*

For the sake of analysis, all regulations about zoning, land use, parceling and development, such as zoning laws, land use laws and building codes, have been gathered under the unique topic "land use regulations" (Table 8).

**Table 8.** Climate action references in city land use regulations.

| City | Adaptation | Cooperation | Mitigation | Resilience |
|---|---|---|---|---|
| Curitiba | 137 | 2 | 23 | 31 |
| Goiânia | 358 | 13 | 61 | 552 |
| Manaus | 22 | 0 | 7 | 4 |
| Salvador | 70 | 0 | 23 | 10 |
| São Paulo | 64 | 0 | 25 | 8 |
| Total | 651 | 15 | 139 | 605 |

In Manaus, Salvador and São Paulo, the most referenced climate action axis was, firstly, Adaptation, followed by Mitigation and Resilience. The cities of Goiânia and Curitiba were the only ones to present references to decentralized cooperation actions in their land use regulations. In the city of Goiânia, the most mentioned axis was Resilience, followed by Adaptation and Mitigation. In the city of Curitiba, the most mentioned axis was Adaptation, followed by Resilience and Mitigation (Figure 9). Considering the overall sum of references, the Adaptation axis was the most mentioned, with 46.2%, followed by Resilience, with 42.9%, Mitigation, with 9.9% and Cooperation with only 1.1% (Figure 9).

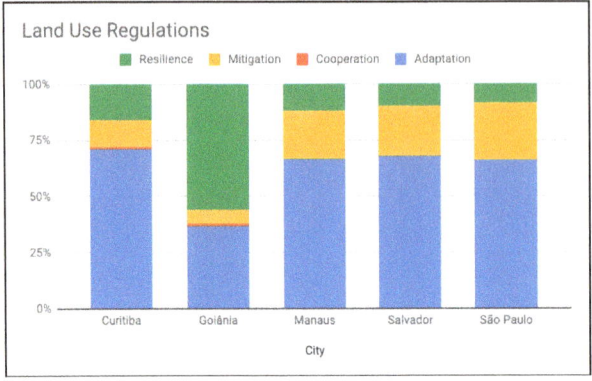

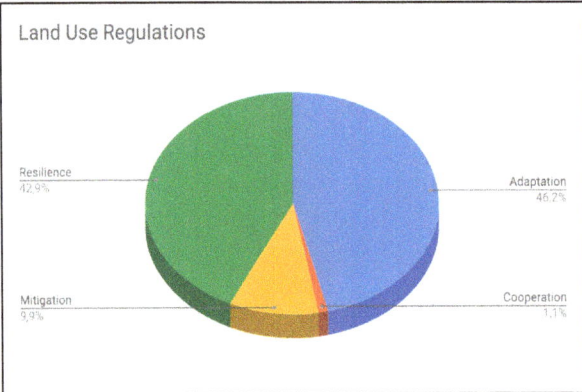

**Figure 9.** Climate action references in city land use regulations.

## 3.5. Sectoral Plans

According to the 1988 Brazilian Constitution, municipalities have the powers to provide for all essential public services at the local level, following both national and state directives. Thus, the data survey involved the following public policies: housing, mobility, sanitation, solid waste, urban drainage, disasters, and health. The goal was to detect the incorporation of local climate actions also in these fields. According to the Brazilian 1988 Constitution and the 2001 Statute of the City, local public policies must be coordinated and integrated, following the city's Master Plan guidelines.

According to national regulations, the edition of local plans on Sanitation, Waste Management, and Urban Mobility is mandatory to Municipalities [61–63]. Drainage plans are not mandatory; however, due to extreme weather events, an increasing number of cities have recently issued, if not proper plans, at least specific regulations on urban drainage and rainwater management. Concerning housing, even though local Housing Plans are as well not mandatory, in order to access National Social Housing Fund resources municipalities must edit Social Housing Plans, focused on housing provision for low-income families [64].

Regarding disasters, the National System of Civil Protection and Defense, regulated by a 2012 National Law [65], expressly integrates the Municipalities as Members, with a set of local attributions, amongst them the powers to incorporate protection and civil defense actions in municipal planning, to map disaster risk areas and to declare emergency situations and public calamity at the local level. Thus, it is incumbent to the Municipalities, if not to draft typical local disaster prevention and mitigation plans, at least to structure their respective Civil Defense at the local level.

Lastly, concerning Public Health, the 1988 Brazilian Constitution has expressly granted Municipalities the powers to provide, with the technical and financial cooperation of the union and the states, health care services for the population at the local level. The Municipalities are thus included in the Unified Health System (SUS), and must, therefore prepare and periodically update their respective Local Health Plans [66]. Tables 9–13 and Figures 10–14 presents the analysis results that are obtained per investigated city.

Table 9. Climate action references in Curitiba's sectoral plans.

| CURITIBA | Adaptation | Cooperation | Mitigation | Resilience |
|---|---|---|---|---|
| Disasters | 9 | 0 | 0 | 24 |
| Drainage | 531 | 1 | 35 | 179 |
| Health | 2 | 0 | 0 | 2 |
| Housing | 112 | 10 | 23 | 4 |
| Mobility | 1 | 0 | 2 | 0 |
| Sanitation | 1318 | 18 | 37 | 64 |
| Waste | 59 | 4 | 37 | 5 |

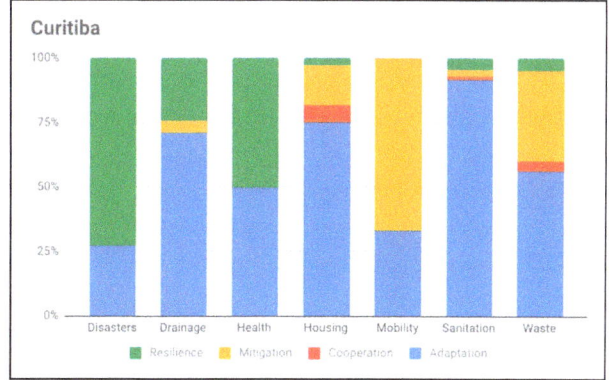

Figure 10. Climate action references in Curitiba's sectoral plans.

Table 10. Climate action references in Goiânia's sectoral plans.

| GOIÂNIA | Adaptation | Cooperation | Mitigation | Resilience |
|---|---|---|---|---|
| Disasters | 20 | 0 | 0 | 2 |
| Drainage | 57 | 0 | 3 | 4 |
| Health | 44 | 0 | 1 | 6 |
| Housing | 0 | 0 | 0 | 0 |
| Mobility | 0 | 0 | 0 | 0 |
| Sanitation | 0 | 0 | 0 | 0 |
| Waste | 76 | 8 | 86 | 23 |

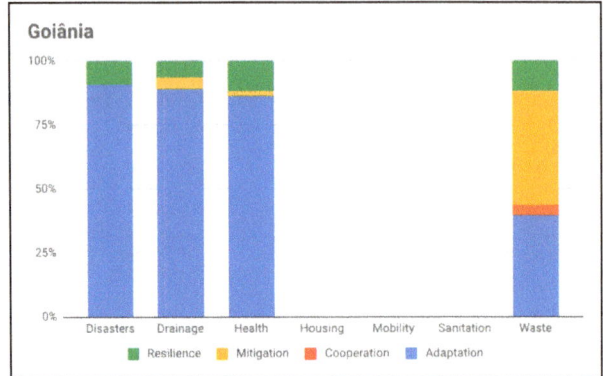

**Figure 11.** Climate action references in Goiânia's sectoral plans.

**Table 11.** Climate action references in Manaus' sectoral plans.

| MANAUS | Adaptation | Cooperation | Mitigation | Resilience |
|---|---|---|---|---|
| Disasters | 39 | 6 | 0 | 2 |
| Drainage | 0 | 0 | 0 | 0 |
| Health | 39 | 6 | 0 | 2 |
| Housing | 0 | 0 | 0 | 0 |
| Mobility | 16 | 0 | 131 | 107 |
| Sanitation | 553 | 3 | 30 | 12 |
| Waste | 40 | 0 | 117 | 3 |

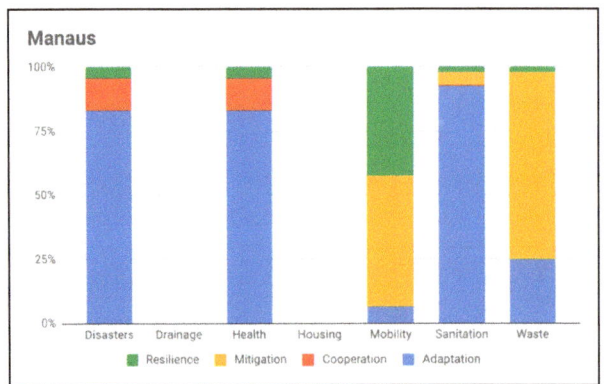

**Figure 12.** Climate action references in Manaus' sectoral plans.

**Table 12.** Climate action references in Salvador's sectoral plans.

| SALVADOR | Adaptation | Cooperation | Mitigation | Resilience |
|---|---|---|---|---|
| Disasters | 99 | 0 | 2 | 118 |
| Drainage | 0 | 0 | 0 | 0 |
| Health | 195 | 13 | 6 | 52 |
| Housing | 0 | 0 | 0 | 0 |
| Mobility | 36 | 9 | 62 | 11 |
| Sanitation | 18 | 0 | 1 | 0 |
| Waste | 1 | 0 | 0 | 0 |

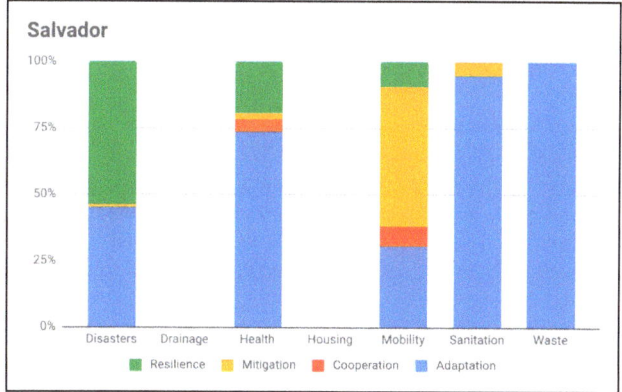

Figure 13. Climate action references in Salvador's sectoral plans.

Table 13. Climate action references in São Paulo's sectoral plans.

| SÃO PAULO | Adaptation | Cooperation | Mitigation | Resilience |
|---|---|---|---|---|
| Disasters | 25 | 0 | 0 | 41 |
| Drainage | 1666 | 4 | 202 | 425 |
| Health | 271 | 0 | 3 | 0 |
| Housing | 103 | 0 | 8 | 12 |
| Mobility | 43 | 7 | 228 | 4 |
| Sanitation | 721 | 10 | 202 | 93 |
| Waste | 84 | 6 | 153 | 48 |

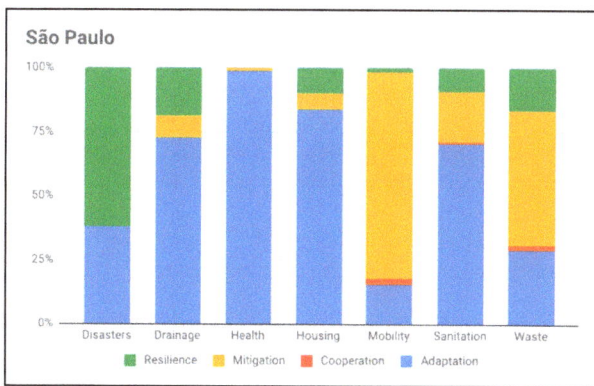

Figure 14. Climate action references in São Paulo's sectoral plans.

Among the five investigated cities, only São Paulo and Curitiba presented references to climate actions in all sectoral policies: Disasters, Drainage, Health, Housing, Mobility, Sanitation, and Waste management. Goiânia did not present references to climate actions in Housing, Mobility, and Sanitation. Manaus and Salvador, in turn, did not present references to climate actions in Drainage and Housing.

Firstly, concerning Disasters Management, the most referenced climate action axis in the cities of Curitiba, Salvador—the latter, the Brazilian city with the most significant number of inhabitants in the risk area [67]—and São Paulo was Resilience, followed by Adaptation. In Goiânia and the axis most commonly referred to was Adaptation. Unlike the other investigated cities, the Municipality of Manaus dealt with the prevention and remediation of disasters in its Municipal Health Plan and not, as expected, in civil defense regulations.

In what concerns Drainage, the most referenced climate action axis in the cities of Curitiba, Goiânia and São Paulo was Adaptation, followed by Resilience, in second, and Mitigation, in third place. Only the cities of São Paulo and Curitiba presented references to cooperation actions in urban drainage. Manaus and Salvador did not present any results in this field, as the document survey did not locate any documents specifically dedicated to urban drainage for both cities.

Only São Paulo and Curitiba presented references to climate action in Housing policy documents. Adaptation was the axis with the highest number of references, followed, in the case of São Paulo, by Resilience, in the second place, and Mitigation, in the third place; in the case of Curitiba, Mitigation occupied the second place, followed respectively by Cooperation, in third, and Resilience, in fourth.

In the field of Urban Mobility, except for Goiânia, all the investigated cities presented a higher number of references to Mitigation, followed by Adaptation in São Paulo, Salvador and Curitiba, and Resilience in Manaus. Only São Paulo and Salvador presented references to the Cooperation axis. The city of Goiânia did not present any references to climate action in its Municipal Urban Mobility Plan, still under development.

Concerning Sanitation, the axis of climate action with the most significant number of references in the cities of São Paulo, Salvador, Manaus, and Curitiba, was Adaptation. Mitigation was the second most mentioned axis in São Paulo, Salvador, and Manaus; in Curitiba, it was the Resilience axis. There are references to Cooperation only in the Sanitation Plans of Curitiba and São Paulo. The Municipal Sanitation Plan of Goiânia (in preparation) did not present references to any climate action axis.

For Waste Management, Adaptation was the most referenced axis in both Curitiba and Salvador Waste Management Plans. In the cities of São Paulo, Manaus and Goiânia, Mitigation was the most referenced axis, followed by Adaptation and Resilience. Only the cities of Curitiba, Goiânia and São Paulo presented references to the Cooperation axis in their Local Waste Management Plans.

Finally, concerning Public Health, Adaptation was the most referenced climate action axis in all five Local Health Plans, followed by Mitigation, in São Paulo, and by Resilience, in Curitiba, Goiânia, and Salvador. References to Coordination actions were found only in the Health Plans of Salvador and Manaus.

A word search focused on the Municipal Public Health Plans pointed out that only Salvador's contains references to the expression "climate change", detected in two contexts: the articulation of health and climate agendas and the prevention of arboviruses such as yellow fever. Furthermore, based on the assumption that health promotion is a necessary component of sustainable urban development, another specific Word Search was carried out, aimed at locating references to "health" in all the investigated Local Climate Plans, Master Plans, Land Use Regulations and other Sectoral Plans (Table 14). The results confirm the initial hypothesis that "health" is an integral dimension of urban planning in the five investigated cities, including concerning climate change issues.

**Table 14.** References to health in city sectoral plans.

| Documents | Curitiba | Goiania | Manaus | Salvador | São Paulo |
|---|---|---|---|---|---|
| Climate Plan | 0 | 5 | 23 | 34 | 8 |
| Master Plan | 43 | 46 | 9 | 69 | 33 |
| Land Use Regulations | 62 | 37 | 7 | 11 | 25 |
| Disasters | 1 | 0 | 948* | 21 | 3 |
| Drainage | 6 | 0 | 0 | 0 | 31 |
| Housing | 84 | 1 | 0 | 0 | 6 |
| Sanitation | 279 | 0 | 18 | 0 | 343 |
| Waste | 110 | 124 | 82 | 0 | 224 |

* Disaster management is a content of the Manaus Municipal Public Health Plan, hence returns the highest number of results.

Figure 15 figure illustrates that there are references to health in the Master Plans and the Land Use Regulations of all the investigated cities. All Local Climate Plans, except Curitiba's, also make references to health. Finally, there are, references to health in most of the sectoral plans of the investigated cities, especially sanitation and waste management plans. Overall results are sorted by sectoral policy and climate action axis presented, in turn, the following findings (Table 15).

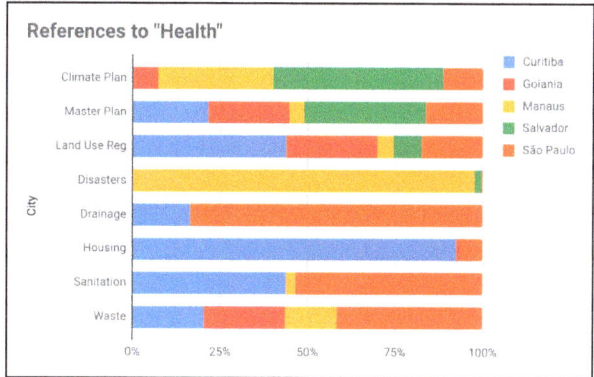

Figure 15. References to health in city sectoral plans.

Table 15. Climate action references in sectoral policies.

| Sectoral Policy | Adaptation | Cooperation | Mitigation | Resilience |
| --- | --- | --- | --- | --- |
| Disasters | 192 | 6 | 2 | 187 |
| Drainage | 2254 | 5 | 24 | 608 |
| Housing | 215 | 10 | 31 | 16 |
| Mobility | 96 | 16 | 423 | 122 |
| Sanitation | 1437 | 31 | 270 | 169 |
| Public Health | 561 | 19 | 10 | 62 |
| Waste | 260 | 18 | 394 | 79 |
| Total | 5015 | 105 | 1370 | 1243 |

Adaptation was the most referenced axis in the majority of the sectoral policies documents: Disasters, Drainage, Housing, Sanitation, and Public Health. Mobility and Waste management Document presented more references to Mitigations. Resilience was the second most referred axis in Drainage, Disaster, and Mobility. Considering the total number of references, the Adaptation axis was the most referenced, with 71.2%, followed by Resilience, with 14.3%, Mitigation, with 13.3% and Cooperation, with 1.2% (Figure 16).

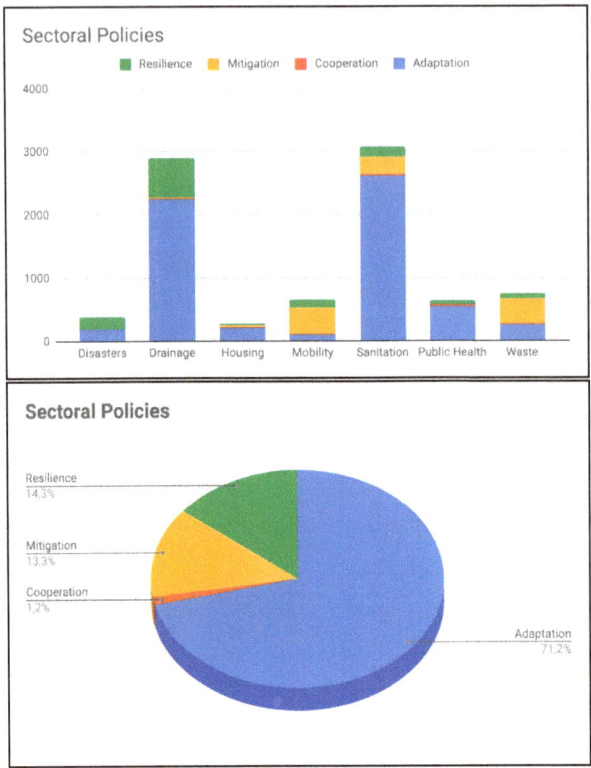

**Figure 16.** Climate action references in sectoral policies.

## 4. Discussion

*4.1. Institutional Frameworks and Climate Action in Brazil*

The 1992 United Nations Conference on Environment and Development (UNCED) fostered the edition of the 1992 United Nations Framework Convention on Climate Change (UNFCCC). As its hosts, Brazil has been for some time, one of the Global South main protagonists in climate change negotiations (the situation has changed dramatically with the election of Brazilian President Jair Bolsonaro in October 2018, a far-right open climate skeptic anti-environmentalist politician. Just before taking on office, the new Presidential Cabinet withdrew Brazil's proposal to host the 25th UNFCCC Conference of Parties (COP 25), to be held in December 2019, and since the very first days of the federal mandate, the appointed Minister of Environment, Ricardo Salles, has been making significant cuts on climate programs and initiatives at the national level. In this context, it is essential to highlight that this paper portrays the institutional framework before President Bolsonaro's inauguration, as comprised in the years of 2017 and 2018. In other words, this study discloses a much more positive standing of Brazil in the climate action than what it is actually today) [30]. Nevertheless, the country has only issued a national climate change policy (PNMC) long after the edition of the 1997 Kyoto Protocol, through the approval of a national law in the year of 2009 [68].

In resonance to the Brazilian Federation legal framework, PNMC expressly contemplates amongst its primary goals the promotion of climate adaptation in all levels—national, regional and local—with the participation and collaboration of all implicated stakeholders. Furthermore, all measures taken at the national level must consider and integrate regional and local actions taken by both public and private entities. PNMC also fosters integration between mitigation and adaptation strategies in all federative levels, as well as the engagement of all sectors in the implementation of climate policies, plans, programs, and strategic actions.

Under the 1992 UNFCCC, all states signatory to the 2015 Paris Agreement, regardless of their level of economic development, are obliged to comply with the voluntary commitments embodied in their respective NDC. The NDC mechanism was formulated to effectively engage and commit the UNFCCC member states in the pursuit of the 1.5-degree Celsius target. As a result, it can be affirmed that unlike its predecessor—the 1997 Kyoto Protocol—the Paris Agreement is a legally binding agreement [69].

Following its adherence to the Paris Agreement in September 2016, Brazil's national government duly presented its NDC in 2017, with an emissions reduction target of 37% below the national emissions average until 2025, and a complementary reduction target of 43% below 2005 levels until 2030 [20], as well as adaptation measures, all based in the 2016 National Adaptation Plan (PNA) [70]. The 2016 PNA main objectives are to promote risk management and reduction at the national level; to guide the expansion and dissemination of knowledge; to foster cooperation and coordination between the public and private sectors, and to identify and to propose adaptation measures. PNA covers a total of 11 strategic sectors: industry, vulnerable social groups, water, infrastructure, coastal zone, food security, agriculture, biodiversity, cities, health, and disaster prevention, and mitigation.

With regard specifically to the linkages between climate change and public health, the Ministry of Health has edited a Sectoral Health Plan for Mitigation and Adaptation to Climate Change in 2013 [71]. This plan's main objective is to organize the public health services response to climate change in terms of both adaptation and mitigation, through four lines of action: health surveillance, health care, health promotion and education, and health research, each pointing out objectives, indicators, expected results and the governmental agency responsible for each action.

The 2013 Sectoral Health Plan appoints the critical challenges of climate change to public health. Firstly, deaths and morbidities due to disasters and communicable diseases such as dengue fever, yellow fever, Zika virus, and leishmaniosis. Secondly, deaths and morbidities due to non-communicable diseases such as cardiovascular and respiratory diseases, linked to air pollution, and also skin cancer, linked to the destruction of ozone layer. The plan highlights the vital role to be played by local authorities regarding resilience, adaptation, and mitigation. In terms of resilience, the plan focuses on natural disasters, to be caused by extreme weather events. In terms of adaptation, the plan points local actions in the fields of housing, sanitation, and green areas. Lastly, for mitigation, the plan highlights the importance of active urban mobility in order to target air pollution and promote physical activity.

The 1988 Brazilian Constitution expressly grants municipalities the powers to act on environmental preservation as well as on urban planning. Thus, climate actions are expected to permeate both environmental and urban policies at the local level, either in local climate plans or in local regulations in general, such as Master Plans, Land Use Regulations, and Sectoral Policy Plans.

Nonetheless, at the national level, the topics "climate change", "sustainable urban development", "public health" and "disasters management" are under the responsibility of separate Federal bodies, respectively, Ministry of Environment, Ministry of Cities (the Ministry of Cities was terminated by President Jair Bolsonaro in January 2019—its former attributions have been partially absorbed by the Ministry of Regional Development, Ministry of Health, and Ministry of Science, Technology, Innovation and Communications). There is roughly no common agenda specifically focused on urban climate policy, which results in the lack of consistent monitoring of the local policies, plans, programs and climate actions currently conducted by the Brazilian municipalities [72,73].

*4.2. Local Emissions Inventories and Climate Plans*

Firstly, four of the five surveyed cities have published GHG emissions inventories in the context of decentralized cooperation agreements: Curitiba, with the support of ICLEI; Goiânia, with support from the Inter-American Development Bank; Salvador, with support from ICLEI and WRI Brazil and São Paulo, with support from the World Bank. All four investigated cities, having compiled GHGs inventories, have subsequently edited their respective Local Climate Plans. Thus, preparation of GHGs inventories seems to be, in fact, a necessary preparatory step for climate plans. Moreover, all the investigated Local Climate Plans came to light as of 2009, year of the edition of the National Policy on Climate Change (PNMC).

Manaus was the only city not to assemble an emissions inventory or a specific Climate Plan. Nevertheless, its 2011 Environmental Code contains references to Adaptation and Mitigation actions, allowing its use as a proxy for a climate plan in this investigation. Of the five cities investigated, only Curitiba edited documents specifically dedicated to Adaptation and Resilience, following the internationally recommended standards. Also, it has contemplated climate actions in its Master Plan, Land Use Regulation, and all Sectoral Plans. Hence, this city is at the forefront of climate action, ahead of the other surveyed cities. Decentralized cooperation and the edition of the PNMC can be pointed out as the two main engines for the structuring of local climate actions, considering, firstly, the participation of international actors on the local emissions inventories and climate plans; and, secondly, the adherence of the investigated cities to national guidelines—PNMC included—while structuring their local public policies.

*4.3. Contribution to Greenhouse Gas Emission Reduction*

The PNMC emphasizes city's role in climate Adaptation, first of all, and in urban Resilience, but not so much in climate Mitigation, what is consistent with the data contemplated in the National GHG inventory. National emissions estimates show that the contribution of the most relevant sectors to urban activities—e.g., Waste and Energy—is relatively low when compared to the Agriculture and Land Use sectors. In line with the PNMC, Adaptation was the climate action axis with the most significant number of references in the Master Plans, Land Use Regulations and Sectoral Plan. Mitigation, however, had the highest number of references in the local Climate Plans, Urban Mobility and Waste Management, what enhances the cities concerns with fossil fuels consumption and sustainable management of waste residues and effluents.

Furthermore, in line with the PNMC, Resilience was the second most referenced climate action axis in Land Use Regulations, Disasters Management, Drainage, Health and Urban Mobility Plans, demonstrating that the investigated cities have made efforts to incorporate the coping of shock and stress factors into their planning and management processes. Decentralized cooperation was overall the least referenced climate action. Nonetheless, it was possible to detect references to cooperation actions in the compiled documents from all investigated cities, confirming their engagement with the international community regarding climate change issues.

*4.4. Climate Action in Public Health and Disasters Management*

Qualitative analysis results reported in this paper revealed that health promotion effectively integrates Local Climate Plans, Master Plans, Land Use Regulations and Sectoral Plans of the five investigated cities, with a more significant number of references to Health promotion actions in Sanitation and Waste Management Plans, what reinforces the critical connection between Sanitation and Public Health policies in urban areas. Municipal Public Health Plans and the Disaster Management regulations of the five surveyed cities do incorporate climate action, mainly in the Adaptation axis, but not so much in Resilience. The analyzed Local Public Health and Disaster Management Policies did present some connections, but not, as it would be desirable, effective coordination of projects and actions in the investigated cities. Municipality of Manaus, at this point, provided innovative treatment to the issue, incorporating Disaster management into the Municipal Public Health Plan.

*4.5. Limitations of the Study*

This study encountered a number of limitations. The following issues should be considered when interpreting the findings of this study: (a) The data collection process, even undertaken thoroughly, may have unintentionally omitted some policy documents—e.g., the ones that are approved after the data is collected; (b) The study does not factor in the maturity of the policy—e.g., whether just released or in place for quite some time; (c) The study does not factor in whether the policy is implemented; and if so how effective results it generated; (d) The content analysis technique, even conducted with the aid of a software, may contain unconscious bias—e.g., selection of keywords and nodes; (e) The results of the content analysis, even reviewed with utmost care, may have interpreted differently—e.g., due to unconscious bias of the researchers; (f) The findings could not be crosschecked with any other empirical research, since this is a pioneering study; (g) The study portrays the institutional framework before President Bolsonaro's inauguration, as comprised in the years of 2017 and 2018; meaning it discloses a much more positive standing of Brazil in the climate action than what it is actually today, and; (h) The study did not investigate the preparedness of the planning and emergency frameworks in combatting major climate-related catastrophes, such as the recent Amazon rainforest fires (Figure 17). In our prospective studies, we will overcome these limitations.

**Figure 17.** A snapshot from August 2019 Amazon rainforest fires [74].

## 5. Conclusions

The study at hand focused on tackling an important question of: "Are, and if yes how, cities considering the climate emergency in their local planning mechanisms?" The results of the study revealed that despite the existing federal presidential level climate crisis denial and efforts towards deforestation of Amazonia, the investigated municipalities have attempted to address the issue at least through municipal legislations, regulations and plans. These municipalities not only incorporate climatic issues in their urban planning and urban management regulations, but also contribute to the fulfillment of Brazil's climate commitments, more in terms of adaptation than in terms of mitigation, consistently with the findings of local and national emissions inventories.

Concerning possible policy improvements at the municipal level, the study recommends greater coordination between local health promotion and disaster management actions by the investigated cities. This is at utmost importance in order to foster urban resilience and sustainability. However, these municipalities' policies concentrate more on adaptation rather than mitigation [75]. Meaning that there is much more room for a better policy for adequately tackle the biggest problem of our time—i.e., climate crisis—further developing strategies in urban adaptation and resilience. According to the obtained results from our analysis, it is possible to draw the following insights and conclusions.

Firstly, decentralized cooperation and the edition of the National Policy on Climate Change (PNMC) were the main drivers for the structuring of local climate actions by the five investigated cities. However, with the new federal administration in office, the national climate change policy has the risk of staying current or not being endorsed and implemented [76].

Secondly, by including climate actions in their Master Plans, Land Use regulations and Sectoral Plans, the investigated cities do contribute to the compliance of Brazil's NDC, more in terms of Adaptation than Mitigation, coherently with the data of national and local GHG emissions inventories and with the guidelines of the PNMC and the National Adaptation Plan (PNA). It was not possible, however, to determine precisely to what extent—i.e., emissions percentages—the investigated cities can contribute to the national target of 37% of GHG emissions. It can only be inferred that local mitigation actions concentrate in Mobility and Waste Management policies.

Thirdly, even though the investigated Brazilian cities have included adaptation and resilience measures in their Master Plans, land use regulations and sectoral policies, there is significant room for improvements, especially considering that mitigation, not adaptation, was by far the most referenced axis in four of the five investigated Local Climate Plans.

Fourthly, even though cooperation was overall the least referenced climate action axis, all investigated cities have presented some references to cooperation actions apart from their respective GHGs inventories, what confirms some degree of engagement with the international community regarding climate change issues.

Fifthly, health, as expected, effectively integrates the Climate Plans, Master Plans, Land Use Regulations and Sectoral Plans of the five investigated cities, especially in Waste Management and Sanitation. Besides the climate emergency, due to inadequate infrastructure and informal housing—so-called Favelas of Brazil—health has been a major policy focus and concern in the nation.

Sixthly, the study findings reveal that Brazilian cities are rich in terms of policy but poor in terms of their implementation. Although the five investigated cities showed some connection points between their climate, public health, and disaster management policies, they should further coordinate and integrate these policies in order to develop urban resilience and foster SDGs fulfillment at the local level. These local level policy and implementation attempts are hoped to showcase promising results on the ground and generate motivation for other Brazilian municipalities, states and the federal government to further invest in tackling the climate crisis—that is upon Brazil [77,78] and the entire globe. The results indicate that the investigated cities have the necessary tools to act on climate mitigation, adaptation and resilience, but assessing the effectiveness of these actions goes beyond the objectives and scope of the proposed research and could be developed by further studies on the matter.

Lastly, the literature clearly indicates that climate change and its catastrophic consequences are the inevitable harsh reality of our time [79,80]. In such dire straits, we need to move urgently from climate policy/agenda to climate action [81,82]. This is to say, the lip service (e.g., rhetoric and/or policy without effective implementation and impact) will not get the desired outcomes. Cities need to urgently lead the way in climate change action [83], however, they need to find ways to successfully implement policies as well. The road to successful implementation is highly bumpy, we need to find ways to transform barriers into enablers of action on climate change and sustainable urban development [84–86]. Citizens, scientist, institutions, policymaking, and multilevel governance structures along with good/honest politics are required for such immediate action and practice turnaround (i.e., changing the attitude, behavior, and choice)—both industrial and household [87,88].

**Author Contributions:** D.S. designed and led the research, under the supervision of A.P.; D.S. and A.P. jointly prepared the first draft of the manuscript; T.Y. and M.K. revised, and increased the rigor, relevance and reach of the manuscript. All authors read and approved the final version of the manuscript.

**Funding:** This research did not receive any specific grant from funding agencies in the public, commercial or not-for-profit sectors.

**Acknowledgments:** The authors thank anonymous referees for their constructive comments on an earlier version of the manuscript.

**Conflicts of Interest:** The authors declare no conflict of interest.

## References

1. Van Loon, A.F.; Gleeson, T.; Clark, J.; Van Dijk, A.I.; Stahl, K.; Hannaford, J.; Hannah, D.M. Drought in the Anthropocene. *Nat. Geosci.* **2016**, *9*, 89–91. [CrossRef]
2. Yigitcanlar, T.; Kamruzzaman, M.; Foth, M.; Sabatini-Marques, J.; Costa, E.; Ioppolo, G. Can cities become smart without being sustainable? A systematic review of the literature. *Sustain. Cities Soc.* **2019**, *45*, 348–365. [CrossRef]
3. Scranton, R. *Learning to Die in the Anthropocene: Reflections on the End of a Civilization*; City Lights Publishers: San Francisco, CA, USA, 2015.
4. Yigitcanlar, T.; Foth, M.; Kamruzzaman, M. Towards post-anthropocentric cities: Reconceptualising smart cities to evade urban ecocide. *J. Urban Technol.* **2019**, *26*, 147–152. [CrossRef]
5. Parmesan, C.; Yohe, G. A globally coherent fingerprint of climate change impacts across natural systems. *Nature* **2003**, *421*, 37. [CrossRef] [PubMed]
6. Burke, M.; Dykema, J.; Lobell, D.B.; Miguel, E.; Satyanath, S. Incorporating climate uncertainty into estimates of climate change impacts. *Rev. Econ. Stat.* **2015**, *97*, 461–471. [CrossRef]
7. Pecl, G.T.; Araújo, M.B.; Bell, J.D.; Blanchard, J.; Bonebrake, T.C.; Chen, I.C.; Falconi, L. Biodiversity redistribution under climate change: Impacts on ecosystems and human well-being. *Science* **2017**, *355*, eaai9214. [CrossRef] [PubMed]
8. Jakob, M.; Steckel, J.C. Implications of climate change mitigation for sustainable development. *Environ. Res. Lett.* **2016**, *11*, 104010. [CrossRef]
9. Yigitcanlar, T. Planning for smart urban ecosystems: Information technology applications for capacity building in environmental decision making. *Theor. Empir. Res. Urban Manag.* **2009**, *4*, 5–21.
10. Aldieri, L.; Carlucci, F.; Vinci, C.P.; Yigitcanlar, T. Environmental innovation, knowledge spillovers and policy implications: A systematic review of the economic effects literature. *J. Clean. Prod.* **2019**, *239*, 118051. [CrossRef]
11. Yigitcanlar, T.; Kamruzzaman, M.; Buys, L.; Ioppolo, G.; Sabatini-Marques, J.; da Costa, E.; Yun, J. Understanding 'smart cities': Intertwining development drivers with desired outcomes in a multidimensional framework. *Cities* **2018**, *81*, 145–160. [CrossRef]
12. Rottz, M.; Sell, D.; Pacheco, R.; Yigitcanlar, T. Digital commons and citizen coproduction in smart cities: Assessment of Brazilian municipal e-government platforms. *Energies* **2019**, *12*, 2813. [CrossRef]
13. Arbolino, R.; Carlucci, F.; Cira, A.; Ioppolo, G.; Yigitcanlar, T. Efficiency of the EU regulation on greenhouse gas emissions in Italy: The hierarchical cluster analysis approach. *Ecol. Indic.* **2017**, *81*, 115–123. [CrossRef]
14. Ingrao, C.; Messineo, A.; Beltramo, R.; Yigitcanlar, T.; Ioppolo, G. How can life cycle thinking support sustainability of buildings? Investigating life cycle assessment applications for energy efficiency and environmental performance. *J. Clean. Prod.* **2018**, *201*, 556–569. [CrossRef]
15. Yigitcanlar, T.; Sabatini-Marques, J.; da-Costa, E.M.; Kamruzzaman, M.; Ioppolo, G. Stimulating technological innovation through incentives: Perceptions of Australian and Brazilian firms. *Technol. Forecast. Soc. Chang.* **2019**, *146*, 403–412. [CrossRef]
16. Arbolino, R.; De Simone, L.; Carlucci, F.; Yigitcanlar, T.; Ioppolo, G. Towards a sustainable industrial ecology: Implementation of a novel approach in the performance evaluation of Italian regions. *J. Clean. Prod.* **2018**, *178*, 220–236. [CrossRef]
17. Zapf, M.; Pengg, H.; Weindl, C. How to Comply with the Paris Agreement Temperature Goal: Global Carbon Pricing According to Carbon Budgets. *Energies* **2019**, *12*, 2983. [CrossRef]
18. Yigitcanlar, T.; Kamruzzaman, M. Does smart city policy lead to sustainability of cities? *Land Use Policy* **2018**, *73*, 49–58. [CrossRef]
19. Manowska, A.; Nowrot, A. The Importance of Heat Emission Caused by Global Energy Production in Terms of Climate Impact. *Energies* **2019**, *12*, 3069. [CrossRef]

20. Brasil. *Sumário Executivo. Documento-Base para Subsidiar os Diálogos Estruturados Sobre a Elaboração de uma Estratégia de Implementação e Financiamento da Contribuição Nacionalmente Determinada do Brasil ao Acordo de Paris*; Ministério do Meio Ambiente: Brasília, Brazil, 2017.
21. Octaviano, C.; Paltsev, S.; Gurgel, A.C. Climate change policy in Brazil and Mexico: Results from the MIT EPPA model. *Energy Econ.* **2016**, *56*, 600–614. [CrossRef]
22. Broto, V.C. Urban governance and the politics of climate change. *World Dev.* **2017**, *93*, 1–15. [CrossRef]
23. Hale, T. "All Hands-on Deck": The Paris Agreement and Nonstate Climate Action. *Glob. Environ. Politics* **2016**, *16*, 12–22. [CrossRef]
24. United Nations, General Assembly. *Resolution Adopted by the General Assembly on 27 July 2012. 66/288. The Future We Want*; United Nations: Washington, DC, USA, 2012.
25. United Nations, General Assembly. *Resolution Adopted by the General Assembly on 25 September 2015*; United Nations: Washington, DC, USA, 2015.
26. United Nations, General Assembly. *Resolution Adopted by the General Assembly on 23 December 2016; 71/256. New Urban Agenda*; United Nations: Washington, DC, USA, 2016.
27. Jacobi, P.; Sulaiman, S. Governança ambiental urbana em face das mudanças climáticas. *Rev. USP* **2016**, *109*, 133–142. [CrossRef]
28. Nobre, C. Mudanças climáticas globais: Possíveis impactos nos ecossistemas do país. *Parcer. Estratégicas* **2001**, *6*, 239–258.
29. Pessini, L.; Sganzerla, A. As mudanças climáticas e seus impactos no reino a vida: Perspectivas para um futuro não apocalíptico. *Revista Iberoamericana de Bioética* **2016**, *2*, 1–13.
30. Lahsen, M. A science–policy interface in the global south: The politics of carbon sinks and science in Brazil. *Clim. Chang.* **2009**, *97*, 339. [CrossRef]
31. Engle, N.L.; Lemos, M.C. Unpacking governance: Building adaptive capacity to climate change of river basins in Brazil. *Glob. Environ. Chang.* **2010**, *20*, 4–13. [CrossRef]
32. Yin, R.K. *Case Study Research*; Sage Publications: Thousand Oaks, CA, USA, 2003.
33. Owen, G.T. Qualitative methods in higher education policy analysis: Using interviews and document analysis. *Qual. Rep.* **2014**, *19*, 1–19.
34. Brasil. Lei 6.938 de 31 de Agosto de 1981. Available online: http://www.planalto.gov.br/ccivil_03/Leis/L6938.htm (accessed on 24 August 2019).
35. Brasil. Lei 10.257 de 10 de Julho de 2001—Estatuto da Cidade. Available online: http://www.planalto.gov.br/ccivil_03/Leis/LEIS_2001/L10257.htm (accessed on 24 August 2019).
36. Yigitcanlar, T.; Metaxiotis, K.; Carrillo, F.J. *Building Prosperous Knowledge Cities: Policies, Plans and Metrics*; Edward Elgar: Cheltenham, UK, 2012.
37. Yigitcanlar, T. Position paper: Benchmarking the performance of global and emerging knowledge cities. *Expert Syst. Appl.* **2014**, *41*, 5549–5559. [CrossRef]
38. Intergovernmental Panel on Climate Change—IPCC. *Global Warming of 1.5 °C: An IPCC Special Report on the Impacts of Global Warming of 1.5 °C above Pre-industrial Levels and Related Global Greenhouse Gas Emission Pathways, in the Context of Strengthening the Global Response to the Threat of Climate Change, Sustainable Development, and Efforts to Eradicate Poverty*; Intergovernmental Panel on Climate Change: Geneva, Switzerland, 2018.
39. Pacheco, R.; Silva, C. Global Warming Potential of Biomass-to-Ethanol: Review and Sensitivity Analysis through a Case Study. *Energies* **2019**, *12*, 2535. [CrossRef]
40. Ministério da Ciência, Tecnologia, Inovação e Comunicação—MCTIC. *Terceira Comunicaçãao Nacional do Brasil à Convenção-Quadro das Nações Unidas sobre Mudança do Clima*; MCTIC: Brasília, Brazil, 2018.
41. Azevedo, T. *Emissões de GEE do Brasil e suas Implicações para Políticas Públicas e a Contribuição Brasileira para o Acordo de Paris*; Documento síntese; Período 1970–2015; Observatório do Clima; SEEG: Brasilia, Brazil, 2017.
42. Malhi, Y.; Roberts, J.T.; Betts, R.A.; Killeen, T.J.; Li, W.; Nobre, C.A. Climate change, deforestation, and the fate of the Amazon. *Science* **2008**, *319*, 169–172. [CrossRef]
43. Caviglia-Harris, J.L. Agricultural innovation and climate change policy in the Brazilian Amazon: Intensification practices and the derived demand for pasture. *J. Environ. Econ. Manag.* **2018**, *90*, 232–248. [CrossRef]
44. Farias, A.R.; Mingoti, R.; Valle, L.D.; Spadotto, C.A.; Lovisi Filho, E. *Comunicado Técnico. Identificação, Mapeamento e Quantificação das Áreas Urbanas do Brasil*; Embrapa: Brasilia, Brazil, 2017.
45. SEEG. Plataforma SEEG Brasil. Available online: http://plataforma.seeg.eco.br/ (accessed on 24 August 2019).

46. De Azevedo, T.R.; Junior, C.C.; Junior, A.B.; dos Santos Cremer, M.; Piatto, M.; Tsai, D.S.; Barreto, P.; Martins, H.; Sales, M.; Galuchi, T.; et al. SEEG initiative estimates of Brazilian greenhouse gas emissions from 1970 to 2015. *Sci. Data* **2018**, *5*, 180045. [CrossRef] [PubMed]
47. Souza, S. *Frota de Manaus em Circulação Supera Marca de 713 mil Veículos*; A Crítica: Manaus, Brazil, 2017. Available online: https://www.acritica.com/channels/manaus/news/frota-de-manaus-superamarca-de-713-mil-veiculos (accessed on 24 August 2019).
48. Suframa. O que é o Projeto ZFM? Available online: http://site.suframa.gov.br/assuntos/modelo-zona-franca-de-manaus/o-que-e-o-projeto-zfm (accessed on 24 August 2019).
49. Nobre, C.A.; Sampaio, G.; Borma, L.S.; Castilla-Rubio, J.C.; Silva, J.S.; Cardoso, M. Land-use and climate change risks in the Amazon and the need of a novel sustainable development paradigm. *Proc. Natl. Acad. Sci. USA* **2016**, *113*, 10759–10768. [CrossRef] [PubMed]
50. Liu, Y.; Brito, J.; Dorris, M.R.; Rivera-Rios, J.C.; Seco, R.; Bates, K.H.; Goldstein, A.H. Isoprene photochemistry over the Amazon rainforest. *Proc. Natl. Acad. Sci. USA* **2016**, *113*, 6125–6130. [CrossRef] [PubMed]
51. Prefeitura de Curitiba. *2o e 3o Inventários de Gases de Efeito Estufa para a Cidade de Curitiba*; Prefeitura Municipal de Curitiba: Curitiba, Brazil, 2016.
52. Abreu, V. *Produção de $CO_2$ em Goiânia é 28.5% acima da média nacional*; O Popular: Goiânia, Brazil, 2017.
53. Prefeitura de Salvador. *Inventário de Emissões dos Gases do Efeito Estufa de Salvador*; Secretaria Cidade Sustentável: Salvador, Brazil, 2016.
54. Instituto Ekos Brasil & Geolock Consultoria e Engenharia Ambiental. *Inventário de Emissões e Remoções Antrópicas de Gases de Efeito Estufa do Município de São Paulo de 2003 a 2009 com Atualização para 2010 e 2011 nos Setores Energia e Resíduos*; ANTP: São Paulo, Brazil, 2013.
55. Governo do Estado da Bahia. *Primeiro Inventário de Emissões Antrópicas de Gases de Efeito Estufa do Estado da Bahia*; SEMA: Salvador, Brazil, 2010. Available online: http://www.consultaesic.cgu.gov.br/busca/dados/Lists/Pedido/Attachments/456145/RESPOSTA_PEDIDO_1%2020100915%20inventario%20emissoes%20BA.pdf (accessed on 24 August 2019).
56. Governo do Estado de São Paulo. *Comunicação Estadual. 1o Inventário de Emissões Antrópicas de Gases de Efeito Estufa Diretos e Indiretos do Estado de São Paulo*; SMA; CETESB: São Paulo, Brazil, 2011.
57. Governo do Estado do Paraná. *Inventário de Emissões Antrópicas Diretas e Indiretas de Gases de Efeito Estufa 2005–2012. Resumo Executivo*; SEMA: Curitiba, Brazil, 2015.
58. World Health Organization—WHO. *Ambient air Pollution: A Global Assessment of Exposure and Burden of Disease*; World Health Organization: Geneva, Switzerland, 2016.
59. Saldiva, P. *Vida Urbana e Saúde. Os Desafios dos Habitantes das Metrópoles*; Contexto: São Paulo, Brazil, 2018.
60. Brasil. Constituição da Republica Federativa do Brasil de 1988. Available online: http://www.planalto.gov.br/ccivil_03/Constituicao/Constituicao.htm (accessed on 24 August 2019).
61. Brasil. Lei 11.445 de 5 de Janeiro de 2007. Available online: http://www.planalto.gov.br/ccivil_03/_Ato2007-2010/2007/Lei/L11445.htm (accessed on 24 August 2019).
62. Brasil. Lei 12.305 de 2 de Agosto de 2010. Available online: http://www2.mma.gov.br/port/conama/legiabre.cfm?codlegi=636 (accessed on 24 August 2019).
63. Brasil. Lei 12.587 de 3 de Janeiro de 2012. Available online: http://www.planalto.gov.br/ccivil_03/_Ato2011-2014/2012/Lei/L12587.htm (accessed on 24 August 2019).
64. Brasil. Lei 11.124 de 16 de Junho de 2005. Available online: http://www.planalto.gov.br/ccivil_03/_Ato2004-2006/2005/Lei/L11124.htm (accessed on 24 August 2019).
65. Brasil. Lei 12.608 de 10 de Abril de 2012. Available online: http://www.planalto.gov.br/ccivil_03/_Ato2011-2014/2012/Lei/L12608.htm (accessed on 24 August 2019).
66. Brasil. Lei 8.080 de 19 de Setembro de 1990. Available online: http://www.planalto.gov.br/ccivil_03/leis/L8080.htm (accessed on 24 August 2019).
67. IBGE. *População em Áreas de Risco no Brasil*; IBGE: Rio de Janeiro, Brazil, 2018.
68. Brasil. Lei 12.187 de 29 de Dezembro de 2009. Available online: http://www.planalto.gov.br/ccivil_03/_Ato2007-2010/2009/Lei/L12187.htm (accessed on 24 August 2019).
69. Rei, F.; Gonçalves, A.F.; de Souza, L.P. Acordo de Paris: Reflexões e Desafios para o Regime Internacional de Mudanças Climáticas. *Veredas Do Direito* **2017**, *14*, 81–99. [CrossRef]
70. Ministério do Meio Ambiente (MMA). *Plano Nacional de Adaptação à Mudança do Clima—Sumário Executivo*; Ministério do Meio Ambiente: Brasília, Brazil, 2016.

71. Ministério da Saúde. *Plano Setorial da Saúde para Mitigação e Adaptação à Mudança do Clima*; Ministério da Saúde: Brasília, Brazil, 2013.
72. Maglio, I.; Philippi, A., Jr. Sustentabilidade Ambiental e Mudanças Climáticas. In *Gestão Urbana e Sustentabilidade*; Philippi, A., Jr., e Bruna, G., Eds.; Manole: Barueri, Brazil, 2018; pp. 428–453.
73. Yigitcanlar, T.; Sabatini-Marques, J.; Lorenzi, C.; Bernardinetti, N.; Schreiner, T.; Fachinelli, A.; Wittmann, T. Towards smart Florianópolis: What does it take to transform a tourist island into an innovation capital? *Energies* **2018**, *11*, 3265. [CrossRef]
74. Sopala, R. A copyright free image by of 2019 Amazon fires. Available online: https://pixabay.com/photos/fire-forest-fire-children-fear-4429478/ (accessed on 3 September 2019).
75. Galán-Martín, A.; Pozo, C.; Azapagic, A.; Grossmann, I.E.; Mac Dowell, N.; Guillén-Gosálbez, G. Time for global action: An optimised cooperative approach towards effective climate change mitigation. *Energy Environ. Sci.* **2018**, *11*, 572–581. [CrossRef]
76. Rochedo, P.R.; Soares-Filho, B.; Schaeffer, R.; Viola, E.; Szklo, A.; Lucena, A.F.; Rathmann, R. The threat of political bargaining to climate mitigation in Brazil. *Nat. Clim. Chang.* **2018**, *8*, 695–698. [CrossRef]
77. Eduardo, V.; Franchini, M. *Brazil and Climate Change: Beyond the Amazon*; Routledge: Abingdon, UK, 2017.
78. Giannini, T.C.; Costa, W.F.; Cordeiro, G.D.; Imperatriz-Fonseca, V.L.; Saraiva, A.M.; Biesmeijer, J.; Garibaldi, L.A. Projected climate change threatens pollinators and crop production in Brazil. *PLoS ONE* **2017**, *12*, e0182274. [CrossRef] [PubMed]
79. Zheng, B.; Xu, Q.; Shen, Y. The relationship between climate change and quaternary glacial cycles on the Qinghai–Tibetan Plateau: Review and speculation. *Quat. Int.* **2002**, *97*, 93–101. [CrossRef]
80. Hartmann, B. Rethinking climate refugees and climate conflict: Rhetoric, reality and the politics of policy discourse. *J. Int. Dev. J. Dev. Stud. Assoc.* **2010**, *22*, 233–246. [CrossRef]
81. Spratt, D.; Sutton, P. *Climate Code Red: The Case for Emergency Action*; Scribe Publications: Sidney, Australia, 2008.
82. Tang, Z.; Brody, S.D.; Quinn, C.; Chang, L.; Wei, T. Moving from agenda to action: Evaluating local climate change action plans. *J. Environ. Plan. Manag.* **2010**, *53*, 41–62. [CrossRef]
83. Rosenzweig, C.; Solecki, W.; Hammer, S.A.; Mehrotra, S. Cities lead the way in climate–change action. *Nature* **2010**, *467*, 909–911. [CrossRef] [PubMed]
84. Burch, S. Transforming barriers into enablers of action on climate change: Insights from three municipal case studies in British Columbia, Canada. *Glob. Environ. Chang.* **2010**, *20*, 287–297. [CrossRef]
85. Yigitcanlar, T. *Rethinking Sustainable Development: Urban Management, Engineering, and Design*; IGI Global: Hershey, PA, USA, 2010.
86. Yigitcanlar, T. *Sustainable Urban and Regional Infrastructure Development: Technologies, Applications and Management: Technologies, Applications and Management*; IGI Global: Hershey, PA, USA, 2010.
87. Selin, H.; VanDeveer, S.D. *Changing Climates in North American Politics: Institutions, Policymaking, and Multilevel Governance*; MIT Press: Cambridge, MA, USA, 2009.
88. Shove, E. Beyond the ABC: Climate change policy and theories of social change. *Environ. Plan. A* **2010**, *42*, 1273–1285. [CrossRef]

 © 2019 by the authors. Licensee MDPI, Basel, Switzerland. This article is an open access article distributed under the terms and conditions of the Creative Commons Attribution (CC BY) license (http://creativecommons.org/licenses/by/4.0/).

*Article*

# The State of Smart Cities in China: The Case of Shenzhen

Richard Hu

Faculty of Business, Government and Law, University of Canberra, Bruce, ACT 2601, Australia; richard.hu@canberra.edu.au

Received: 2 October 2019; Accepted: 12 November 2019; Published: 17 November 2019

**Abstract:** China is at the midpoint of its urbanisation—the largest scale in human history. The recent smart city movement is influencing the discourse and practice of China's urbanisation, with numerous cities claiming to build smart cities and/or adopting some forms of smart city strategies and initiatives. A so-called 'latecomer's advantage' is being exploited to advance their pursuit for a smart city status, not only to catch up with overseas counterparts, but to overtake them and become international leaders. This local-level enthusiasm strikes a chord with the central government's strategy of building an 'innovative nation' to drive its economic transformation towards a knowledge economy. This converging central-local interest is creating a 'smart city mania' across the nation, which, however, has not received due attention in the international literature, and thus deserves critical examination and reflection to inform policy debates. To address this gap, this study investigates the state of smart cities in China, based on a case study of Shenzhen, China's fastest-growing, experimental city. Shenzhen grew from a fishing village into an international metropolis in 40 years, and has now won a nickname of 'China's Silicon Valley' or 'China's smartest city'. This study analyses the state of Chinese smart cities and the pursuit for a smart Shenzhen from the perspectives of the smart city as a concept, as an urban development paradigm, and as an urban regime, drawing upon the international smart city literature. It concludes that a technology-centric approach to smart cities in China, as illustrated by the Shenzhen case, have advanced innovation capacity and economic growth through capitalising on a 'latecomer's advantage'. However, this 'latecomer's advantage' may translate into a 'latecomer's disadvantage' for this approach's lack of institutional adaptation, and for its insufficient attention to social and environmental problems covered under the shiny economic boom. This latecomer's disadvantage is likely to impact the long-term sustainability of Chinese cities.

**Keywords:** smart cities; Shenzhen; Chinese cities; latecomer's advantage; sustainability

## 1. Introduction

The recent decades have witnessed the surge of a smart city movement, a globalised urban discourse and development paradigm, which had been initially facilitated by technological advancement, but has quickly progressed beyond a technological dimension to articulate with economic, social and environmental goals in contemporary urban governance and policy [1–4]. The smart city movement, in the forms of growing numbers of smart city programs across the world, has attracted a burgeoning body of literature on its conceptualisation, policy approaches, desired outcomes, and mythification and demythification [4–7]. These practical and intellectual efforts, coupled with the industry's pursuit of market opportunities, the media's reportage and debates, and the community's general favour for a 'smart' vision for their cities, have created 'the smart city' a new buzzword. This buzzword was diffused into China to shape the policy making—both top-down and bottom-up—of the largest urbanisation process in human history there. China, a latecomer in the smart city movement, has now the largest number of smart city programs, and has set ambitious targets of constructing smart cities and a smart society of international leadership [8]. Despite these booming smart cities in China, they are not fully engaged in the international scholarship, and there is no critical literature on the state of their development to draw insights and to inform an understanding of their trajectory.

Set against these above backgrounds, this study is centred on answering this question: to what extent do the smart cities in China conform to or differ from their international counterparts? In order to answer this question, this study comprises three major components. First, it deconstructs 'the smart city', through a review of literature, as a concept, as an urban development paradigm, and as an urban regime, so as to establish an analytical framework for this study. Second, it applies the analytical framework to provide an overview on the state of smart cities in China. Third, it further applies the analytical framework to investigate Shenzhen, 'China's Silicon Valley' or 'China's smartest city' [9]. Discussing these findings sheds light on the characteristics that are unique to Chinese smart cities, and the latecomer's advantage and disadvantage associated the Chinese approach to smart cities.

This article reports this study and its findings. This introductory section sets the contexts, asks the research question and introduces the organisation of the article. Section 2 is a literature review to deconstruct the smart city. Section 3 explains the methods used. Sections 4 and 5 present findings on an overview of Chinese smart cities and the case of smart Shenzhen, respectively. Section 6 discusses the characteristics of Chinese smart cities. It further discusses the latecomer's advantage that has enabled these Chinese characteristics, and critically points out the inherited disadvantage in the long run. Section 7 concludes the article and suggests limitations of this study and thematic strands of further research.

## 2. Deconstructing the Smart City

This literature review approaches the smart city from three perspectives. First, it approaches the conceptualisation of smart city to clarify what the concept *is*, or *should be*, so as to capture the core of the concept and to differentiate an external layer of conceptualisation attached to this core. Second, it situates the understanding of the smart city movement in a longitudinal perspective of urban development in the recent decades, shifting from urban competitiveness, to urban sustainability, and to urban smartness. Third, it identifies a smart city coalition of key interest groups and stakeholders involved in the making and promotion of the smart city movement, and the rationale behind, which have formed a new urban regime. These three perspectives forge an integrative framework for approaching the smart city as a concept, as an urban development paradigm, and as an urban regime, to underpin this study's examination of Chinese smart cities and Shenzhen. Each of the three perspectives is explicated below.

### 2.1. Conceptual Decoupling

Smart city is the new buzzword in urban scholarship, policy and practice. A vast body of smart city literature is burgeoning in the recent decades, contributing to as well as riding on the concept's popularity, and its 'fuzziness' [6]. This body of literature helps to spread, or 'brand', the concept generally, but is limited in clarifying what the concept is about, and what it is not about. In policy making, the smart city means very different things to different cities, varying by local economic, geographic and technological settings [10]. Several systematic and comprehensive literature reviews converge on an understanding that the smart city concept is multi-faceted, complex, confusing, ambiguous, contingent, fragmented, and lacks cohesion [2–7]. This conceptual blurring, while likely to trigger more and further literature debates, is impacting a shared understanding and a consensus on theory building among researchers. Furthermore, this will be likely to impact the effectiveness of the concept's applicability in making policies and shaping contemporary urban development, and to confuse the general readership.

The literature both diverges and converges, however, on defining the smart city. They diverge on delineating the conceptual scope, and on the extent to which the smart city concept can be expanded to incorporate contemporary urban or planetary challenges, in particular those wicked problems of climate change, urbanisation, quality of life, and social polarisation. There have been efforts to integrate the smart city thesis with the sustainable city thesis to explore possible synergies for sustainable smart cities. However, a large gap still exists between the smart city and the sustainable city frameworks: while the smart city performance measurements focus on the efficiency of smart solutions, they should also include ultimate goals of environmental, economic and social sustainability [1]. This gap is empirically attested that city smartness and environmental performances are not positively correlated [11]. A sociotechnical framework highlights the interactions between the social and the technological dimensions of smart cities that incorporate not only technological and symbolic value, but also institutional collaboration and instrumental value, to achieve successful technological innovation [12]. Along this sociotechnical line, smart people and smart economy are incorporated into innovation ecosystems for smart cities [13]. These theoretical propositions and associated empirical studies make contributions through situating the smart city movement in the broader urban agenda for research and policy thinking. However, they tend to fall into a pitfall of creating the smart city an umbrella term, diluting its conceptual core.

On the other hand, the literature converges on a conceptual core of the smart city—technology, in the context of the rise of the knowledge economy. Technological advancement and the knowledge economy have been mutually facilitating in the recent decades, and have interacted to position innovation onto an unprecedented, important place in urban development, especially in economic development [14]. In spite of the diversity of directions to align the smart city to, technology remains in the core to be the defining factor of the smart city. Then, the applications of the technology in urban infrastructure and services to achieve desired outcomes in economic, social and environmental dimensions constitute an outer layer of the smart city conceptualisation. The range of these desired outcomes for the smart city may include productivity, sustainability, wellbeing, liveability and governance, while technology is one of the major drivers [4]. Technology has always been a major shaping force of urbanisation and cities [15]. However, the most salient attribute of the recent technology development is the exponential advancement of information technology from the 1990s, along with the accelerating development of the knowledge economy [16]. The major technological enabler of the smart city movement is the ubiquitous and instant access to information and information processing capability [17], in such forms as big data and internet-of-things (IoT).

Combining the diverging and converging literature points out a two-layer, in a broad sense, conceptualisation of the smart city: an internal layer of information technological advancement combined with the knowledge economy; and an external layer of articulating the smart city with various urban concepts and challenges, such as sustainability, climate change, and social inclusion. These two layers of conceptualisation encompass the major debates on the smart city concepts, applications, and urban policy goals, as captured in the several major systematic literature reviews [2,4,5,7].

*2.2. Urban Smartness Paradigm*

The recent smart city movement did not emerge out of context or abruptly. It has evolved along the paradigm shifts in urban development from the late 1970s, which can be broadly classified into competitive city, sustainable city, and smart city (Table 1). The conceptual division and the temporal sequence of these three paradigms are not as clearly cut as they are illustrated in Table 1, which serves an illustrative purpose only. These paradigms have been evolutionary, along a generally linear path mixed with cyclicality, in the recent urban development history. They each have defining attributes, but are often interlinked in conceptualisation and informing urban policy and planning. The sustainable turn—a shift from urban competitiveness to urban sustainability—in urban development approaches and policy goals has been driven by an imperative to address the intricate environmental and social problems in cities, which have been attributed, in part at least, to a focus on economic growth only in urban policy [14]. Urban competitiveness, while still a paramount policy goal in urban development, has tended to employ an integrative approach to incorporate non-economic dimensions such as environmental protection and social inclusion [18]. This trend has even led to a re-conceptualisation of urban competitiveness as urban sustainability, that is, sustainability makes a city's competitiveness [19]. Fusing sustainability and competitiveness, which are intrinsically contradictory in some way, has been a major policy aspiration as well as policy challenge in the strategies for many global cities [20,21]. Sustainability has become a new competitive edge, which is marketable and brandable, in a still neoliberalism-dominated urban policy discourse [14].

Table 1. Paradigm shifts in urban development.

| Paradigm Shifts | Timeframe | Driving Factors | Policy Priorities | Defining Features |
|---|---|---|---|---|
| Competitive city | 1970s- | Globalisation; Neoliberalism | Urban competitiveness; Economic growth | Economic-centrism |
| Sustainable city | 1990s- | Environmental challenges; Social problems | Urban sustainability; Balanced development with environmental protection and social equity | Environmentalism |
| Smart city | 2010s- | Technological disruption; Knowledge economy | Urban smartness; Innovation | Technology-enabling |

Source: Partially adapted from [14].

The smart city has come as one of the several models of urban development in the context of the new economy, which 'is directly based on the production, distribution, and use of knowledge and technology' [14] (p. 317). Each of these models—knowledge city, creative city, and smart city—has attracted tremendous attention in scholarship and practice. While overlapping to various degrees, they 'focus on different aspects of the new economy ... and each has its limitations' [14] (p. 317). The smart city, as discussed above, is essentially enabled by disruptive technology, which defines the internal layer of its conceptualisation. However, in the external layer of its conceptualisation, the smart city is often articulated into dialogues with the preceding urban development paradigms to achieve competitive city or sustainable city goals. These conceptual articulations between the major urban development paradigms create mixed outcomes even contradictions. For example, there is a mismatch between sustainable city and smart city in policies: while sustainability city policies fully take into account technology, sustainability is insufficiently addressed in smart city initiatives [1]. Empirically, smart cities do not always achieve environmental sustainability outcomes [11]. While it will take more practice and research, and the test of time, to clarify the articulation of smart city with a comprehensive set of policy goals, differentiation of the two layers of the smart city's conceptualisation and identification of a consensus-based internal conceptual core helps to focus on the concept that is technology-enabled and contextualised in the knowledge economy. This conceptual decoupling, as discussed above, also informs an understanding of how the smart city fits into the urban development paradigms.

### 2.3. The Smart City Regime

A politic-economic perspective helps to unpack the smart city coalition that has been acting to push the surge of the smart city movement, and thus has formed an urban regime. The smart city movement is a recent paradigm of the neoliberal urbanism that has been in place since the 1970s, proceeding along the trajectory of competitive city, sustainability city, and now smart city (Table 1). The smart city movement, as a form of neoliberal urbanism, joined the urban imaginaries of sustainable development, smart growth, and new urbanism—a new consensus on incorporating economic prosperity, ecological integrity and social equity into urban sustainability—in an era of market triumphalism [22]. Such neoliberal urbanism concepts as smart growth, new regionalism, new urbanism and sustainable development are all put under an umbrella smart city agenda, with a focus on the notion of 'smartness', to address contemporary urban imaginaries of competitiveness and sustainability [20,22]. 'The power of the smart city imaginary to capture the minds of corporations, policymakers and average citizens makes it an important means through which cities are being (re)constructed in the 21st century' [23] (p. 22).

The smart city movement is often criticised for a technology-centrism to argue that a corporate-led smart urbanism should shift from being technology-intensive to being knowledge-intensive, governed by a more socially just use of digital technology [24]. However, this normative proposition is often used to disguise its neoliberal nature. The smart city acts to 'sell' a city in the global economy, and masks entrepreneurial governance and strategies, oriented to a utility to foster globalised business enterprises and further economic development [25,26]. In understanding the formation of such a smart city coalition, a focus needs to be placed on the relationalities through which the smart city idea has taken root in territories [23]. Global technology firms—IBM, Siemens and Cisco—have constructed a market for cities, as a scalable community, for their knowledge through reducing, standardising and simplifying urban problems for the sale of their proprietary software and hardware, and consultancy services [27]. However, the global technology firms' agency in driving the smart city discourse is overemphasised in critical literature; city governments are key actors advancing the smart city paradigm, in a rhetoric of city-wide benefits, but geared to attracting businesses in a globalised economy [25]. The relationalities between the governments and the global technology firms have forged the leadership of the smart city coalition and the functioning of the urban regime, which has right fitted into the neoliberal urbanism contexts for entrepreneurial governance pursuing the urban imaginaries of competitiveness, sustainability and smartness that have been shaping urban development paradigms in the recent decades (Table 1).

## 3. Methods

The above literature review has deconstructed the smart city phenomenon as a concept, as an urban development paradigm, and as an urban regime. These perspectives establish an analytical framework for approaching the smart city: the smart city is centred on technological advancement and the knowledge economy; it is articulated into contemporary urban development paradigms comprising urban competitiveness, sustainability and smartness; and it is led and advocated by a neoliberal urban regime pursuing new forms of innovation-led growth in an increasingly competitive global knowledge economy. This study employs this analytical framework to investigate the Shenzhen case study to illustrate the state of smart cities in China, and to test to what extent smart cities in China, as observed in the Shenzhen experience, conform with or differ from it.

This study was undertaken through two major phases. In phase I, it synthesised an overview of the smart city movement in China. This phase I included three steps: step 1—collecting urbanisation data from the World Bank and economic composition data from the National Bureau of Statistics of China, both in 1978–2018, to analyse China's economic and urban transformations; step 2—analysing three milestone central government policies in 2012, 2014 and 2017, respectively, which have outlined China's smart city strategy, to identify thematic evolutions; step 3—drawing upon data from several consultancy reports on Chinese smart cities in general, and the latest smart city vision for Xiong'an, China's newest city making, to map out the smart city movement from bottom-up.

In Phase II, this study moved on to unpack smart Shenzhen. This phase II included three steps: step 1—collecting the 1980–2017 time-series data on Shenzhen's economic composition and hi-tech sectors from the Shenzhen Statistical Bureau, to analyse the city's economic transformation towards a knowledge economy and its rise as a global knowledge city; Step 2—making a content analysis of Shenzhen's master plan in 2010, and smart city plan in 2018, to present transformative planning thinking; Step 3—a fieldtrip by the author in 19–25 September 2019 to Shenzhen to observe its smart cityscapes and engage local residents to have their perceptions of their city, so as to obtain first-hand experience. The author also visited Shenzhen's neighbouring city Hong Kong to draw some experiential knowledge on the comparisons of these two cities. These multiple sources of data and information were triangulated and synthesised, through the smart city analytical framework established through the above literature review.

The selection of Shenzhen as a case study for smart cities in China is based upon the city's unique status in the Chinese urban system. Shenzhen was largely a rural area 40 years ago, and was designated as an experimental city to spearhead China's modernisation agenda of 'reform and opening-up' [28]. Shenzhen's rapid growth into an international metropolis—population growth by 40-fold, employment growth by 68-fold, and gross domestic product (GDP) growth by 11,452-fold, in 1979–2017—is most representative of China's urbanisation and economic recovery [28]. Shenzhen is dubbed as 'China's Silicon Valley' for its global innovation hub status. The tech giant Huawei, now the world's leader of 5G technology, is based in Shenzhen, forging a world-class hi-tech cluster together with other hi-tech firms, large and small. Partly because of the presence of these hi-tech firms and cluster, and partly because of its status as a new planned city, the Shenzhen Government is proactive in embracing the smart city concept and branding it as a leading smart city in China. Shenzhen is ranked as China's top 1 smart city, beating Shanghai and Beijing [29]. In this study, the case of Shenzhen provides a prism to interrogate the state of smart cities in China.

## 4. The Smart City Movement in China

Deloitte [29] estimated that there were over 1000 smart city projects worldwide, and half of them were in China as of 2016. These figures are indicative of the spreading smart cities—claimed, planned, or being constructed—since the actual, accurate numbers are hard to collect, depending on how the concept is defined and what criteria are used to label a smart city. It is also reported that more than 700 Chinese cities have proposed or claimed to construct smart cities in government reports or development strategies as of 2019 [30]. Europe, North America, Japan and South Korea have been leading regions in smart city development, but China, a latecomer, is catching up and is leading in terms of numbers of smart cities [29]. In the recent decade, a smart city movement has been emerging across China, riding on the international trend, but at a faster speed. This section below will elaborate on the coming of this Chinese smart city movement in the context of China's transformative economic growth and urbanisation, driven by a national strategy, and pursued by a coalition comprising entrepreneurial governments and market-sought technology and service firms. To further expose the popularity of and zeal for the smart city in China, this section will then illustrate how this concept has shaped the imaginary of Xiong'an, China's latest grand new city being planned right now.

The smart city movement has emerged when China's economic growth and urbanisation entered a transformative stage since around 2010. China's modernisation agenda of 'reform and opening-up', led by Deng Xiaoping, commenced in 1978 when it was a rural society in grave poverty. Since then, China has created the second largest economy in the world and the largest urbanisation process in human history. However, these achievements did not come without problems, among which pollution, resource consumption and environmental degradation have been hallmarks of China's urban revolution [31]. Discussion on changing the nation's development path, which was deemed as unsustainable, started from the mid-1990s. However, this discussion did not translate into effective policy making and implementation until the 2010s when the tertiary sectors of services exceeded the secondary sectors of industries in China's GDP composition, and China's urbanisation rate surpassed 50% (Figure 1). China's transformation into an urban society of the knowledge economy has propelled a policy discourse on pursuing an innovative nation and a new-type urbanisation with sustainability, and even a so-called 'ecological civilisation', to shift the nation's rapid development from quantitative growth to qualitative upgrade [31]. The smart city, an imported concept, has proven to be a right fit with this policy discourse marked by innovation, sustainability and the knowledge economy.

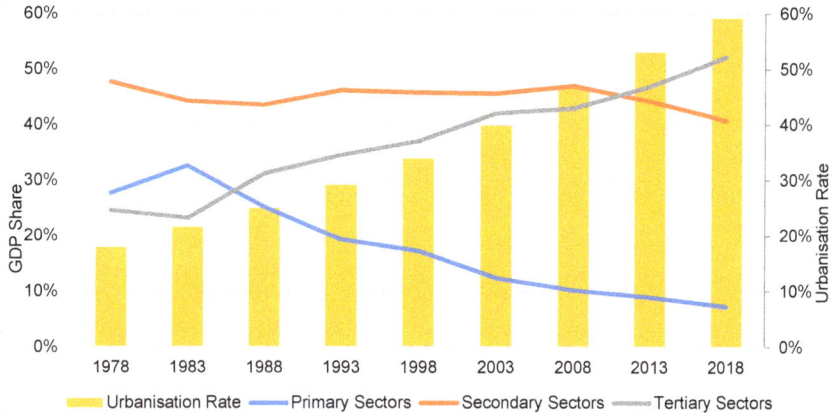

**Figure 1.** China's GDP composition and urbanisation rate, 1978–2018. Source: GDP data—National Bureau of Statistics of China [32]; urbanisation data—The World Bank [33], created by the author.

The smart city movement has been institutionalised into a national strategy by several top-down central government policies, in 2012, 2014 and 2017, respectively. In November 2012, the Ministry of Housing and Urban-Rural Development (MOHURD) released *A Notice on Conducting National Pilot Smart Cities*. This notice defined the smart city as a new model of strengthening urban planning, construction, and management through comprehensively utilising modern science and technology to integrate information resources and to coordinate application systems; it also regarded the development of smart cities as a key measure to achieve the national goals of innovation-driven development, new-type urbanisation, and a comprehensive well-off society [34]. This notice attached a set of pilot indicators for smart cities, comprising four major dimensions—support system and infrastructure, smart construction and liveability, smart management and services, and smart industry and economy [34]. To apply for becoming 'national pilot smart cities', local cities should have in place an economic and social development plan incorporating the smart city, a completed smart city development outline, secured funding source (e.g., government budget), and leadership arrangement. Consequently, three groups of pilot smart city programs were announced in 2013–2015, amounting to a total of 277, with 112 in the eastern coastal region, 91 in the middle region, and 74 in the western region [35]. In 2014, the State Council released the *National New-Type Urbanisation Strategy (2014–2020)*, China's first national urban plan, to re-orient its urbanisation process at a critical time of its economic and urban transformations as discussed above. This plan contained one section on the smart city, outlining dimensions of construction in digital networks, planning management, urban infrastructure, public services, hi-tech industry and social governance [36]. In his report *Secure a Decisive Victory in Building a Moderately Prosperous Society in All Aspects and Strive for the Great Success of Socialism with Chinese Characteristics for a New Era* addressing the 19th Communist Party of China's National Congress, the most important platform for making national strategic guidelines, in October 2017, Chinese President Xi Jinping used the term 'smart society' in his elaboration on 'making China a country of innovators' [37]. The smart society concept represents a conceptual expansion of and thus a more ambitious aspiration than the smart city, as a national strategy. In the official propaganda, 'the smart society' is described as a theoretical innovation, a developed Chinese version of 'the smart city', and a new way of China's development in a 'new era' headed by Xi Jinping [8].

The mushrooming smart cities in China, advocated by the state, presents unusually lucrative market opportunities for technology and consultancy firms. Behind the booming smart city movement is a coalition of the public sector and the private sector—the entrepreneurial governments seeking new ways of economic and urban development and the global technology firms seeking market profits—join forces in capitalising on the smart city. It is estimated that the market value for building the Chinese smart cities increased from RMB 740 billion in 2014 to RMB 10,500 billion in 2019, and is forecast to reach RMB 25,000 in 2022 [30]. This booming market has attracted the established global firms such as IBM and Cisco, and those locally-grown firms that have rapidly developed to achieve global competitiveness and reputation. Several China-based hi-tech giants, including Huawei, Baidu, Alibaba, and Tencent, are more advantaged than the overseas competitor firms in securing the smart city opportunities. It is reported that these Chinese technology firms have signed strategic collaboration frameworks with 300 Chinese cities to construct smart cities [38]. Business consultancy firms and lobby organisations have been no less enthusiastic in capturing the smart city market opportunities [39]. They publish reports, organise events, and engage governments and the industry to promote their services. For example, Deloitte [29] repackaged a term 'super smart cities' and built a 'China super smart cities index' to seek consultancy service opportunities. On Deloitte's website, it has a service section exclusively on strategic planning for smart cities in China. Numerous start-up firms and websites have also been created to provide services on smart cities in recent years. These actors, both public and private, have pushed the smart city movement in China, and made it a trendy policy agenda for Chinese cities.

Xiong'an represents the latest smart city imaginary in China. Located in Beijing-Tianjin-Hebei mega-city region in north China, Xiong'an, a semi-urban area with a vast body of water and rural land, is 120 km from Beijing and 110 km from Tianjin. On 1 April 2017, the State Council announced the establishment of Xiong'an New Area, a new city to be planned and developed. An urban project to alleviate the increasing crowdedness problems in Beijing and to better integrate Beijing-Tianjin-Hebei mega-city region's development, Xiong'an is also a political project of Xi Jinping who wishes to imprint his name in a new city, just like Deng Xiaoping who orchestrated the creation of Shenzhen and Pudong, Shanghai from scratch [40]. Labelled as a 'millennium plan' and a 'national strategy', Xiong'an is expected to create a new Chinese model of a 'green, smart city' [41]. Xiong'an is a grand plan to be implemented in the coming decades to become a new city of five million population by the mid-21st century. Its 'green, smart city' vision, however, captures the lynchpin of the Chinese smart city movement, and represents a re-orientation, through fusing the aspirations of urban sustainability and urban smartness, of the nation's four decades of rapid urbanisation.

## 5. Smart Shenzhen: Imaginary and Reality

Smart Shenzhen is a new concept but has achieved instant popularity and importance in local policy and daily discourse in recent years, along with the nation-wide smart city movement. This section explains the emergence of the smart Shenzhen discourse and unpacks its nature, employing the analytical framework developed in Section 2. It situates the understanding of smart Shenzhen along the trajectories of the city's economic transformation towards a knowledge economy and the city's planning transformation towards a prioritisation of sustainability and innovation. These economic and planning transformations have paved the way for the Shenzhen's smart city plan in 2018, which formally set 'smart Shenzhen' as a strategic goal for the city's future development.

In 1980, when Shenzhen was first planned as a 'special economic zone' (SEZ) to test and develop experience for China's modernisation agenda of 'reform and opening-up', it was largely a rural area of agriculture and fishing—the primary sectors in China's economic classifications. The city was selected as an SEZ for its proximity to Hong Kong, so as to catch the spill-overs of Hong Kong's investment, knowhow, industries, and international trade. Until the beginning of the 21st century, Shenzhen's economy, like the national economy, had been industrialisation-led and foreign-oriented through attracting foreign direct investment and exporting manufactured products. As illustrated in Figure 2, the secondary sectors of industrial productions had a steady growth in Shenzhen's GDP composition from the 1980s on and reached the climax of 54% in 2005. The primary sectors' share in GDP sharply declined in the early 1980s, and became negligible in the mid-1990s. The year 2005 is a dividing line in Shenzhen's economic structure: from then on, the GDP share of the secondary sectors started to decline; meanwhile, that of the tertiary sectors kept growing, and surpassed that of the secondary sectors in 2008 to reach 58% in 2017 (Figure 2). These statistics indicate that Shenzhen started its economic transformation from an industrial base to a knowledge base from the turn of the century, and established its knowledge economy in the second decade of the 21st century.

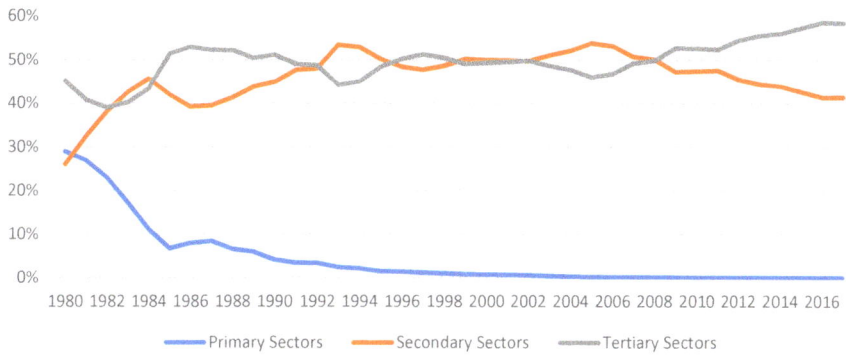

**Figure 2.** Shenzhen's GDP composition, 1980–2017. Source: Shenzhen Statistical Bureau [42], created by the author.

To further illustrate the establishment of the knowledge economy in Shenzhen in the 2010s, Table 2 presents the growth measured by several indicators. These selected indicators reflect the city's progress in hi-tech industries, which have won it the nickname of 'China's Silicon Valley'. These indicators may be broadly categorised into outputs and inputs of hi-tech development: outputs—added value of new industries, international trade value of hi-tech industries and patents certified; inputs—research and development (R&D) expenditure and R&D personnel. They all experienced massive growth to various degrees in 2010–2017, and their growths are clearly interrelated (Table 2). The impressive growth by 226% for added value of new industries in 2010–2017 contributed to the city's economic transformation towards an established knowledge economy, for which an increase of 193% for R&D expenditure was also an explanatory factor in the same period. China's robust investment in R&D in the recent decade has put the nation in an increasingly advantageous position in the global innovation race [19]. Shenzhen is a leader in investing in innovation-led development among the Chinese cities.

**Table 2.** Growth of the knowledge economy in Shenzhen, 2010–2017.

| Years | Added Value of New Industries (RMB Million) | International Trade Value of Hi-Tech Industries (USD 10,000) | Patents Certified | R&D Expenditure (RMB 10,000) | R&D Personnel (Persons) |
| --- | --- | --- | --- | --- | --- |
| 2010 | 282,051 | 19,770,075 | 34,951 | 3,333,102 | 177,756 |
| 2011 | 334,134 | 22,416,000 | 39,363 | 4,161,363 | 176,107 |
| 2012 | 398,244 | 25,206,532 | 48,662 | 4,883,738 | 218,090 |
| 2013 | 513,777 | 30,784,842 | 49,756 | 5,846,115 | 213,641 |
| 2014 | 585,595 | 24,762,288 | 53,687 | 6,400,662 | 192,600 |
| 2015 | 720,540 | 25,424,844 | 72,120 | 7,323,851 | 206,327 |
| 2016 | 809,167 | 22,764,476 | 75,043 | 8,429,693 | 233,927 |
| 2017 | 918,719 | 22,775,570 | 94,250 | 9,769,377 | 281,369 |
| Change (2010–2017) | 226% | 15% | 170% | 193% | 58% |

Note: New industries include broad hi-tech industries and cultural and creative industries. Source: Shenzhen Statistical Bureau [42], created by the author.

Shenzhen's economic transformation has influenced, as well as has been influenced by, the city's planning transformation. A dozen of plans and strategies have ever been made to guide and to respond to the city's rapid development. Of them, three master plans in 1986, 1996 and 2010 are the most important for their strategic and statutory status in the Chinese planning system. The latest 2010 plan is a 'transformative plan' since it was a planning effort to 'transform' the city's development approach and planning direction that had been in place since the city's genesis [28]. This transformative planning thinking has been influencing the city's transformation in many dimensions including the above-discussed economic transformation in the recent decade. This plan's making was built upon a critical reflection on the city's rapid, expansive development which had generated environmental degradation, resource and land waste, and social disparity; and was thus deemed as 'unsustainable' [28]. The 2010 plan identified four challenges—constraints of fundamental resources, structural contradiction in urban development, fragility of social development model, and low efficiently in strategic space uses—confronting the city, and sought a transformative and sustainable pathway to achieve an integration of economic and social development with environmental protection [43].

The 2010 plan's sustainable development goal marked a fundamental departure from the city's previous development paradigm that had followed an industrialisation-led growth trajectory. Instead, it called for economic transformation through (1) strengthening the pivotal industries of hi-tech, finance, logistics, and culture; and developing new industries; (2) increasing the share of R&D expenditure in GDP, and enhancing the local enterprises' capacity of indigenous innovation; (3) innovation-led upgrade of traditional industries; and (4) developing circular economy and green industries [43]. It is hard to conclude a cause-and-effect relationship between these economic transformation policies and Shenzhen's growth in the knowledge economy as indicated in Table 2. However, they demonstrate that the city's economic transformation and planning transformation have been converging to the same direction—an international metropolis of the knowledge economy and innovation. The 2010 plan has contextualised the development paradigms and planning goals for a smart city, as we have often observed elsewhere, without using the term 'smart city', which did not appear in the local and broader Chinese planning discourses in 2006–2010 when the plan was made.

Smart Shenzhen is the latest urban imaginary under the afore-mentioned contexts of economic transformation and planning transformation. Pingshan New District, a local administrative unit in Shenzhen, was listed among the first group of China's pilot smart city programs in 2013. In the recent years, the smart city has taken a prominent position in the local discourse of innovation-led urban development, and Shenzhen is generally regarded as a leader of the China's smart city movement, or the smartest Chinese city, in various smart city rankings, media reportage, and local government's branding. Under these atmosphere and aspiration for a leading smart city, the Shenzhen Government promulgated the *Shenzhen Municipal New-Type Smart City Construction Master Plan* on 30 July 2018. This smart city plan, as admitted in the document, responded to the national strategy of 'a smart society', and aimed to enable Shenzhen's new vision of 'a modern, international, and innovative city' [44]. The plan offered a 'city-wide', 'integrative' approach to the smart city, with a focus on enhancing 'services for people's livelihood and urban governance capacity' [44], two popular terms in the Chinese urban policy discourse in the recent decade. The plan presented a smart city structure, including platforms, systems, domains, operations, and supports, as outlined in Figure 3. This smart city structure for Shenzhen essentially reflects the latest perception of, and approach to, the smart city in China, given Shenzhen's leading position in constructing a smart city among Chinese cities.

Shenzhen has two giant telecom firms—Huawei and Tencent—based in the city. These two firms are playing pivotal roles in advancing the smart city movement in China, and are spreading their global business outreach aggressively. Huawei's hardware and Tencent's software put their home city in an edgy position of constructing the smart city, nationally and internationally. At an operational level, what is outlined in the smart city structure (Figure 3) is much materialised in Shenzhen. In 2016, Shenzhen became the first Chinese city to propose a 'gigaband city', through the deployment of next-generation networks by Huawei in collaboration with telecom service provider China Telecom, to deliver 100% gigabit coverage for communities [45]. Huawei's 'smart city' project is seeking to make its home city 'smarter, safer and more efficient', and is also being experimented in more than 160 cities across 40 countries [46]. In the public domain of Shenzhen today, 'the smart city' permeates, symbolically and operationally (Figure 4). The use of smart technology in Shenzhen has seen tangible results of improvement in areas of public security, telemedicine and transport [47]. Shenzhen, and other Chinese cities, has the most widespread use of smartphones in daily transactions, information access, and mobility, nearly rendering cash and bankcards obsolete. The 'city-wide' deployment of sensors and cameras, in the public spaces especially, has the whole city almost under monitoring and surveillance (Figure 4). Data from these multiple channels are collected, centralised, and integrated into data sharing to inform urban management, planning, and individual decisions. Ping An Financial Centre, the city's highest tower (599 m), was completed in 2016 to claim Shenzhen's rise as an international metropolis. The Ping An Smart City Operations Command Centre established within the tower showcases Shenzhen's utilisation of big data to inform urban management and decision-making in multiple dimensions of governance, transport, safety, and social, economic and environmental development (Figure 5). Shenzhen's application of new technologies in urban infrastructure and operation is more advanced than its neighbouring city across the border—Hong Kong, as acknowledged by residents in both cities. Shenzhen and Hong Kong are virtually one conurbation but are administratively separated under the governance arrangement of 'one country two systems'. Shenzhen was a new city created for its proximity to Hong Kong. Now, Shenzhen is catching up and even overtaking Hong Kong, an established leading global city, in terms not only of economic power—Shenzhen's GDP surpassed Hong Kong's in 2018 [48]—but also of urban smartness. A latecomer's advantage seems to be testified by the Shenzhen–Hong Kong relationality and comparison, purely measured by economic growth and observed through the smart city operationalisation.

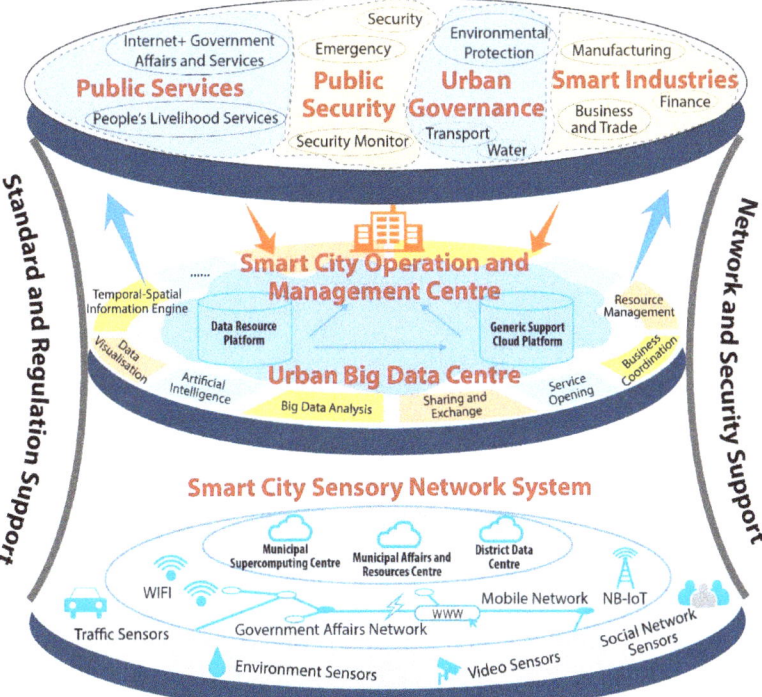

**Figure 3.** Shenzhen's smart city structure. Source: Shenzhen Government [44], recreated by the author.

**Figure 4.** Smart Shenzhen in public spaces. Note: 1—a resident riding a sharing bike passing by an E-police station; 2—surveillance camera in the civic plaza; 3—a smart community stand supported by Chinese hi-tech giant Alibaba; 4—public display of commercial credit information. Source: The author's photography.

**Figure 5.** Ping An smart city operations command centre. Source: The author's photography.

## 6. Smart Cities with Chinese Characteristics?

This section discusses the uniqueness, if there is any, of the smart cities in China, as illustrated by the case of Shenzhen. It is discussed from the three perspectives drawn from the literature review in Section 2—smart city as a concept, as an urban development paradigm, and as an urban regime—to reveal to what extent the Chinese smart cities conform to the international smart city framework, or they have emerged following a pathway with Chinese characteristics. This section finally discusses the latecomer's advantage and disadvantage of Chinese smart cities: a latecomer's advantage in capitalising on technological advancement and economic growth in a short term may turn into a long-term disadvantage if institutional adaptation and sustainable concerns in social and environmental dimensions are not in place in China's current smart city movement. Each of these points are discussed below.

First, the conceptualisation of smart cities in China is strongly technology-centric. It builds, almost exclusively, upon the latest advancement in the information technology, such as IoT, big data, 5G technology, artificial intelligence, and cloud computing, to explore their utility in urban data collection, analysis, and sharing to enable more efficient urban management and services. This technology-centrism, despite its 'people-oriented' rhetoric [49], exactly conforms to the internal layer of the smart city conceptualisation discussed in Section 2, and has underscored the Chinese smart city discourse, including smart city policies and initiatives at both central government and local government levels. The smart city movement has arisen against the backdrop of China's rapid urbanisation and its multiple dimensions of progress and problems—economic growth, environmental pressure, social polarisation and urban governance challenge, which are awaiting smart solutions [50]. However, the smart city movement hardly goes beyond the technology-centrism to concern those non-technological dimensions, which constitute the external layer of the smart city conceptualisation, except for an innovation-led economic transformation towards a knowledge economy. To sum up, the Chinese smart city captures the internal layer but does not fully embrace the external layer, of the conceptualisation of the smart city as discussed in Section 2. Consequently, the smart city policies and initiatives in China have focused on the instrumentality of the smart city to achieve efficiency and effectiveness in urban management and economic upgrade. It has not established a direct dialogue with the sustainability challenges confronting Chinese cities. This technology-centrism links the Chinese smart city movement with, as well as differentiates it from, the international smart city movement that has attempted to cross-fertilise urban smartness and urban sustainability [1]. Furthermore, a narrow focus on technology and infrastructure—the hardware of smart cities—has been made at the expense of involving different stakeholders and public participation into an inclusive urbanisation [51,52].

Second, the smart city movement right fits in with the urban development paradigm being pursued in China. The Chinese cities are at a critical moment of transitioning from an industrial economy to a knowledge economy, and from quantitative expansion to qualitative upgrade, after four decades of rapid economic growth and urbanisation. While 'innovation' became the buzzword in policy making at all levels of government in order to build an 'innovative nation' or 'innovative city', the smart city term was immediately incorporated into the 'innovation' discourse after it was imported from overseas through 'urban policy mobilities' [53]. The smart city was even conceptually broadened as 'the smart society' and was endorsed as a national strategy. However, aligned to the first point discussed on the technology-centrism, the Chinese smart city does not necessarily translate into a new urban development paradigm itself. Rather, the smart city offers a new urban imaginary about enabling innovation capacity and enhancing edgy competitiveness for Chinese cities and their economy in the inter-city competitions and international competitions. Urban smartness is contributory to urban competitiveness, and is thus brought onto a neoliberal urbanism pathway that the Chinese cities have followed for four decades. New technologies and industries are encouraged, alongside innovations and entrepreneurship, to achieve desired outcomes of economic growth and better livelihood [52]. In neoliberalising the smart city as an urban development approach, the Chinese cities are not significantly different from cities elsewhere; they are even more explicit and focused on promoting the smart city as a mode of urban development and growth.

Third, the urban regime for the Chinese smart cities has probably the most imprinted Chinese characteristic. The smart city initiatives in China were driven not only by technological rationalities, but also by political rationalities [54]. The political regime's formation combines top-down and bottom-up interactions: governments at both central and local levels converge on accepting and advocating smart cities, even creating a 'smart city mania'. This shared and concerted inter-governmental understanding and action do not apply to the smart city movement only; they apply to almost every major public initiative that is of the central government's favour and is then immediately spread across the nation. China's unitary, centralised political system has enabled the formation of this urban regime, and explains its difference from the Western nations. The central government directly intervened in the smart cities through setting a national strategy and participating in policies and programs for them. At the local level, the smart city policies do not significantly stray from the top-down, outcome-oriented line of the national approach' [52] (p. 5). This interventionist approach by the central government is in stark contrast to the many smart cities in the West where local city governments have taken the lead and the central government's role is obscure or absent. In the developed countries, Australia is an unusual case in that the federal government has pursued the smart city as a national urban policy since 2016, but it has insisted on a coordinative role for itself while the operation and implementation rest with stakeholders—both public and private—at local levels [55].

The technology firms are also key actors in the Chinese smart city regime. Those China-based giant firms (e.g., Huawei, Baidu, Alibaba, Tencent), which have quickly achieved international competitiveness, are ready smart city services providers at home and overseas [56,57]. They are often favoured by the Chinese governments in deploying smart cities projects under a policy culture of growing 'indigenous innovation'. The smart city movement is also favourably received by the city residents, who have enjoyed a new way of working and living facilitated by the smart technologies, even though their role of participating or being included in its making is limited [51]. The residents often take pride in the rapid progress of their cities, especially in the technology-enabled convenience, efficiency and safety, despite a common concern of losing privacy and being under surveillance, which is deemed as a potential, if not imminent, menace. The Chinese smart city regime, like those elsewhere, also comprises the government, the technology firms, and the community, but the government prevails over the other parties in shaping the regime and determining how it functions. At an operational level, however, the mode of financing and operation is becoming more diverse, and is likely to become more market-oriented, while the government's role will focus on setting standardisation, and making laws and plans [58]. Some Chinese pilot smart cities have indicated citizen engagement mechanisms, although they are designed and implemented in less sophisticated and effective forms than their counterparts, such as in European Union (EU) [10].

Fourth, the latecomer's advantage in enabling the Chinese smart cities seems to embody a latecomer's disadvantage in the long run. As discussed above, the Chinese smart cities have demonstrated a strong technology-centrism to advance innovation capacity and economic transformation for the cities and the nation through top-down and bottom-up consensus and collaboration. Behind this 'Chinese' approach is a logic of so-called 'latecomer's advantage' which was first used to justify the nation's modernisation which started only 40 years ago: China was a latecomer; it may, however, catch up and even surpass the Western developed economies through learning their best practice and avoiding their mistakes [31]. This logic, rightly or wrongly, has informed China's rapid industrialisation and urbanisation in the past four decades, and its campaign for an 'innovative nation' in the recent decade. In some way, this logic has also explained China's rapid leap in the global smart city movement from a latecomer—the development of Chinese smart cities remained in a preliminary stage in the late 2000s compared with EU and North America [59]—to a leader, in numerical terms at least, largely attributed to the cooperation between the government and the technology firms [58]. The smart city strategy is also being discussed to inject new dynamics into the 'belt and road initiative', an ambitious global infrastructure program to outreach China's economic influence overseas [49].

However, some caution should be required in comprehending the booming smart cities in China. The rapidity of promoting a nation-wide smart city movement through its unique urban regime has 'succeeded' in applying the latest technology and advancing economic growth, with outcomes that are measurable, tangible and worth celebrating. This approach and achieved outcomes present a contrast to its neighbouring country India, which shared many similar societal situations but is now lagging behind in urbanisation and smart cities initiatives [60]. However, it should also be noted that the Chinese smart cities have 'failed' to engage the important contemporary urban challenge of sustainable development, which is covered under the shiny urban images of modernisation, prosperity and smartness. A latecomer's advantage in achieving rapid technological and economic outcomes could lead to a latecomer's disadvantage in achieving sustainable development in the end, if a technology-centric pathway would continue to be followed without institutional adaptation to engage public participation, and to incorporate important non-economic challenges confronting Chinese cities.

## 7. Conclusions, Limitations and Further Research

This study has tried to address the question of to what extent the smart cities in China conform to or differ from their international counterparts. By doing so, not only has it identified the conformity and disconformity between them, but also has addressed the lack of critical literature on Chinese smart cities and established a dialogue between them and the international smart city movement. This study has included an overview of the Chinese smart cities and a case study of Shenzhen—a leader in the Chinese smart city movement—analysed through perspectives of the smart city as a concept, as an urban development paradigm, and as an urban regime, drawn from the international literature. These analyses have revealed a general accordance between the Chinese smart cities and this analytical framework; they have also identified several explicit characteristics that mark the Chinese smart cities, and differentiate them from international smart cities: a technology-centric conceptualisation, an instrumentality of the smart city in driving an urban development paradigm for innovation and transformation towards a knowledge economy, and a government-dominated urban regime to pursue not only smart cities but also a smart society. China, a latecomer in the smart city movement, is rapidly catching up and has now the largest number of smart cities, claimed or being constructed, in the world. A latecomer's advantage in utilising the latest technology and driving economic growth and transformation may, however, translate into a latecomer's disadvantage in the long run if the urban challenge of sustainable development—in the sense of a balanced development between economic growth, environmental protection and social equity—continues to be excluded from the smart city discourse without making institutional adaptation and engaging public participation.

This preliminary study has limitations that need to be considered in comprehending the findings and conclusions. The single case of Shenzhen could be compared with other smart cities in China and overseas, to draw deeper insights into the state of smart cities in China, especially in comparison with international smart cities. These limitations inform further research to be undertaken possibly by the author and other interested researchers in the future. This study partially originated from an intention to set the scene and provoke more attention and research on the mushrooming smart cities in China. Further research may be undertaken along two strands, through building upon this study's analytical framework and findings, and addressing the limitations identified. One strand is a comprehensive comparison of the numerous smart cities in China, which is empirically grounded, to investigate their policy initiatives and outcomes. The other strand is comparative case studies of representative Chinese smart cities, such as this case of Shenzhen, with international counterpart smart cities to verify or falsify those characteristics identified by this study for the smart cities in China.

**Funding:** This research received funding from the Faculty of Business, Government and Law, University of Canberra.

**Acknowledgments:** Coco Liu assisted with the design of Figures 3 and 4.

**Conflicts of Interest:** The author declares no conflict of interest.

## References

1. Ahvenniemi, H.; Huovila, A.; Pinto-Seppä, I.; Airaksinen, M. What are the differences between sustainable and smart cities? *Cities* **2017**, *60*, 234–245. [CrossRef]
2. Albino, V.; Berardi, U.; Dangelico, R.M. Smart cities: Definitions, dimensions, performance, and initiatives. *J. Urban Technol.* **2015**, *22*, 3–21. [CrossRef]
3. Ruhlandt, R.W.S. The governance of smart cities: A systematic literature review. *Cities* **2018**, *81*, 1–23. [CrossRef]
4. Yigitcanlar, T.; Kamruzzaman, M.; Buys, L.; Ioppolo, G.; Sabatini-Marques, J.; da Costa, E.M.; Yun, J.J. Understanding 'smart cities': Intertwining development drivers with desired outcomes in a multidimensional framework. *Cities* **2018**, *81*, 145–160. [CrossRef]
5. Angelidou, M. Smart cities: A conjuncture of four forces. *Cities* **2015**, *47*, 95–106. [CrossRef]
6. Anthopoulos, L. Smart utopia vs. smart reality: Learning by experience from 10 smart city cases. *Cities* **2017**, *63*, 128–148. [CrossRef]
7. Mora, L.; Bolici, R.; Deakin, M. The first two decades of smart-city research: A bibliometric analysis. *J. Urban Technol.* **2017**, *24*, 3–27. [CrossRef]
8. Shan, Z. The Smart Society Points out a Direction for Societal Informationalisation [Zhi Hui She Hui Wei She Hui Xin Xi Hua Zhi Ming Fang Xiang]. 2018. Available online: opinion.people.com.cn/n1/2018/0124/c1003-29782429.html (accessed on 16 September 2019).
9. Nylander, J. *Shenzhen Superstars: How China's Smartest City is Challenging Silicon Valley*; CreateSpace Independent Publishing Platform: Scotts Valley, CA, USA, 2017.
10. Kang, Y.; Zang, L.; Chen, C.; Ge, Y.; Li, H.; Cui, Y.; Hart, T. *Comparative Study of Smart Cities in Europe and China*; EU Commission and China Academy of Telecommunications Research: Beijing, China, 2014.
11. Yigitcanlar, T.; Kamruzzaman, M. Does smart city policy lead to sustainability of cities? *Land Use Policy* **2018**, *73*, 49–58. [CrossRef]
12. Meijer, A.; Thaens, M. Urban technological innovation: Developing and testing a sociotechnical framework for studying smart city projects. *Urban Aff. Rev.* **2018**, *54*, 363–387. [CrossRef]
13. Appio, F.P.; Lima, M.; Paroutis, S. Understanding smart cities: Innovation ecosystems, technological advancements, and societal challenges. *Technol. Forecast. Soc. Chang.* **2019**, *142*, 1–14. [CrossRef]
14. Hu, R. Planning for economic development. In *The Routledge Handbook of Planning History*; Hein, C., Ed.; Routledge: London, UK; New York, NY, USA, 2018; pp. 313–324.
15. Yigitcanlar, T. *Technology and the City: Systems, Applications and Implications*; Routledge: London, UK; New York, NY, USA, 2016.
16. Hu, R. Clustering: Concentration of the knowledge-based economy in Sydney. In *Building Prosperous Knowledge Cities: Policies, Plans and Metrics*; Yigitcanlar, T., Metaxiotis, K., Carrillo, F.J., Eds.; Edward Elgar: Cheltenham, UK, 2012; pp. 195–212.
17. Shin, D.H. Ubiquitous city: Urban technologies, urban infrastructure and urban informatics. *J. Inf. Sci.* **2009**, *35*, 515–526. [CrossRef]
18. Hu, R.; Blakely, E.; Zhou, Y. Benchmarking the competitiveness of Australian global cities: Sydney and Melbourne in the global context. *Urban Policy Res.* **2013**, *31*, 435–452. [CrossRef]
19. Blakely, E.J.; Hu, R. *Crafting Innovative Places for Australia's Knowledge Economy*; Palgrave Macmillan: Singapore, 2019.
20. Herrschel, T. Competitivenes. and sustainability: Can 'smart city regionalism' square the circle? *Urban Stud.* **2013**, *50*, 2332–2348. [CrossRef]
21. Hu, R. Sustainable development strategy for the global city: A case study of Sydney. *Sustainability* **2015**, *7*, 4549–4563. [CrossRef]
22. Gibbs, D.; Krueger, R.; MacLeod, G. Grappling with smart city politics in an era of market triumphalism. *Urban Stud.* **2013**, *50*, 2151–2157. [CrossRef]
23. Shelton, T.; Zook, M.; Wiig, A. The 'actually existing smart city'. *Camb. J. Reg. Econ. Soc.* **2015**, *8*, 13–25. [CrossRef]

24. McFarlane, C.; Soderstrom, O. On alternative smart cities: From a technology-intensive to a knowledge-intensive smart urbanism. *City* **2017**, *21*, 312–328. [CrossRef]
25. Wiig, A. IBM's smart city as techno-utopian policy mobility. *City* **2015**, *19*, 258–273. [CrossRef]
26. Wiig, A. The empty rhetoric of the smart city: From digital inclusion to economic promotion in Philadelphia. *Urban Geogr.* **2016**, *37*, 535–553. [CrossRef]
27. McNeill, D. Global firms and smart technologies: IBM and the reduction of cities. *Trans. Inst. Br. Geogr.* **2015**, *40*, 562–574. [CrossRef]
28. Hu, R. *The Shenzhen Phenomenon: From Fishing Village to Global Knowledge City*; Routledge: London, UK; New York, NY, USA, 2020.
29. Deloitte. *Super Smart City: Happier Society with Higher Quality*; Deloitte China: Beijing, China, 2018.
30. Qianzhan. Market Analysis for Chinese Smart Cities in 2019 [2019 Nian Zhong Guo Zhi Hui Cheng Shi Hang Ye Shi Chang Fen Xi]. 2019. Available online: https://bg.qianzhan.com/report/detail/300/190226-6493a8ba.html (accessed on 17 August 2019).
31. Hu, R.; Chen, W. *Global Shanghai Remade: The Rise of Pudong New Area*; Routledge: London, UK; New York, NY, USA, 2019.
32. National Bureau of Statistics of China. Annual Data. 2019. Available online: http://data.stats.gov.cn/easyquery.htm?cn=C01 (accessed on 13 September 2019).
33. The World Bank. Urban Population. 2019. Available online: https://data.worldbank.org/indicator/SP.URB.TOTL.IN.ZS?locations=CN (accessed on 6 August 2019).
34. MOHURD (Ministry of Housing and Urban-Rural Development). A Notice on Conducting Naitonal Pilot Smart Cities [Guan Yu Kai Zhan Guo Jia Zhi Hui Cheng Shi Shi Dian Gong Zuo De Tong Zhi]. 2012. Available online: www.mohurd.gov.cn/wjfb/201212/t20121204_212182.html (accessed on 15 September 2019).
35. Guo, M.; Liu, Y.; Yu, H.; Hu, B.; Sang, Z. An overview of smart city in China. *China Commun.* **2016**, *13*, 203–211. [CrossRef]
36. State Council. *National New-Type Urbanisation Strategy (2014–2020) [Guo Jia Xin Xing Cheng Shi Hua Gui Hua (2014–2020)]*; State Council: Beijing, China, 2014.
37. Xi, J. Secure a Decisive Victory in Building a Moderately Prosperous Society in All Aspects and Strive for the Great Success of Socialism with Chinese Characteristics for a New Era. 2017. Available online: http://www.xinhuanet.com/english/download/Xi_Jinping\T1\textquoterights_report_at_19th_CPC_National_Congress.pdf (accessed on 2 November 2019).
38. Ren, R. China has the Largest Number of Smart Cities under Construction in the Globe [Zhong Guo Zai Jian Zhi Hui Cheng Shi Quan Qiu Di Yi]. 2018. Available online: https://www.leiphone.com/news/201812/anRj0WUF6c7zJxT9.html (accessed on 12 September 2019).
39. Yang, K.; Clery, A.; Liello, D.D. *Smart Cities in China*; The EU SME Centre: Beijing, China, 2015.
40. Hu, R. Xiong'an, Xi Jinping's New City-Making Machine Turned on. 2018. Available online: https://theconversation.com/xiongan-xi-jinpings-new-city-making-machine-turned-on-95442 (accessed on 16 September 2019).
41. Xiong'an Management Commission. Xiong'an China [Zhong Guo Xiong An]. 2019. Available online: http://www.xiongan.gov.cn (accessed on 16 September 2019).
42. Shenzhen Statistical Bureau. *Shenzhen Statistical Yearbook*; China Statistics Press: Beijing, China, 2018.
43. Shenzhen Government. *Shenzhen Municipal Master Plan (2010–2020) [Shen Zhen Shi Cheng Shi Zong Ti Gui Hua (2010–2020)]*; Shenzhen Government: Shenzhen, China, 2010.
44. Shenzhen Government. Shenzhen Municipal New-Type Smart City Construction Master Plan [Shen Zhen Shi Xin Xing Zhi Hui Cheng Shi Jian She Zong Ti Fang An]. 2018. Available online: www.sz.gov.cn/zfgb/2018/gb1062/201807/t20180730_13798766.htm (accessed on 19 July 2019).
45. Tomás, J.P. Turning Shenzhen into a Smart City. 2016. Available online: https://enterpriseiotinsights.com/20160809/smart-cities/turning-shenzhen-smart-city-tag23 (accessed on 3 September 2019).
46. Chen, F. A Look at Shenzhen and Huawei's 'Smart City' Project. *Asia Times*. 11 July 2019. Available online: https://www.asiatimes.com/2019/07/article/a-look-at-shenzhen-and-huaweis-smart-city-project/ (accessed on 1 October 2019).
47. GovInsider. Shenzhen's 'Maslow Model' for Smart Cities. 2018. Available online: https://govinsider.asia/connected-gov/shenzhens-maslow-model-for-smart-cities/?print=true (accessed on 3 September 2019).

48. Cheong, A. Shenzhen Economy Overtakes HK's to Rank First in Bay Area. *China Daily*. 1 March 2018. Available online: https://www.chinadailyhk.com/articles/190/97/169/1519914449127.html (accessed on 3 May 2019).
49. Zhang, L.; Zhang, Z.; Xiang, Q.; Liu, B. Opportunities and challenges for smart city development in China. *J. Civ. Eng. Arch.* **2018**, *12*, 273–287. [CrossRef]
50. Appleyard, B.; Zheng, Y.; Watson, R.; Bruce, L.; Sohmer, R.; Li, X.; Qian, J. *Smart Cities: Solutions for China's Rapid Urbanization*; The Natural Resources Defense Council: New York, NY, USA, 2007.
51. Chan, J.K.S.; Anderson, S. *Rethinking Smart Cities: ICT for New-type Urbanization and Public Participation at the City and Community Level in China*; United Nations Development Programme China: Beijing, China, 2015.
52. Ganot, S. *Smart City Policies in China: National and Local Goals*; Hebrew University of Jerusalem: Jerusalem, Israel, 2016.
53. McCann, E. Urban policy mobilities and global circuits of knowledge: Toward a research agenda. *Ann. Assoc. Am. Geogr.* **2011**, *101*, 107–130. [CrossRef]
54. Yu, W.; Xu, C. Developing smart cities in China: An empriical analysis. *Int. J. Public Adm. Digit. Age* **2018**, *5*, 76–91. [CrossRef]
55. Hu, R. City deals: Old wine in new bottles? In *From Turnbull to Morrison: The Trust Divide*; Evans, M., Grattan, M., McCaffrie, B., Eds.; Melbourne University Press: Melbourne, Australia, 2019; pp. 242–255.
56. Artigas, Á. *Surveillance, Smart Technologies and the Development of Safe City Solutions: The Case of Chinese ICT Firms and Their International Expansion to Emerging Markets*; Institut Barcelona d'Estudis Internacionals: Barcelona, Spain, 2017.
57. Cave, D.; Hoffman, S.; Joske, A.; Ryan, F.; Thomas, E. *Mapping China's Technology Giants*; The Australian Strategic Policy Institute: Sydney, Australia, 2019.
58. Li, Y.; Lin, Y.; Geertman, S. The development of smart cities in China. In Proceedings of the 14th International Conference on Computers in Urban Planning and Urban Management 2015, Cambridge, MA, USA, 7–10 July 2015.
59. Lu, D.; Tian, Y.; Liu, V.Y.; Zhang, Y. The performance of the smart cities in China—A comparative study by means of self-organizing maps and social networks analysis. *Sustainability* **2015**, *7*, 7604–7621. [CrossRef]
60. Chandrasekar, K.S.; Bajracharya, B.; O'Hare, D. A comparative analysis of smart city initiatives by China and India—Lessons for India. In Proceedings of the 9th International Urban Design Conference: Smart Cities for 21st Century Australia: How Urban Design Innovation Can Change Our Cities 2016, Canberra, Australia, 7–9 November 2016.

 © 2019 by the author. Licensee MDPI, Basel, Switzerland. This article is an open access article distributed under the terms and conditions of the Creative Commons Attribution (CC BY) license (http://creativecommons.org/licenses/by/4.0/).

MDPI  
St. Alban-Anlage 66  
4052 Basel  
Switzerland  
Tel. +41 61 683 77 34  
Fax +41 61 302 89 18  
www.mdpi.com  

*Energies* Editorial Office  
E-mail: energies@mdpi.com  
www.mdpi.com/journal/energies

www.ingramcontent.com/pod-product-compliance
Lightning Source LLC
LaVergne TN
LVHW071938080526
838202LV00064B/6630